SYMMETRIC CYCLES

SYMMETRIC CYCLES

Andrey O. Matveev

JENNY STANFORD
PUBLISHING

Published by

Jenny Stanford Publishing Pte. Ltd.
101 Thomson Road
#06-01, United Square
Singapore 307591

Email: editorial@jennystanford.com
Web: www.jennystanford.com

British Library Cataloguing-in-Publication Data
A catalogue record for this book is available from the British Library.

Symmetric Cycles

ISBN 978-981-4968-81-2 (Hardcover)
ISBN 978-1-003-43832-8 (eBook)

To the memory of my parents

Irina M. Matveeva

1934–2014

Oleg S. Matveev

1931–2022

To the memory of

Dr. Damir N. Gainanov

1954–2022

Contents

Preface

The present monograph begins, in a sense, where the book *Pattern Recognition on Oriented Matroids*, published with *De Gruyter* in 2017, left off.

Given an ordered *two-letter alphabet* (θ, α) and an integer *dimensional* parameter t, the *discrete hypercube* $\{\theta, \alpha\}^t$ is the *vertex set* of the *hypercube graph* $\boldsymbol{\Gamma}(t, 2; \theta, \alpha)$, a central discrete mathematical construct with countless applications in theoretical computer science, Boolean function theory, combinatorics, combinatorial optimization, coding theory, discrete and computational geometry, etc.

Let $E_t := [t] := \{1, \ldots, t\}$ denote a *ground set* of indices. A pair $\{X, Y\}$ of vertices $X := (X(1), \ldots, X(t))$ and Y of the hypercube $\{\theta, \alpha\}^t$ by definition is an *edge* of the graph $\boldsymbol{\Gamma}(t, 2; \theta, \alpha)$ if the *Hamming distance* $|\{e \in E_t\colon\ X(e) \neq Y(e)\}|$ between the ordered tuples X and Y is 1.

A vertex $Z \in \{\theta, \alpha\}^t$ of the hypercube graph $\boldsymbol{\Gamma}(t, 2; \theta, \alpha)$ has its *'positive part'* $\mathfrak{p}(Z) := Z^+$, defined to be the subset $\{e \in E_t\colon\ Z(e) = \theta\} \subseteq E_t$, and its *'negative part'* $\mathfrak{n}(Z) := Z^- := \{e \in E_t\colon\ Z(e) = \alpha\} = E_t - \mathfrak{p}(Z)$.

Recall that the *power set* $\mathbf{2}^{[t]}$ of the set E_t is defined to be the family $\{A\colon\ A \subseteq E_t\}$ of all subsets of E_t. The partial ordering of this family by inclusion turns the power set $\mathbf{2}^{[t]}$ into the *Boolean lattice* $\mathbb{B}(t)$ of rank t. Throughout the monograph, by convention we interpret the power set

$$\mathbf{2}^{[t]} = \left\{\mathfrak{n}(Z)\colon\ Z \in \{\theta, \alpha\}^t\right\}$$

of the ground set E_t as the family of the negative parts of vertices of a particular *discrete hypercube*, $\{1, -1\}^t \subset \mathbb{R}^t$, or $\{0, 1\}^t \subset \mathbb{R}^t$, associated with the two-letter alphabet $(\theta, \alpha) := (1, -1)$, or $(\theta, \alpha) := (0, 1)$, respectively, where the letters θ and α are regarded as *real numbers*.

In the monograph, we consider both the *hypercube graphs*

$$\boldsymbol{H}(t, 2) := \boldsymbol{\Gamma}(t, 2; 1, -1)\,, \quad \text{and} \quad \widetilde{\boldsymbol{H}}(t, 2) := \boldsymbol{\Gamma}(t, 2; 0, 1)\,,$$

though our main interest lies in the graph $\boldsymbol{H}(t, 2)$ on its vertex set $\{1, -1\}^t$.

A *symmetric cycle* $\boldsymbol{D} := (D^0, D^1, \ldots, D^{2t-1}, D^0)$ in the hypercube graph $\boldsymbol{H}(t, 2)$ is defined to be a $2t$-cycle, with its *vertex set* $\mathrm{V}(\boldsymbol{D}) := \{D^0, D^1, \ldots, D^{2t-1}\}$, such that

$$D^{k+t} = -D^k\,, \quad 0 \le k \le t-1\,.$$

The vertex set $\mathrm{V}(\boldsymbol{D})$ of the symmetric cycle $\boldsymbol{D}$ is a *maximal positive basis* of the space $\mathbb{R}^t$. For any vertex $T \in \{1, -1\}^t$ of the graph $\boldsymbol{H}(t, 2)$ there exists a *unique inclusion-minimal* and *linearly independent* subset (of *odd* cardinality) $\boldsymbol{Q}(T, \boldsymbol{D}) \subset \mathrm{V}(\boldsymbol{D})$ such that

$$T = \sum_{Q \in \boldsymbol{Q}(T, \boldsymbol{D})} Q\,.$$

In particular, that linear algebraic decomposition describes how the members of the family of subsets $\{\mathfrak{n}(Q) \colon Q \in \boldsymbol{Q}(T, \boldsymbol{D})\} \subset \mathbf{2}^{[t]}$ *vote* for, or against, the elements of the ground set E_t, thus arriving at their *collective decision* on the subset $\mathfrak{n}(T) \subseteq E_t$.

Informally speaking, we discuss in the monograph various aspects of how (based on the above decomposition) all the vertices of the discrete hypercube $\{1, -1\}^t$, as well as the power set of the ground set E_t, emerge from a *rank* 2 *oriented matroid*, from an underlying *rank* 2 *system of linear inequalities*, and thus literally from an *arrangement* of t distinct *straight lines* crossing a common point on a piece of paper.

In the introductory Chapter 2, we briefly recall basic properties of vertex decompositions in hypercube graphs $\boldsymbol{H}(t, 2)$ with respect to their symmetric cycles. We then establish a connection between *coherent decompositions* in the hypercube graphs $\boldsymbol{H}(t, 2)$ and $\widetilde{\boldsymbol{H}}(t, 2)$.

In Chapter 3, we recall some enumerative results on rank 2 infeasible systems of linear inequalities related to arrangements of oriented lines in the plane. *Dehn–Sommerville type relations* are presented that concern the numbers of faces of abstract simplicial complexes associated with large-size decomposition sets for vertices of hypercube graphs.

Chapter 4 concerns additional *Dehn–Sommerville type relations* that are valid for large-size decomposition sets for vertices in hypercube graphs. We present a common *orthogonality relation* that establishes a connection between enumerative properties of large-size decomposition sets in the graphs $\boldsymbol{H}(s, 2)$ and $\boldsymbol{H}(t, 2)$ with specific dimensional parameters s and t.

In Chapter 5, we give a few comments on certain *distinguished* symmetric cycles in hypercube graphs. The main results of the chapter relate to the *interval* structure of the negative parts of vertices of hypercube graphs, to *computation-free decompositions* with respect to the distinguished symmetric cycles, and to statistics on decompositions. We also discuss *equinumerous decompositions* of vertices. We conclude the chapter by mentioning that vector descriptions of vertex decompositions with respect to arbitrary symmetric cycles are *valuations* on the Boolean lattices of subsets of the vertex sets of hypercube graphs.

In Chapter 6, we touch on the question on a structural connection between the decomposition sets for vertices whose negative parts are *comparable* by inclusion. Further enumerative results concern statistics on *partitions* of the negative parts of vertices of hypercube graphs and on decompositions of vertices. The key computational tool that allows us to present quite fine statistics is an approach to enumeration of *ternary Smirnov words* (i.e., words over a three-letter alphabet, such that adjacent letters in the words never coincide) discussed in Appendix A.

An even more involved analysis, based on enumeration of *Smirnov words* over *four-letter* alphabets (also discussed in Appendix A), leads us in Chapter 7 to statistics on *unions* of the negative parts of vertices of hypercube graphs and on decompositions of vertices.

An innocent-looking transformation (with serious applications given later in Chapter 9) of vertices of a discrete hypercube $\{1, -1\}^t$, that turns a vertex of the hypercube graph $\boldsymbol{H}(t, 2)$ into its '*relabeled opposite*', is presented and discussed in Chapter 8.

Recall that a nonempty family of nonempty subsets $\mathcal{A} := \{A_1, \ldots, A_\alpha\} \subset \mathbf{2}^{[t]}$ of the ground set E_t is called a *clutter* (*Sperner family*), if no set from the family $\mathcal{A}$ contains another. One says that a subset $B \subseteq E_t$ is a *blocking set* of the clutter $\mathcal{A}$ if the set B has a nonempty intersection with each member A_i of $\mathcal{A}$. The *blocker* $\mathfrak{B}(\mathcal{A})$ of the clutter $\mathcal{A}$ is defined to be the family of all inclusion-minimal blocking sets of $\mathcal{A}$. In Chapter 9, we drastically change the dimensionality of our research constructs from t to 2^t, since we indirectly represent the families of blocking sets of clutters as the negative parts of relevant vertices of the hypercube graphs $\boldsymbol{H}(2^t, 2)$ and $\widetilde{\boldsymbol{H}}(2^t, 2)$ associated with the discrete hypercubes $\{1, -1\}^{2^t}$ and $\{0, 1\}^{2^t}$, respectively. We describe in detail a '*blocking/voting*' connection between the families of blocking sets of the clutters $\mathcal{A}$ and $\mathfrak{B}(\mathcal{A})$.

In Appendix A, Smirnov words over three-letter and four-letter alphabets are enumerated.

In Appendix B we investigate the enumerative properties of the subset families generated by the *self-dual clutters* $\mathcal{A} = \mathfrak{B}(\mathcal{A})$.

Andrey O. Matveev
February 2023

Chapter 1

Preliminaries and Notational Conventions

Throughout the monograph, ':=' means equality by definition.

– Consider an *arrangement* of distinct *straight lines*

$$\{\mathbf{x} \in \mathbb{R}^2 \colon \langle \boldsymbol{a}_e, \mathbf{x} \rangle = 0,\ e \in E_t\} \tag{1.1}$$

in the plane $\mathbb{R}^2$ (throughout the monograph, $\langle \boldsymbol{v}, \boldsymbol{w} \rangle$ means the standard scalar product $\sum_e v_e w_e$ of real vectors $\boldsymbol{v}$ and $\boldsymbol{w}$ of relevant dimension), with the set of corresponding normal vectors

$$\boldsymbol{A} := \{\boldsymbol{a}_e \colon e \in E_t\} .$$

One associates with the arrangement (1.1) the *rank* 2 *system* of *homogeneous strict linear inequalities*

$$\{\langle \boldsymbol{a}_e, \mathbf{x} \rangle > 0 \colon \mathbf{x} \in \mathbb{R}^2,\ \boldsymbol{a}_e \in \boldsymbol{A}\} .$$

The lines of the arrangement are *oriented*: a vector $\boldsymbol{v} := (v_1, v_2) \in \mathbb{R}^2 - \{\mathbf{0}\}$ lies on the *positive side* of a line $\boldsymbol{L}_e := \{\mathbf{x} \in \mathbb{R}^2 \colon \langle \boldsymbol{a}_e, \mathbf{x} \rangle = 0\}$ if we have $\langle \boldsymbol{a}_e, \boldsymbol{v} \rangle > 0$. Similarly, a *region* $\boldsymbol{T}$ of the arrangement (1.1), that is, a *connected component* of the *complement* $\mathbb{R}^2 - \bigcup_{e \in E_t} \boldsymbol{L}_e$, lies on the *positive side* of the line $\boldsymbol{L}_e$ if we have $\langle \boldsymbol{a}_e, \boldsymbol{v} \rangle > 0$, for an arbitrary vector $\boldsymbol{v} \in \boldsymbol{T}$.

Symmetric Cycles
Andrey O. Matveev

ISBN 978-981-4968-81-2 (Hardcover), 978-1-003-43832-8 (eBook)
www.jennystanford.com

The arrangement (1.1) *realizes* a *rank* 2 *oriented matroid* $\mathcal{N}$ in the following manner (see, e.g., Example 7.1.7 in Ref. [1]): For a vector $\boldsymbol{v} \in \mathbb{R}^2$, the ordered tuple of *signs* $X := (\mathrm{sign}(\langle \boldsymbol{a}_e, \boldsymbol{v}\rangle) : e \in E_t) \in \{1, 0, -1\}^t := \{\text{'+'}, \text{'0'}, \text{'−'}\}^t$ is called a *covector* of $\mathcal{N}$.

The covectors $T \in \mathcal{T} \subset \{1, -1\}^t$ of the oriented matroid $\mathcal{N}$ are called its *topes* (*maximal covectors*), and the set of topes $\mathcal{T}$ is in a one-to-one correspondence with the set of *regions* of the arrangement (1.1). The *cocircuits* $C^* \in \{1, 0, -1\}^t$ of the oriented matroid $\mathcal{N} := (E_t, \mathcal{T})$, on the *ground set* E_t, and with its set of topes $\mathcal{T}$, are the covectors that correspond to the *rays* emanating from the origin; *one* sign component of each cocircuit is 0.

- Given a vector $\mathbf{z} := (z_1, \ldots, z_t) \in \mathbb{R}^t$, we denote its *support* $\{e \in E_t : z_e \neq 0\}$ by $\mathrm{supp}(\mathbf{z})$.

- Now consider the *rank t oriented matroid* $\mathcal{H} := (E_t, \{1, -1\}^t)$ on the *ground set* E_t, and with its set of *topes* (*maximal covectors*) $\mathcal{T} := \{1, -1\}^t$, *realizable* (see, e.g., Example 2.1.4 in Ref. [1]) as the *arrangement* of *coordinate hyperplanes*

$$\big\{\{\mathbf{x} := (\mathrm{x}_1, \ldots, \mathrm{x}_t) \in \mathbb{R}^t : |\mathrm{supp}(\mathbf{x})| = t-1,\ \mathrm{x}_e = 0\} : e \in E_t\big\} \tag{1.2}$$

in the space $\mathbb{R}^t$.

The hyperplanes of the arrangement are *oriented*: a vector $\boldsymbol{v} := (v_1, \ldots, v_t) \in \mathbb{R}^t - \{\mathbf{0}\}$ lies on the *positive side* of a hyperplane $\boldsymbol{H}_e := \{\mathbf{x} \in \mathbb{R}^t : |\mathrm{supp}(\mathbf{x})| = t-1,\ \mathrm{x}_e = 0\}$ if $v_e > 0$. Similarly, a *region* $\boldsymbol{T}$ of the arrangement (1.2), that is, a *connected component* of the *complement* $\mathbb{R}^t - \bigcup_{e \in E_t} \boldsymbol{H}_e$, lies on the *positive side* of the hyperplane $\boldsymbol{H}_e$ if $v_e > 0$, for an arbitrary vector $\boldsymbol{v} \in \boldsymbol{T}$. For a vector $\boldsymbol{v} \in \mathbb{R}^t$, the *sign* tuple $X := (\mathrm{sign}(v_e) : e \in E_t) \in \{1, 0, -1\}^t := \{\text{'+'}, \text{'0'}, \text{'−'}\}^t$ is a *covector* of the oriented matroid $\mathcal{H}$. The *cocircuits* $C^* \in \{1, 0, -1\}^t$ of $\mathcal{H}$ are the covectors that have *one* sign component different from 0.

- If $\mathcal{M}$ is one of the above oriented matroids $\mathcal{N} := (E_t, \mathcal{T})$ and $\mathcal{H} := (E_t, \mathcal{T} := \{1, -1\}^t)$, then a sign tuple $S := (S(1), \ldots, S(t)) \in \{1, 0, -1\}^t$, with exactly *one* zero component $S(i) = 0$, is called a *subtope* of $\mathcal{M}$ if there are two topes, $T' := (T'(1), \ldots, T'(t)) \in \mathcal{T}$, and $T'' \in \mathcal{T}$, such that the *Hamming distance* between the tuples T' and T'' is 1, that is, $|\{e \in E_t : T'(e) \neq T''(e)\}| = 1$, and $T'(i) \neq T''(i)$.

Note that the subtopes of the rank 2 oriented matroid $\mathcal{N}$ are its cocircuits.

The *separation set* $\mathbf{S}(T', T'')$ of topes T' and T'' is defined by

$$\mathbf{S}(T', T'') := \{e \in E_t\colon T'(e) \neq T''(e)\}\,,$$

and the *graph distance* between them is

$$d(T', T'') := |\mathbf{S}(T', T'')|\,,$$

that is, $d(T', T'')$ is the *Hamming distance* between the tuples T' and T''. Recall that

$$\begin{aligned} d(T', T'') &= t - \frac{1}{4}\|T'' + T'\|^2 = \frac{1}{4}\|T'' - T'\|^2 \\ &= \frac{1}{2}(t - \langle T'', T'\rangle)\,, \end{aligned}$$

where $\|X\|^2 := \langle X, X\rangle$. If t is *even*, then we have[1]

$$\langle T'', T'\rangle = 0 \iff d(T', T'') = \frac{t}{2}\,.$$

- The vertex set of the *tope graph* of an oriented matroid $\mathcal{M}$ by definition is its set of topes $\mathcal{T}$. Topes T' and T'' are *adjacent* in the tope graph if they have a *common subtope*.

Note that the tope graph of the oriented matroid $\mathcal{N}$ is a *cycle* on its vertex set $\mathcal{T}$. The tope graph of the oriented matroid $\mathcal{H}$ is the *hypercube graph* on its vertex set $\{1, -1\}^t$.

The tope graph turns into the *Hasse diagram* of the *tope poset* of the oriented matroid $\mathcal{M}$, *based* at a tope $B \in \mathcal{T}$, if we partially order the set of topes as follows:

$$T' \preceq T'' \iff \mathbf{S}(B, T') \subseteq \mathbf{S}(B, T'')\,.$$

The graph distance $d(B, \cdot)$ becomes in this situation the *rank function* of the tope poset of the oriented matroid $\mathcal{M}$.

- In the present monograph, we use almost everywhere the nonstandard representation of *sign components* '+', '0' and '−' of covectors/vertices by the *real numbers* 1, 0 and −1, respectively. Nevertheless, most figures involve the traditional notation.

[1]Interesting subsets of vertices of discrete hypercubes $\{1, -1\}^t$, with *zero* pairwise *scalar products*, are the rows of *Hadamard matrices*, see, e.g., Refs. [2, 3, 4].

As a common rule, no matter what the notation is used, covectors/vertices are always thought of as elements of the *real Euclidean space* $\mathbb{R}^t$ (or $\mathbb{R}^{2^t}$, in Chapter 9) of *row vectors*.

- For readability, we prefer the nonstandard notation $\mathfrak{p}(X)$ and $\mathfrak{n}(X)$ for the *positive parts* and *negative parts* of tuples $X \in \{1, -1\}^t$, instead of the traditional notation X^+ and X^-, respectively. Thus, by convention we define the parts by

$$\mathfrak{p}(X) := \{e \in E_t \colon X(e) = \quad 1\} =: X^+ ,$$

and

$$\mathfrak{n}(X) := \{e \in E_t \colon X(e) = -1\} =: X^- .$$

- We denote by $\mathrm{T}^{(+)}$ the row tuple of all one's

$$\mathrm{T}^{(+)} := (+, \ldots, +) := (1, \ldots, 1) \in \mathbb{R}^t .$$

If $\mathrm{T}^{(+)}$ is a (maximal) covector of our oriented matroid $\mathcal{M}$, then it is called the *positive tope* of $\mathcal{M}$. We say that $\mathrm{T}^{(+)}$ is the *positive vertex* of the discrete hypercube $\{1, -1\}^t$, and that it is the *positive vertex* of the hypercube graph $\boldsymbol{H}(t, 2)$. The row tuple $\mathrm{T}^{(-)}$ is the element $\mathrm{T}^{(-)} := -\mathrm{T}^{(+)} := (-, \ldots, -) := (-1, \ldots, -1)$ of the space $\mathbb{R}^t$; if the tuple $\mathrm{T}^{(-)}$ is a (maximal) covector of the oriented matroid $\mathcal{M}$, then $\mathrm{T}^{(-)}$ is called the *negative tope* of $\mathcal{M}$.

- A subset $A \subseteq E_t$ of the ground set of the oriented matroid $\mathcal{N}$ is called *acyclic* if there is a tope $F \in \mathcal{T}$ of $\mathcal{N}$ such that $\mathfrak{p}(F) \supseteq A$.

- Given a tuple $T \in \{1, -1\}^t$, and a subset $A \subseteq E_t$, the notation ${}_{-A}T$ is used to denote the tuple obtained from T by *sign reversal* or *reorientation* on the set A:

$$({}_{-A}T)(e) := \begin{cases} -T(e) , & \text{if } e \in A, \\ T(e) , & \text{if } e \notin A. \end{cases}$$

Thus, by convention we have $\mathfrak{n}({}_{-A}\mathrm{T}^{(+)}) := A$.

One says that the oriented matroid ${}_{-A}\mathcal{N}$, whose set of topes by definition is the set ${}_{-A}\mathcal{T} := \{{}_{-A}T : T \in \mathcal{T}\}$, is obtained from the oriented matroid $\mathcal{N}$ by *reorientation* on the subset A.

Since we have ${}_{-A}\{1, -1\}^t = \{1, -1\}^t$ and, as a consequence, ${}_{-A}\mathcal{H} = \mathcal{H}$, the oriented matroid $\mathcal{H}$ is insensitive to reorientations.

– The unit vectors of the standard basis of the space $\mathbb{R}^t$ are denoted by $\boldsymbol{\sigma}(i) := (0, \ldots, \underset{\substack{\uparrow \\ i}}{1}, \ldots, 0)$, $1 \leq i \leq t$.

– We denote by $\#\mathcal{A}$ the number α of sets in a family $\mathcal{A} := \{A_1, \ldots, A_\alpha\}$. The cardinality of a finite set A is denoted by $|A|$.

– An *abstract simplicial complex* Δ on its *vertex* set E_t is defined to be a family $\Delta \subseteq \mathbf{2}^{[t]}$, such that $\{e\} \in \Delta$, for every vertex $e \in E_t$, and the following implications hold:

$$A, B \subseteq E_t \,, \quad A \subseteq B \in \Delta \quad \Longrightarrow \quad A \in \Delta \,.$$

Members of the family Δ are called *faces*. The inclusion-maximal faces are the *facets* of the complex Δ.

Chapter 2

A 2D Perspective on Higher Dimensional Discrete Hypercubes and the Power Sets of Finite Sets

The *power set* $\mathbf{2}^{[t]}$ of the t-element set $E_t := [t] := \{1, \ldots, t\}$, that is, the family of all subsets of the set E_t, emerges in a somewhat miraculous manner, in a 2D context,[1] from t distinct straight lines crossing a common point on a piece of paper.

Indeed, on the one hand, the sets composing the family $\mathbf{2}^{[t]}$ can be seen as the *negative parts* of vertices $T \in \{1, -1\}^t$ of the hypercube graph $\boldsymbol{H}(t, 2)$.

On the other hand, the sequence of *sign tuples* describing the *regions* of our *arrangement* of *oriented lines* in the plane $\mathbb{R}^2$ can be interpreted as the *vertex sequence* $\vec{\mathrm{V}}(\boldsymbol{D})$ of a *symmetric cycle* $\boldsymbol{D}$ in the graph $\boldsymbol{H}(t, 2)$. Since the sequence $\vec{\mathrm{V}}(\boldsymbol{D}) \subset \{1, -1\}^t$ is an ordered *maximal positive basis* of the space $\mathbb{R}^t$, any vertex $T \in \{1, -1\}^t$ of the graph $\boldsymbol{H}(t, 2)$ admits its expansion into a unique *inclusion-minimal* and *linearly independent* subset (of *odd* cardinality) of the vertex set $\mathrm{V}(\boldsymbol{D})$.

---

[1] 2D means 2-*dimensional*.

Symmetric Cycles
Andrey O. Matveev

ISBN 978-981-4968-81-2 (Hardcover), 978-1-003-43832-8 (eBook)
www.jennystanford.com

In Section 2.1 of this introductory chapter we briefly recall basic properties of vertex decompositions in hypercube graphs $\boldsymbol{H}(t, 2)$ with respect to their symmetric cycles. In Section 2.2 we establish a connection between coherent decompositions in the hypercube graphs $\widetilde{\boldsymbol{H}}(t, 2)$ and $\boldsymbol{H}(t, 2)$ whose vertex sets are the sets $\{0, 1\}^t$ and $\{1, -1\}^t$, respectively.

2.1 Symmetric Cycles in Hypercube Graphs, and Vertex Decompositions

Given the hypercube graph $\boldsymbol{H}(t, 2)$ on its vertex set $\{1, -1\}^t$, a *symmetric cycle* $\boldsymbol{D} := (D^0, D^1, \ldots, D^{2t-1}, D^0)$ in the graph $\boldsymbol{H}(t, 2)$, with its *vertex sequence*[2]

$$\vec{\mathrm{V}}(\boldsymbol{D}) := (D^0, D^1, \ldots, D^{2t-1}), \tag{2.1}$$

and with its *vertex set*

$$\mathrm{V}(\boldsymbol{D}) := \{D^0, D^1, \ldots, D^{2t-1}\}, \tag{2.2}$$

is defined to be a $2t$-cycle such that

$$D^{k+t} = -D^k, \quad 0 \le k \le t-1. \tag{2.3}$$

This section gives an overview of basic properties of the vertex decompositions in hypercube graphs $\boldsymbol{H}(t, 2)$ with respect to their symmetric cycles.

2.1.1 The Vertex Set of a Symmetric Cycle Is the Set of Topes of a Rank 2 Oriented Matroid

As noted in Remark 1.7 of Ref. [5], the vertex set $\mathrm{V}(\boldsymbol{D})$ of a symmetric cycle $\boldsymbol{D}$ in a hypercube graph $\boldsymbol{H}(t, 2)$ is the set of *topes* (i.e., *maximal covectors*) of a *rank* 2 *oriented matroid* denoted as $\mathcal{H}_{\boldsymbol{D}}$.

The *ground set* of $\mathcal{H}_{\boldsymbol{D}}$ is the ground set E_t of the oriented matroid $\mathcal{H} := (E_t, \{1, -1\}^t)$.

[2]Figure 2.6 on page 19 explains why we should make a distinction between vertex sequences and vertex sets.

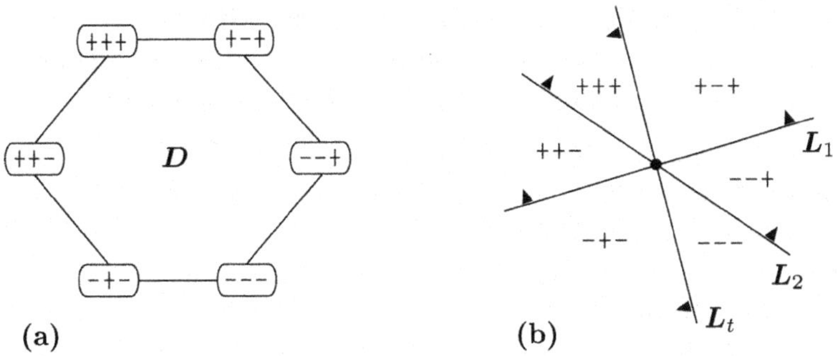

Figure 2.1 (a) A *symmetric cycle* $\boldsymbol{D}$ in the hypercube graph $\boldsymbol{H}((t := 3), 2)$. The graph $\boldsymbol{H}(t, 2)$ is the tope graph of the oriented matroid $\mathcal{H} := (E_t, \{1, -1\}^t)$, on the ground set E_t, with its set of topes $\{1, -1\}^t$.
(b) A central arrangement of $t := 3$ oriented lines $\{\boldsymbol{L}_1, \boldsymbol{L}_2, \boldsymbol{L}_t\}$ that realizes the rank 2 oriented matroid $\mathcal{H}_{\boldsymbol{D}} := (E_t, \mathrm{V}(\boldsymbol{D}))$, on the ground set E_t, with its set of topes $\mathrm{V}(\boldsymbol{D})$. The *positive sides* of lines are marked by black triangles. The *regions* of the arrangement are labeled with the vertices of the cycle $\boldsymbol{D}$, that is, with the *topes* of the oriented matroid $\mathcal{H}_{\boldsymbol{D}}$.

The set of *cocircuits* of the oriented matroid $\mathcal{H}_{\boldsymbol{D}}$ is the subset of *subtopes* of $\mathcal{H}$, each of which is *covered* in the *big face lattice* of the oriented matroid $\mathcal{H}$ by a pair of *adjacent* vertices of the cycle $\boldsymbol{D}$.

The rank 2 oriented matroid $\mathcal{H}_{\boldsymbol{D}}$ can be represented by a *central line arrangement* in the plane.[3] The set $\mathrm{V}(\boldsymbol{D})$ of *topes* of $\mathcal{H}_{\boldsymbol{D}}$ corresponds to the centrally symmetric set of *regions* of the arrangement. The set of *cocircuits* of $\mathcal{H}_{\boldsymbol{D}}$ corresponds to the centrally symmetric set of *rays* emanating from the origin; see Example 7.1.7 in Ref. [1].

Example 2.1. Suppose $t := 3$, and consider a symmetric cycle $\boldsymbol{D}$ in the hypercube graph $\boldsymbol{H}(t, 2)$, depicted in Figure 2.1(a).

The *vertex set* $\mathrm{V}(\boldsymbol{D})$ of the cycle $\boldsymbol{D}$ is the *set* of *topes* of a rank 2 *oriented matroid* $\mathcal{H}_{\boldsymbol{D}}$ realized in the plane as a *central arrangement* of *oriented lines* depicted in Figure 2.1(b).

[3] By convention, each line of a *central* arrangement crosses the origin $\mathbf{0} := (0, 0)$ of the plane $\mathbb{R}^2$.

2.1.2 The Vertices of a Symmetric Cycle with Inclusion-Maximal Positive Parts Are Detected via the Local Maxima of the Cardinalities of Parts

Given a symmetric cycle $\boldsymbol{D}$ in a hypercube graph $\boldsymbol{H}(t, 2)$, we denote by $\mathbf{max}^{+}(\mathrm{V}(\boldsymbol{D}))$ the subset of all vertices $T \in \mathrm{V}(\boldsymbol{D})$ that have *inclusion-maximal positive parts*. Note that the family $\{\mathfrak{p}(T)\colon T \in \mathbf{max}^{+}(\mathrm{V}(\boldsymbol{D}))\} \subset \mathbf{2}^{[t]}$ is a *clutter*, that is, its members are pairwise incomparable by inclusion.

Let $(T^{k_1}, T^{k_2}, T^{k_3})$ be a 2-path in the symmetric cycle $\boldsymbol{D}$ of the graph $\boldsymbol{H}(t, 2)$, where $T^{k_1} \neq T^{k_3}$. We have

$$T^{k_2} \in \mathbf{max}^{+}(\mathrm{V}(\boldsymbol{D})) \iff |\mathfrak{p}(T^{k_1})| = |\mathfrak{p}(T^{k_3})| = |\mathfrak{p}(T^{k_2})| - 1\,;$$

see Remark 1.8(ii) in [5]. In other words, let $\{f\} := \mathbf{S}(T^{k_1}, T^{k_2})$, and $\{g\} := \mathbf{S}(T^{k_2}, T^{k_3})$. Then the inclusion $T^{k_2} \in \mathbf{max}^{+}(\mathrm{V}(\boldsymbol{D}))$ holds if and only if $T^{k_2}(f) = T^{k_2}(g) = 1$.

Example 2.2. Suppose $t := 3$, and consider a symmetric cycle $\boldsymbol{D}$ in the hypercube graph $\boldsymbol{H}(t, 2)$, depicted in Figure 2.2. The subset of vertices of the cycle $\boldsymbol{D}$ that have *locally maximal* cardinalities of their positive parts is precisely the subset $\mathbf{max}^{+}(\mathrm{V}(\boldsymbol{D}))$ of vertices of the cycle $\boldsymbol{D}$ with *inclusion-maximal* positive parts.

Note that for a vertex $T^{k_2} \in \mathrm{V}(\boldsymbol{D})$, the related separation sets are $\mathbf{S}(T^{k_1}, T^{k_2}) = \{1\}$, and $\mathbf{S}(T^{k_2}, T^{k_3}) = \{3\}$. Since $T^{k_2}(1) = T^{k_2}(3) = 1$, we have the inclusion $T^{k_2} \in \mathbf{max}^{+}(\mathrm{V}(\boldsymbol{D}))$.

2.1.3 The Subset of Vertices of a Symmetric Cycle with Inclusion-Maximal Positive Parts Is a Critical Committee for the Oriented Matroid $\mathcal{H} := (E_t, \{1, -1\}^t)$

According to Definitions 1.1(i) and 1.3(ii) of Ref. [5], an *inclusion-minimal* subset $\mathcal{K}^{*} \subset \{1, -1\}^t$, such that

$$\sum_{K \in \mathcal{K}^{*}} K = \mathrm{T}^{(+)}\,, \tag{2.4}$$

is called a *critical committee* for the oriented matroid $\mathcal{H} := (E_t, \{1, -1\}^t)$.

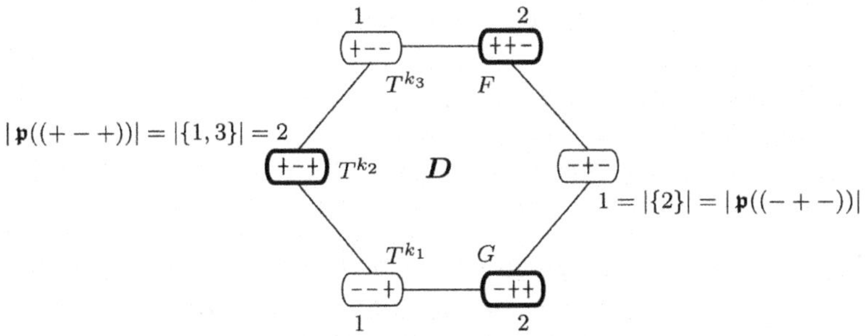

Figure 2.2 A symmetric cycle $\boldsymbol{D}$ in the hypercube graph $\boldsymbol{H}((t := 3), 2)$. The vertices of the cycle are labeled with the cardinalities of their positive parts. Since the cardinalities of the positive parts of vertices T^{k_2}, F and G are *locally maximal* (e.g., we have $|\mathfrak{p}(T^{k_1})| = |\mathfrak{p}(T^{k_3})| = |\mathfrak{p}(T^{k_2})| - 1$), this guarantees that these three vertices are the only vertices that belong to the vertex set of the cycle $\boldsymbol{D}$ and have *inclusion-maximal positive parts*, that is, $\mathbf{max}^+(\mathrm{V}(\boldsymbol{D})) = \{T^{k_2}, F, G\}$.
The set $\mathbf{max}^+(\mathrm{V}(\boldsymbol{D}))$ is an inclusion-minimal subset of the set $\{1, -1\}^t$, with the property $\mathrm{T}^{(+)} := (1, 1, 1) = \sum_{K \in \mathbf{max}^+(\mathrm{V}(\boldsymbol{D}))} K = (1, -1, 1) + (1, 1, -1) + (-1, 1, 1)$, that is, $\mathbf{max}^+(\mathrm{V}(\boldsymbol{D}))$ by definition is a *critical committee* for the oriented matroid $\mathcal{H} := (E_t, \{1, -1\}^t)$.

Returning to our symmetric cycle $\boldsymbol{D}$ of the hypercube graph $\boldsymbol{H}(t, 2)$, described in (2.1)(2.3), recall that the subset $\mathbf{max}^+(\mathrm{V}(\boldsymbol{D})) \subset \mathrm{V}(\boldsymbol{D})$ is a *critical committee* for $\mathcal{H}$; see Proposition 1.27 in Ref. [5]. The set $\mathbf{max}^+(\mathrm{V}(\boldsymbol{D}))$ contains an *odd* number of vertices.

For each element $e \in E_t$ of the ground set of the oriented matroid $\mathcal{H}$ we have

$$|\{K \in \mathbf{max}^+(\mathrm{V}(\boldsymbol{D})) \colon K(e) = 1\}| = \left\lceil \frac{|\,\mathbf{max}^+(\mathrm{V}(\boldsymbol{D}))|}{2} \right\rceil .$$

Example 2.3. Suppose $t := 3$, and consider again the symmetric cycle $\boldsymbol{D}$ in the hypercube graph $\boldsymbol{H}(t, 2)$, depicted in Figure 2.2. Since the set $\mathcal{K}^* := \mathbf{max}^+(\mathrm{V}(\boldsymbol{D}))$ is an inclusion-minimal subset of the set $\{1, -1\}^t$ with the property (2.4), the set $\mathbf{max}^+(\mathrm{V}(\boldsymbol{D}))$ by definition is a critical committee for the oriented matroid $\mathcal{H} := (E_t, \{1, -1\}^t)$.

As noted in Remark 1.28 of Ref. [5], a simple linear algebraic argument given in Section 11.1 of Ref. [5] and recalled below in Subsection 2.1.4 guarantees that the set $\mathbf{max}^+(\mathrm{V}(\boldsymbol{D}))$ is the *only* critical committee whose topes *all* belong to the vertex set $\mathrm{V}(\boldsymbol{D})$ of the symmetric cycle $\boldsymbol{D}$.

Transformations of critical tope committees under one-element reorientations of the vertex sets of symmetric cycles are discussed in Section 1.3 of Ref. [5].

2.1.4 Vertex Decompositions in a Hypercube Graph with Respect to a Symmetric Cycle: Linear Algebraic Decompositions with Respect to a (Maximal Positive) Basis of $\mathbb{R}^t$

For a symmetric cycle $\boldsymbol{D}$ in a hypercube graph $\boldsymbol{H}(t, 2)$, described in (2.1)(2.3), the square matrix

$$\mathbf{M} := \mathbf{M}(\boldsymbol{D}) := \begin{pmatrix} D^0 \\ D^1 \\ \vdots \\ D^{t-2} \\ D^{t-1} \end{pmatrix} \in \mathbb{R}^{t \times t} \tag{2.5}$$

is nonsingular, since we have

$$0 \neq |\det(\mathbf{M})| = 2^{t-1} ;$$

see Section 11.1 of Ref. [5]. The vertex sequence $(D^0, D^1, \ldots, D^{t-1})$ of the corresponding $(t-1)$-path in the graph $\boldsymbol{H}(t, 2)$ is an ordered *basis* of the space $\mathbb{R}^t$. This means that the vertex sequence $\vec{\mathrm{V}}(\boldsymbol{D})$ of the symmetric cycle $\boldsymbol{D}$ is an ordered *maximal positive basis* of $\mathbb{R}^t$; cf. Theorem 6.4 in Ref. [6].

Given an arbitrary vertex $T \in \{1, -1\}^t$ of the graph $\boldsymbol{H}(t, 2)$, we thus have

$$T = \boldsymbol{x} \cdot \mathbf{M} ,$$

for the row vector

$$\boldsymbol{x} := \boldsymbol{x}(T) := \boldsymbol{x}(T, \boldsymbol{D}) = (x_1, \ldots, x_t) := T \cdot \mathbf{M}^{-1} . \tag{2.6}$$

If $T := {}_{-A}\mathrm{T}^{(+)}$, for some subset $A \subseteq E_t$, then, instead of $\boldsymbol{x}({}_{-A}\mathrm{T}^{(+)}, \boldsymbol{D})$, we often use the notation

$$\boldsymbol{x}(A) := \boldsymbol{x}(A, \boldsymbol{D}) .$$

While the vertex ${}_{-A}\mathrm{T}^{(+)}$ of the graph $\boldsymbol{H}(t, 2)$ can be viewed as the '*characteristic covector*' of the subset $A \subseteq E_t$, the '$\boldsymbol{x}$-*vector*' ${}_{-A}\mathrm{T}^{(+)} \cdot \mathbf{M}^{-1}$ is yet another *linear algebraic* portrait of the set A. Note that

$$\boldsymbol{x} \in \{-1, 0, 1\}^t .$$

For any vertex $T \in \{1, -1\}^t$ of the hypercube graph $\boldsymbol{H}(t, 2)$, there exists a *unique* inclusion-minimal subset $\boldsymbol{Q}(T, \boldsymbol{D}) \subset \mathrm{V}(\boldsymbol{D})$, such that

$$T = \sum_{Q \in \boldsymbol{Q}(T, \boldsymbol{D})} Q . \tag{2.7}$$

Sometimes we let $\mathfrak{q}(T) := \mathfrak{q}(T, \boldsymbol{D})$ denote the cardinality of the set $\boldsymbol{Q}(T, \boldsymbol{D})$:

$$\mathfrak{q}(T) := |\boldsymbol{Q}(T, \boldsymbol{D})| .$$

Moreover, if $T := {}_{-A}\mathrm{T}^{(+)}$, for some subset $A \subseteq E_t$, we often use for the quantity $|\boldsymbol{Q}(T, \boldsymbol{D})|$ the notation

$$\mathfrak{q}(A) := \mathfrak{q}(A, \boldsymbol{D}) .$$

The subset

$$\boldsymbol{Q}(T, \boldsymbol{D}) = \{x_i(T) \cdot D^{i-1} \colon x_i(T) \neq 0\} \subset \mathbb{R}^t , \tag{2.8}$$

of *odd* cardinality, is *linearly independent*. We certainly have $\sum_{Q \in \boldsymbol{Q}(T, \boldsymbol{D})} \langle T, Q \rangle = t$.

In the particular case, where T is the positive vertex $\mathrm{T}^{(+)}$, the set

$$\begin{aligned} \boldsymbol{Q}(\mathrm{T}^{(+)}, \boldsymbol{D}) &= \{x_i(\mathrm{T}^{(+)}) \cdot D^{i-1} \colon x_i(\mathrm{T}^{(+)}) \neq 0\} \\ &= \mathbf{max}^{+}(\mathrm{V}(\boldsymbol{D})) \end{aligned}$$

is a *critical committee* for the oriented matroid $\mathcal{H} := (E_t, \{1, -1\}^t)$.

Example 2.4. Let us return to the symmetric cycle $\boldsymbol{D}$ in the hypercube graph $\boldsymbol{H}((t := 3), 2)$, considered earlier in Example 2.3; see Figure 2.3.

The $t \times t$ matrix $\mathbf{M}(\boldsymbol{D})$ defined by (2.5) is the nonsingular matrix

$$\mathbf{M} := \mathbf{M}(\boldsymbol{D}) = \begin{pmatrix} -1 & 1 & -1 \\ 1 & 1 & -1 \\ 1 & 1 & 1 \end{pmatrix} .$$

The sequence of its rows, that is, the vertex sequence of the $(t-1)$-path (D^0, D^1, D^{t-1}) in the cycle $\boldsymbol{D}$, is an ordered basis of $\mathbb{R}^t$.

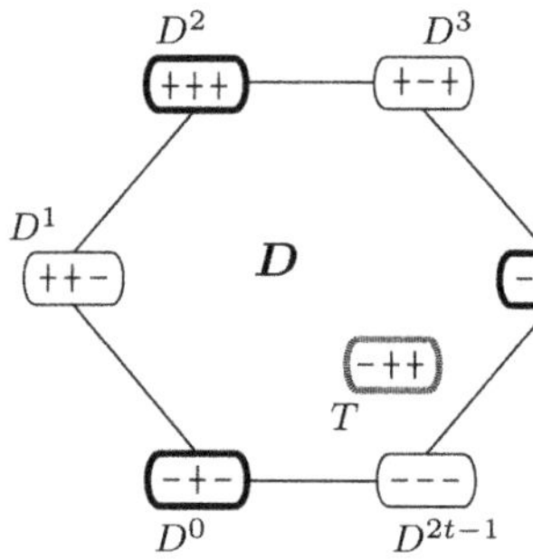

Figure 2.3 A symmetric cycle $\boldsymbol{D}$ in the hypercube graph $\boldsymbol{H}((t := 3), 2)$. The only inclusion-minimal subset of vertices of the cycle $\boldsymbol{D}$, such that their sum is the vertex T of the graph $\boldsymbol{H}(t, 2)$, is the set $\boldsymbol{Q}(T, \boldsymbol{D})$ given in (2.8). Here $\boldsymbol{Q}(T, \boldsymbol{D}) = \{D^0, D^2, D^4\}$.

The inverse matrix is

$$\mathbf{M}^{-1} = \frac{1}{2}\begin{pmatrix} -1 & 1 & 0 \\ 1 & 0 & 1 \\ 0 & -1 & 1 \end{pmatrix} .$$

Pick the vertex $T := (-1, 1, 1)$ of the graph $\boldsymbol{H}(t, 2)$. The corresponding vector $\boldsymbol{x}$ defined by (2.6) is

$$\boldsymbol{x} := T \cdot \mathbf{M}^{-1} = (1, -1, 1) .$$

The subset

$$\boldsymbol{Q}(T, \boldsymbol{D}) = \left\{ \underbrace{x_1 D^0}_{D^0}, \underbrace{x_2 D^1}_{D^4}, \underbrace{x_t D^{t-1}}_{D^2} \right\} = \{D^0, D^2, D^4\} \subset \mathrm{V}(\boldsymbol{D})$$

of the vertex set of the cycle $\boldsymbol{D}$ is a linearly independent subset (a basis, for our particular vertex T under consideration) of the space $\mathbb{R}^t$ such that

$$\begin{aligned} T := (-1, 1, 1) &= \sum_{Q \in \boldsymbol{Q}(T, \boldsymbol{D})} Q = D^0 + D^2 + D^4 \\ &:= (-1, 1, -1) + (1, 1, 1) + (-1, -1, 1) . \end{aligned}$$

2.1.5 Vertex Decompositions in a Hypercube Graph with Respect to a Symmetric Cycle: A Poset-Theoretic Interpretation

Pick a vertex $T \in \{1, -1\}^t$, and regard the vertex set $\mathrm{V}(\boldsymbol{D})$ of a symmetric cycle $\boldsymbol{D}$ in the hypercube graph $\boldsymbol{H}(t, 2)$ as a *subposet* of

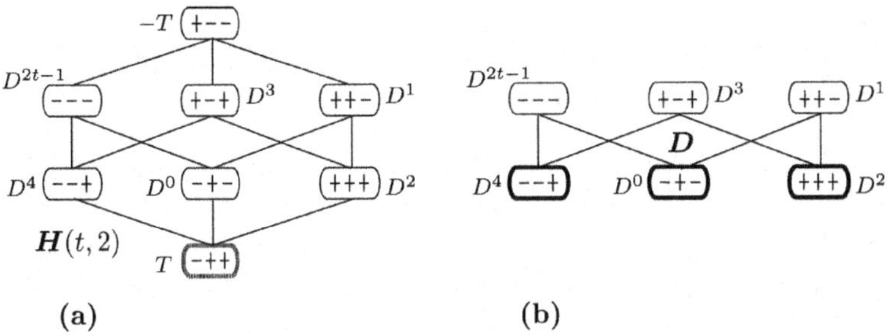

Figure 2.4 **(a)** This hypercube graph $\boldsymbol{H}((t := 3), 2)$ is in fact the Hasse diagram of the tope poset (based at a tope T, and isomorphic to the Boolean lattice of rank t) of the oriented matroid $\mathcal{H} := (E_t, \{1, -1\}^t)$.
(b) A symmetric cycle $\boldsymbol{D} := (D^0, D^1, \ldots, D^{2t-1}, D^0)$ in the graph $\boldsymbol{H}(t, 2)$, whose vertex set $\mathrm{V}(\boldsymbol{D})$ is regarded as a subposet of the tope poset of the oriented matroid $\mathcal{H}$, based at T. The inclusion-minimal subset of vertices in the set $\mathrm{V}(\boldsymbol{D})$, such that their sum is the vertex T, is the set $\boldsymbol{Q}(T, \boldsymbol{D}) = \mathbf{min}\,\mathrm{V}(\boldsymbol{D}) = \{D^0, D^2, D^4\}$ of *minimal* elements of the poset $\mathrm{V}(\boldsymbol{D})$.

the *tope poset* of the oriented matroid $\mathcal{H} := (E_t, \{1, -1\}^t)$, based at the tope T. According to Theorem 11.1 of Ref. [5], the *unique inclusion-minimal* subset $\boldsymbol{Q}(T, \boldsymbol{D}) \subset \mathrm{V}(\boldsymbol{D})$ with the property (2.7) is the set

$$\boldsymbol{Q}(T, \boldsymbol{D}) = \mathbf{min}\,\mathrm{V}(\boldsymbol{D})$$

of *minimal elements* of the subposet $\mathrm{V}(\boldsymbol{D})$.

In the particular case, where the *tope poset* of the oriented matroid $\mathcal{H}$ is based at the positive vertex $\mathrm{T}^{(+)}$, the set

$$\boldsymbol{Q}(\mathrm{T}^{(+)}, \boldsymbol{D}) = \mathbf{min}\,\mathrm{V}(\boldsymbol{D}) = \mathbf{max}^{+}(\mathrm{V}(\boldsymbol{D}))$$

is a *critical committee* for $\mathcal{H}$.

Example 2.5. Suppose $t := 3$. Pick the vertex $T := (-1, 1, 1)$ of the hypercube graph $\boldsymbol{H}(t, 2)$, and consider again the symmetric cycle $\boldsymbol{D} := (D^0, D^1, \ldots, D^{2t-1}, D^0)$ in the graph $\boldsymbol{H}(t, 2)$, depicted in Figure 2.3.

Let us transform the graph $\boldsymbol{H}(t, 2)$ in the way such that it becomes the *Hasse diagram* of the *tope poset* of the oriented matroid $\mathcal{H} := (E_t, \{1, -1\}^t)$, based at the tope T; see Figure 2.4(a).

The unique inclusion-minimal subset $\boldsymbol{Q}(T, \boldsymbol{D}) \subset \mathrm{V}(\boldsymbol{D})$ with the property (2.7), detected in Subsection 2.1.4 by means of elementary linear algebra, is now seen in Figure 2.4(b) as the set $\mathbf{min}\,\mathrm{V}(\boldsymbol{D}) = \{D^0, D^2, D^4\}$ of *minimal* elements of the subposet $\mathrm{V}(\boldsymbol{D})$, where $D^0 := (-1, 1-1)$, $D^2 := (1, 1, 1)$, and $D^4 := (-1, -1, 1)$.

Since the *poset rank* of a vertex in the tope poset is essentially its *Hamming distance* from the bottom vertex T, we see that the set $\boldsymbol{Q}(T, \boldsymbol{D})$ is precisely the subset of vertices in the set $\mathrm{V}(\boldsymbol{D})$ that are *locally nearest* to the vertex T. That is, if $(T^{k_1}, T^{k_2}, T^{k_3})$ is a 2-path in the cycle $\boldsymbol{D}$, where $T^{k_1} \neq T^{k_3}$, then we have

$$T^{k_2} \in \boldsymbol{Q}(T, \boldsymbol{D}) \iff d(T, T^{k_1}) = d(T, T^{k_3}) = d(T, T^{k_2}) + 1\,.$$

2.1.6 Vertex Decompositions in a Hypercube Graph with Respect to a Symmetric Cycle: Decompositions by Means of the Tuples With Inclusion-Maximal Positive Parts in the Set of Reoriented Vertices of the Cycle

If $T \in \{1, -1\}^t$ is a vertex of the hypercube graph $\boldsymbol{H}(t, 2)$, then the inclusion-minimal subset $\boldsymbol{Q}(T, \boldsymbol{D}) \subset \mathrm{V}(\boldsymbol{D})$, with the property (2.7), is the set

$$\boldsymbol{Q}(T, \boldsymbol{D}) = {}_{-\mathfrak{n}(T)}\left(\mathbf{max}^{+}\left({}_{-\mathfrak{n}(T)}\mathrm{V}(\boldsymbol{D})\right)\right)\;; \tag{2.9}$$

see Corollary 11.2 in Ref. [5]. Here ${}_{-\mathfrak{n}(T)}\mathrm{V}(\boldsymbol{D})$ is the sequence of tuples obtained from the vertex sequence $\mathrm{V}(\boldsymbol{D})$ of the cycle $\boldsymbol{D}$ by *reorientation* on the *negative part* $\mathfrak{n}(T)$ of the vertex T; $\mathbf{max}^{+}(\cdot)$ is the subset of all tuples from the resulting sequence that have *inclusion-maximal positive parts*, and the outermost operation ${}_{-\mathfrak{n}(T)}(\cdot)$ means the reverse *reorientation* of the tuples with inclusion-maximal positive parts on the negative part $\mathfrak{n}(T)$ of the vertex T.

Example 2.6. Suppose $t := 3$. We would like to find the decomposition of the vertex $T := (-1, 1, 1)$ of the hypercube graph $\boldsymbol{H}(t, 2)$ with respect to its symmetric cycle $\boldsymbol{D} := (D^0, D^1, \ldots, D^{2t-1}, D^0)$ depicted in Figure 2.5(a).

First, we *reorient* the vertex T, as well as the vertex set $\mathrm{V}(\boldsymbol{D})$ of the cycle $\boldsymbol{D}$, on the *negative part* $\mathfrak{n}(T) = \{1\}$ of the vertex T;

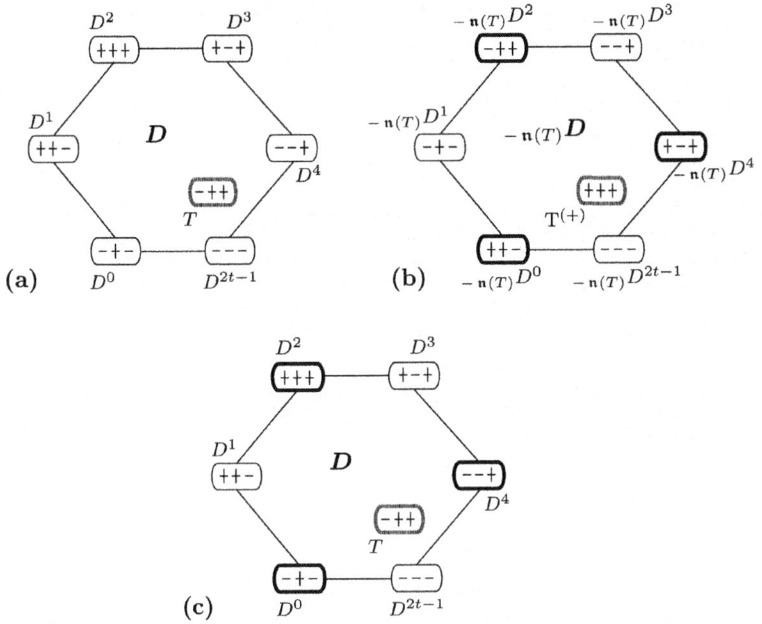

Figure 2.5 (a) A symmetric cycle $\boldsymbol{D} := (D^0, D^1, \ldots, D^{2t-1}, D^0)$ in the hypercube graph $\boldsymbol{H}((t := 3), 2)$, and a vertex T of the graph $\boldsymbol{H}(t, 2)$.
(b) The *reorientation* of the vertex T on its *negative part* $\mathfrak{n}(T)$ turns T into the *positive* sign tuple $\mathrm{T}^{(+)}$. The set $\mathbf{max}^+({}_{-\mathfrak{n}(T)}\mathrm{V}(\boldsymbol{D}))$ of tuples with *inclusion-maximal positive parts* in the reoriented symmetric cycle is the set $\{{}_{-\mathfrak{n}(T)}D^0, {}_{-\mathfrak{n}(T)}D^2, {}_{-\mathfrak{n}(T)}D^4\}$.
(c) The inclusion-minimal subset of vertices $\boldsymbol{Q}(T, \boldsymbol{D}) \subset \mathrm{V}(\boldsymbol{D})$, such that their sum is the vertex T, is the *reoriented* set ${}_{-\mathfrak{n}(T)}\{{}_{-\mathfrak{n}(T)}D^0, {}_{-\mathfrak{n}(T)}D^2, {}_{-\mathfrak{n}(T)}D^4\}$, that is, $\boldsymbol{Q}(T, \boldsymbol{D}) = \{D^0, D^2, D^4\}$.

see Figure 2.5(b). As a result, T turns into the *positive* tuple $\mathrm{T}^{(+)}$. The set $\mathbf{max}^+(\{{}_{-\mathfrak{n}(T)}D\colon D \in \mathrm{V}(\boldsymbol{D})\}) = \{{}_{-\mathfrak{n}(T)}D^0, {}_{-\mathfrak{n}(T)}D^2, {}_{-\mathfrak{n}(T)}D^4\}$, where $D^0 := (-1, 1 - 1)$, $D^2 := (1, 1, 1)$, and $D^4 := (-1, -1, 1)$, is the unique inclusion-minimal subset of tuples in the cycle ${}_{-\mathfrak{n}(T)}\boldsymbol{D}$ such that their sum is $\mathrm{T}^{(+)}$, as explained in Subsection 2.1.6.

We see in Figure 2.5(c) that the subset $\boldsymbol{Q}(T, \boldsymbol{D}) \subset \mathrm{V}(\boldsymbol{D})$, built up in (2.9), is the set ${}_{-\mathfrak{n}(T)}\{{}_{-\mathfrak{n}(T)}D^0, {}_{-\mathfrak{n}(T)}D^2, {}_{-\mathfrak{n}(T)}D^4\}$, that is, $\boldsymbol{Q}(T, \boldsymbol{D}) = \{D^0, D^2, D^4\}$.

2.1.7 The Sizes of Vertex Decompositions: Distance Signals

Let $\boldsymbol{D} := (D^0, D^1, \ldots, D^{2t-1}, D^0)$ be a symmetric cycle in the hypercube graph $\boldsymbol{H}(t, 2)$, and let $T \in \{1, -1\}^t$ be a vertex of the graph $\boldsymbol{H}(t, 2)$. According to Definition 12.2 of Ref. [5], the (row) *distance vector* of the cycle $\boldsymbol{D}$ with respect to the vertex T is the sequence

$$\boldsymbol{z}_{T,\boldsymbol{D}} := (z_{T,\boldsymbol{D}}(0), \ldots, z_{T,\boldsymbol{D}}(2t-1)) := \big(d(T, D) \colon D \in \vec{\mathrm{V}}(\boldsymbol{D})\big)$$

of the graph distances between T and the vertices of $\boldsymbol{D}$.

We denote by $\mathbf{W}$ the $2t \times 2t$ *Fourier matrix* whose rows and columns are indexed starting with zero. Recall that the (m, n)th entry of the matrix $\mathbf{W}$ by definition is $\mathrm{e}^{-\pi \imath mn/t}$, $0 \leq m, n \leq 2t-1$. The *Discrete Fourier Transform* $\hat{\boldsymbol{z}_{T}}_{,\boldsymbol{D}}$ of the distance vector $\boldsymbol{z}_{T,\boldsymbol{D}}$ is defined to be the vector

$$\hat{\boldsymbol{z}_{T}}_{,\boldsymbol{D}} := \boldsymbol{z}_{T,\boldsymbol{D}} \cdot \mathbf{W} \,.$$

Consider the unique inclusion-minimal subset $\boldsymbol{Q}(T, \boldsymbol{D})$ of the vertex set $\mathrm{V}(\boldsymbol{D})$ of the cycle $\boldsymbol{D}$, such that $T = \sum_{Q \in \boldsymbol{Q}(T,\boldsymbol{D})} Q$. It is shown in Chapter 12 of Ref. [5] that

$$\mathfrak{q}(T, \boldsymbol{D}) := |\boldsymbol{Q}(T, \boldsymbol{D})| = t - \frac{1}{4t} \sum_{k=0}^{2t-1} |\hat{z_{T}}_{,\boldsymbol{D}}(k)|^2 \cdot \sin^2 \frac{\pi k}{t} \,,$$

that is,

$$\mathfrak{q}(T, \boldsymbol{D}) = t - \frac{1}{2t} \sum_{\substack{1 \leq k \leq t-1, \\ k \text{ odd}}} |\hat{z_{T}}_{,\boldsymbol{D}}(k)|^2 \cdot \sin^2 \frac{\pi k}{t} \,,$$

since

$$\hat{z_{T}}_{,\boldsymbol{D}}(k) = \begin{cases} t^2 \,, & \text{if } k = 0 \,, \\ 0 \,, & \text{if } k \text{ is even, } k \neq 0 \,, \\ 2\big(-t \cdot (1 - \mathrm{e}^{-\pi \imath k/t})^{-1} \\ \qquad + \sum_{j=0}^{t-1} z_{T,\boldsymbol{D}}(j) \mathrm{e}^{-\pi \imath kj/t}\big) \,, & \text{if } k \text{ is odd} \,; \end{cases}$$

see Proposition 12.3 in Ref. [5].

Now let $\boldsymbol{a}_{T,\boldsymbol{D}} := (a_{T,\boldsymbol{D}}(0), \ldots, a_{T,\boldsymbol{D}}(2t-1))$ denote the *autocorrelation* of the *distance vector* $\boldsymbol{z}_{T,\boldsymbol{D}}$, defined by

$$a_{T,\boldsymbol{D}}(m) := \sum_{n=0}^{2t-1} z_{T,\boldsymbol{D}}(n) \cdot z_{T,\boldsymbol{D}}((n+m) \mod 2t) \,, \quad 0 \leq m \leq 2t-1 \,.$$

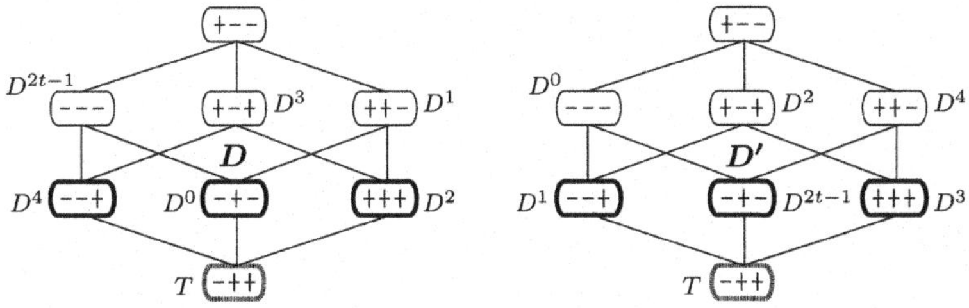

(a) **(b)**

Figure 2.6 Two symmetric cycles $\boldsymbol{D} := (D^0, D^1, \ldots, D^{2t-1}, D^0)$ and $\boldsymbol{D}' := (D^0, D^1, \ldots, D^{2t-1}, D^0)$ in the hypercube graph $\boldsymbol{H}((t := 3), 2)$. The *vertex sets* of the cycles are the same, that is, $\mathrm{V}(\boldsymbol{D}) = \mathrm{V}(\boldsymbol{D}')$, but the names of identical vertices in the cycles do not coincide. In other words, the *vertex sequences* of the cycles are different: $\vec{\mathrm{V}}(\boldsymbol{D}) \neq \vec{\mathrm{V}}(\boldsymbol{D}')$. For example, we have $\vec{\mathrm{V}}(\boldsymbol{D}) \ni D^4 := (-1, -1, 1) =: D^1 \in \vec{\mathrm{V}}(\boldsymbol{D}')$.

By Eq. (13.5) of Ref. [5] we have

$$\mathfrak{q}(T, \boldsymbol{D}) = t - \frac{1}{4}\big(a_{T,\boldsymbol{D}}(0) - a_{T,\boldsymbol{D}}(2)\big) .$$

Example 2.7. (i) Suppose $t := 3$. Consider symmetric cycles $\boldsymbol{D} := (D^0, D^1, \ldots, D^{2t-1}, D^0)$ and $\boldsymbol{D}' := (D^0, D^1, \ldots, D^{2t-1}, D^0)$ in the hypercube graph $\boldsymbol{H}(t, 2)$, depicted in Figures 2.6(a) and (b), respectively. The *vertex sets* $\mathrm{V}(\boldsymbol{D})$ and $\mathrm{V}(\boldsymbol{D}')$ of the cycles are the same, that is, $\mathrm{V}(\boldsymbol{D}) = \mathrm{V}(\boldsymbol{D}')$, but the names of identical vertices in the cycles do not coincide. Thus, the *vertex sequences* $\vec{\mathrm{V}}(\boldsymbol{D})$ and $\vec{\mathrm{V}}(\boldsymbol{D}')$ of the cycles differ. For example, we have $\vec{\mathrm{V}}(\boldsymbol{D}) \ni D^4 := (-1, -1, 1) =: D^1 \in \vec{\mathrm{V}}(\boldsymbol{D}')$.

According to Definition 12.2 of Ref. [5], the distance vectors $\boldsymbol{z}_{T,\boldsymbol{D}}$ and $\boldsymbol{z}_{T,\boldsymbol{D}'}$ of the cycles $\boldsymbol{D}$ and $\boldsymbol{D}'$ with respect to the vertex $T := (-1, 1, 1)$ are

$$\boldsymbol{z}_{T,\boldsymbol{D}} = \left(\underset{\substack{\uparrow \\ d(T,D^0)}}{1}, \overset{\substack{d(T,D^1) \\ \downarrow}}{2}, \underset{\substack{\uparrow \\ d(T,D^2)}}{1}, \overset{\substack{d(T,D^3) \\ \downarrow}}{2}, \underset{\substack{\uparrow \\ d(T,D^4)}}{1}, \overset{\substack{d(T,D^{2t-1}) \\ \downarrow}}{2} \right),$$

$$\boldsymbol{z}_{T,\boldsymbol{D}'} = \left(\underset{\substack{\uparrow \\ d(T,D^0)}}{2}, \overset{\substack{d(T,D^1) \\ \downarrow}}{1}, \underset{\substack{\uparrow \\ d(T,D^2)}}{2}, \overset{\substack{d(T,D^3) \\ \downarrow}}{1}, \underset{\substack{\uparrow \\ d(T,D^4)}}{2}, \overset{\substack{d(T,D^{2t-1}) \\ \downarrow}}{1} \right).$$

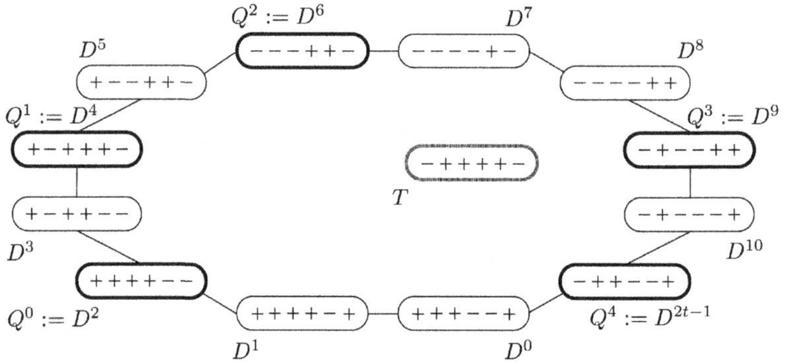

Figure 2.7 A symmetric cycle $\boldsymbol{D} := (D^0, D^1, \ldots, D^{2t-1}, D^0)$ in the hypercube graph $\boldsymbol{H}((t := 6), 2)$, and a vertex $T \in \{1, -1\}^t$ of the graph. The set $\boldsymbol{Q}(T, \boldsymbol{D}) := \{Q^0, Q^1, Q^2, Q^3, Q^4\}$ is the unique inclusion-minimal subset of the vertex set $\mathrm{V}(\boldsymbol{D})$ of the cycle $\boldsymbol{D}$, such that $T = \sum_{Q \in \boldsymbol{Q}(T,\boldsymbol{D})} Q$.

We have

$$a_{T,\boldsymbol{D}}(0) := \sum_{n=0}^{2t-1} \underbrace{z_{T,\boldsymbol{D}}(n)}_{d(T,D^n)}{}^2 = 1^2 + 2^2 + 1^2 + 2^2 + 1^2 + 2^2 = 15\,,$$

$$a_{T,\boldsymbol{D}}(2) := \sum_{n=0}^{2t-1} \underbrace{z_{T,\boldsymbol{D}}(n)}_{d(T,D^n)} \cdot \underbrace{z_{T,\boldsymbol{D}}((n+2) \mod 2t)}_{d(T,D^{(n+2) \mod 2t})}$$
$$= 1 \cdot 1 + 2 \cdot 2 + 1 \cdot 1 + 2 \cdot 2 + 1 \cdot 1 + 2 \cdot 2 = 15\,.$$

We obtain

$$\mathfrak{q}(T, \boldsymbol{D}) = t - \frac{1}{4}\big(a_{T,\boldsymbol{D}}(0) - a_{T,\boldsymbol{D}}(2)\big) = 3 - \frac{1}{4}(15 - 15) = 3\,,$$

and we certainly have $\mathfrak{q}(T, \boldsymbol{D}') = \mathfrak{q}(T, \boldsymbol{D}) = 3$.

(ii) Suppose $t := 6$. Given a symmetric cycle $\boldsymbol{D} := (D^0, D^1, \ldots, D^{2t-1}, D^0)$ in the hypercube graph $\boldsymbol{H}(t, 2)$, depicted in Figure 2.7, the distance vector of the cycle $\boldsymbol{D}$ with respect to the vertex $T := (-1, 1, 1, 1, 1, -1)$ is

$$\boldsymbol{z}_{T,\boldsymbol{D}} = \left(\begin{array}{cccccccccccc} & \overset{1}{\downarrow} & & \overset{3}{\downarrow} & & \overset{5}{\downarrow} & & \overset{7}{\downarrow} & & \overset{9}{\downarrow} & & \overset{2t-1}{\downarrow} \\ 4, & 3, & 2, & 3, & 2, & 3, & 2, & 3, & 4, & 3, & 4, & 3 \\ \underset{0}{\uparrow} & & \underset{2}{\uparrow} & & \underset{4}{\uparrow} & & \underset{6}{\uparrow} & & \underset{8}{\uparrow} & & \underset{10}{\uparrow} & \end{array} \right).$$

We have

$$a_{T,\mathbf{D}}(0) := \sum_{n=0}^{2t-1} \underbrace{z_{T,\mathbf{D}}(n)}_{d(T,D^n)}{}^2$$
$$= 4^2 + 3^2 + 2^2 + 3^2 + 2^2 + 3^2 + 2^2 + 3^2 + 4^2 + 3^2$$
$$+ 4^2 + 3^2 = 114\,,$$
$$a_{T,\mathbf{D}}(2) := \sum_{n=0}^{2t-1} \underbrace{z_{T,\mathbf{D}}(n)}_{d(T,D^n)} \cdot \underbrace{z_{T,\mathbf{D}}((n+2) \mod 2t)}_{d(T,D^{(n+2) \mod 2t})}$$
$$= 4\cdot 2 + 3\cdot 3 + 2\cdot 2 + 3\cdot 3 + 2\cdot 2 + 3\cdot 3$$
$$+ 2\cdot 4 + 3\cdot 3 + 4\cdot 4 + 3\cdot 3 + 4\cdot 4 + 3\cdot 3 = 110\,.$$

Thus, we have

$$\mathfrak{q}(T, \mathbf{D}) = t - \frac{1}{4}\big(a_{T,\mathbf{D}}(0) - a_{T,\mathbf{D}}(2)\big) = 6 - \frac{1}{4}(114 - 110) = 5\,.$$

Indeed, this is the cardinality of the vertex subset $\boldsymbol{Q}(T, \mathbf{D}) = \{Q^0, Q^1, \ldots, Q^4\} \subset \mathrm{V}(\mathbf{D})$; see Figure 2.7.

2.1.8 Basic Metric Properties of Vertex Decompositions in a Hypercube Graph with Respect to a Symmetric Cycle

In view of Remark 13.2 of Ref. [5], for any vertex $T \in \{1, -1\}^t$ of the hypercube graph $\boldsymbol{H}(t, 2)$, and for any symmetric cycle $\mathbf{D}$ in the graph $\boldsymbol{H}(t, 2)$, we have

$$\sum_{Q \in \boldsymbol{Q}(T,\mathbf{D})} d(T, Q) = \frac{1}{2}(\mathfrak{q}(T) - 1)t\,, \tag{2.10}$$

and

$$\mathfrak{q}(T) = 1 + \frac{2}{t} \sum_{Q \in \boldsymbol{Q}(T,\mathbf{D})} d(T, Q)\,. \tag{2.11}$$

As noted in Remark 13.3 of Ref. [5], if $T \notin \mathrm{V}(\mathbf{D})$, then we have

$$\sum_{0 \le i < j \le \mathfrak{q}(T)-1} d(Q^i, Q^j) = \frac{1}{4}\big(\mathfrak{q}(T)^2 - 1\big)t\,, \tag{2.12}$$

$$\mathfrak{q}(T) = \sqrt{1 + \frac{4}{t}\sum\nolimits_{i<j} d(Q^i, Q^j)}\,, \tag{2.13}$$

and

$$\sum_{0\le i<j\le \daleth(T)-1} d(Q^i, Q^j) = \frac{1}{2}\big(\daleth(T)+1\big) \sum_{Q\in\boldsymbol{Q}(T,\boldsymbol{D})} d(T, Q)\,. \tag{2.14}$$

Example 2.8. For a symmetric cycle $\boldsymbol{D}$ of the hypercube graph $\boldsymbol{H}((t := 6), 2)$, and for the vertex $T := (-1, 1, 1, 1, 1, -1) \in \{1, -1\}^t$ of the graph $\boldsymbol{H}(t, 2)$, depicted in Figure 2.7, we have

$$\begin{aligned}
\sum_{Q\in\boldsymbol{Q}(T,\boldsymbol{D})} d(T, Q) &= \sum_{Q\in\{Q^0, Q^1,\ldots, Q^4\}} d(T, Q)\\
&= \underbrace{d(T, D^2)}_{d(T,Q^0)} + \underbrace{d(T, D^4)}_{d(T,Q^1)} + \underbrace{d(T, D^6)}_{d(T,Q^2)}\\
&\quad + \underbrace{d(T, D^9)}_{d(T,Q^3)} + \underbrace{d(T, D^{2t-1})}_{d(T,Q^4)}\\
&= 2+2+2+3+3 = 12\\
&= \frac{1}{2}(\daleth(T)-1)t = \frac{1}{2}(5-1)6\,,
\end{aligned}$$

and

$$\daleth(T) = 1 + \frac{2}{t}\sum_{Q\in\boldsymbol{Q}(T,\boldsymbol{D})} d(T, Q) = 1 + \frac{2}{6}\cdot 12 = 5\,.$$

Further,

$$\begin{aligned}
\sum_{0\le i<j\le \daleth(T)-1} d(Q^i, Q^j) &= \underbrace{d(Q^0, Q^1)}_{2} + \underbrace{d(Q^0, Q^2)}_{4} + \underbrace{d(Q^0, Q^3)}_{5}\\
&\quad + \underbrace{d(Q^0, Q^4)}_{3} + \underbrace{d(Q^1, Q^2)}_{2} + \underbrace{d(Q^1, Q^3)}_{5}\\
&\quad + \underbrace{d(Q^1, Q^4)}_{5} + \underbrace{d(Q^2, Q^3)}_{3}\\
&\quad + \underbrace{d(Q^2, Q^4)}_{5} + \underbrace{d(Q^3, Q^4)}_{2} = 36\\
&= \frac{1}{4}\big(\daleth(T)^2-1\big)t = \frac{1}{4}\big(5^2-1\big)6\,,
\end{aligned}$$

and

$$\daleth(T) = \sqrt{1 + \frac{4}{t}\sum\nolimits_{i<j} d(Q^i, Q^j)} = \sqrt{1 + \frac{4}{6}\cdot 36} = 5\,,$$

and

$$\sum_{0 \le i < j \le \mathfrak{q}(T)-1} d(Q^i, Q^j) = 36$$
$$= \frac{1}{2}(\mathfrak{q}(T)+1) \sum_{Q \in \boldsymbol{Q}(T,\boldsymbol{D})} d(T, Q) = \frac{1}{2}(5+1)12 .$$

2.1.9 The Negative Parts of Vertices, Graph Distances, and the Scalar Products of Vertices

Let $\boldsymbol{D}$ be a symmetric cycle in a hypercube graph $\boldsymbol{H}(t, 2)$. Looking at relations (2.10) and (2.11) for vertices $T \in \{1, -1\}^t$ of the graph $\boldsymbol{H}(t, 2)$, note that the following implications hold:

$$e \in E_t , \quad T(e) = -1$$
$$\Longrightarrow \quad |\{Q \in \boldsymbol{Q}(T, \boldsymbol{D}) \colon Q(e) = -1\}| = \left\lceil \frac{\mathfrak{q}(T)}{2} \right\rceil .$$

We see that

$$\sum_{Q \in \boldsymbol{Q}(T,\boldsymbol{D})} |\mathfrak{n}(Q)| = \sum_{e \in \mathfrak{n}(T)} \left\lceil \frac{\mathfrak{q}(T)}{2} \right\rceil + \sum_{e \in \mathfrak{p}(T)} \left\lfloor \frac{\mathfrak{q}(T)}{2} \right\rfloor$$
$$= |\mathfrak{n}(T)| \cdot \left\lceil \frac{\mathfrak{q}(T)}{2} \right\rceil + (t - |\mathfrak{n}(T)|) \cdot \left\lfloor \frac{\mathfrak{q}(T)}{2} \right\rfloor .$$

Remark 2.9. For any vertex $T \in \{1, -1\}^t$ of the hypercube graph $\boldsymbol{H}(t, 2)$, and for any symmetric cycle $\boldsymbol{D}$ in the graph $\boldsymbol{H}(t, 2)$, we have

(i)

$$\sum_{Q \in \boldsymbol{Q}(T,\boldsymbol{D})} |\mathfrak{n}(Q)| = |\mathfrak{n}(T)| + \frac{1}{2}(\mathfrak{q}(T) - 1)t ,$$

and

$$\mathfrak{q}(T) = 1 - \frac{2}{t}|\mathfrak{n}(T)| + \frac{2}{t} \sum_{Q \in \boldsymbol{Q}(T,\boldsymbol{D})} |\mathfrak{n}(Q)| . \tag{2.15}$$

(ii)

$$\sum_{Q \in \boldsymbol{Q}(T,\boldsymbol{D})} d(T, Q) = -|\mathfrak{n}(T)| + \sum_{Q \in \boldsymbol{Q}(T,\boldsymbol{D})} |\mathfrak{n}(Q)| . \tag{2.16}$$

Example 2.10. Suppose $t := 6$. For a symmetric cycle $\boldsymbol{D}$ of the hypercube graph $\boldsymbol{H}(t, 2)$, and for the vertex $T := (-1, 1, 1, 1, 1, -1) \in \{1, -1\}^t$ of the graph $\boldsymbol{H}(t, 2)$, depicted in Figure 2.7, we have

$$\begin{aligned}\sum_{Q \in \boldsymbol{Q}(T,\boldsymbol{D})} |\mathfrak{n}(Q)| &= \sum_{Q \in \{Q^0, Q^1, \ldots, Q^4\}} |\mathfrak{n}(Q)| \\ &= \underbrace{|\mathfrak{n}(D^2)|}_{2} + \underbrace{|\mathfrak{n}(D^4)|}_{2} + \underbrace{|\mathfrak{n}(D^6)|}_{4} + \underbrace{|\mathfrak{n}(D^9)|}_{3} \\ &\quad + \underbrace{|\mathfrak{n}(D^{2t-1})|}_{3} \\ &= 14 = |\mathfrak{n}(T)| + \frac{1}{2}(\mathfrak{q}(T) - 1)t \\ &= 2 + \frac{1}{2}(5 - 1)6 \,,\end{aligned}$$

and

$$\mathfrak{q}(T) = 5 = 1 - \frac{2}{t}|\mathfrak{n}(T)| + \frac{2}{t} \sum_{Q \in \boldsymbol{Q}(T,\boldsymbol{D})} |\mathfrak{n}(Q)| = 1 - \frac{2}{6} \cdot 2 + \frac{2}{6} \cdot 14 \,.$$

Further, we have

$$\sum_{Q \in \boldsymbol{Q}(T,\boldsymbol{D})} d(T, Q) = 12 = -|\mathfrak{n}(T)| + \sum_{Q \in \boldsymbol{Q}(T,\boldsymbol{D})} |\mathfrak{n}(Q)| = -2 + 14 \,.$$

Let us now turn to relations (2.12)–(2.14). Let $T \in \{1, -1\}^t$ be a vertex of the hypercube graph $\boldsymbol{H}(t, 2)$ with its symmetric cycle $\boldsymbol{D}$, such that $T \notin \mathrm{V}(\boldsymbol{D})$. Relations (2.16) and (2.14) yield

$$\begin{aligned}\sum_{0 \le i < j \le \mathfrak{q}(T)-1} d(Q^i, Q^j) &= \frac{1}{2}(\mathfrak{q}(T) + 1) \sum_{Q \in \boldsymbol{Q}(T,\boldsymbol{D})} d(T, Q) \\ &= \frac{1}{2}(\mathfrak{q}(T) + 1)\Big(-|\mathfrak{n}(T)| + \sum_{Q \in \boldsymbol{Q}(T,\boldsymbol{D})} |\mathfrak{n}(Q)|\Big) \,.\end{aligned}$$

Using (2.15), we obtain

$$\begin{aligned}\sum_{0 \le i < j \le \mathfrak{q}(T)-1} d(Q^i, Q^j) &= \frac{1}{2}(\mathfrak{q}(T) + 1) \\ &\quad \times \Big(-|\mathfrak{n}(T)| + \sum_{Q \in \boldsymbol{Q}(T,\boldsymbol{D})} |\mathfrak{n}(Q)|\Big) \\ &= \frac{1}{2}\Big(1 - \frac{2}{t}|\mathfrak{n}(T)| + \frac{2}{t} \sum_{Q \in \boldsymbol{Q}(T,\boldsymbol{D})} |\mathfrak{n}(Q)| + 1\Big) \\ &\quad \times \Big(-|\mathfrak{n}(T)| + \sum_{Q \in \boldsymbol{Q}(T,\boldsymbol{D})} |\mathfrak{n}(Q)|\Big) \,.\end{aligned}$$

Remark 2.11. Let $\boldsymbol{D}$ be a symmetric cycle in a hypercube graph $\boldsymbol{H}(t,2)$. Let $T \in \{1,-1\}^t$ be a vertex of the graph $\boldsymbol{H}(t,2)$, such that $T \not\in \mathrm{V}(\boldsymbol{D})$. We have

$$\sum_{0\leq i<j\leq \mathfrak{q}(T)-1} d(Q^i,Q^j) = \Big(1-\frac{1}{t}|\mathbf{n}(T)| + \frac{1}{t}\sum_{Q\in\boldsymbol{Q}(T,\boldsymbol{D})}|\mathbf{n}(Q)|\Big) \times \Big(-|\mathbf{n}(T)| + \sum_{Q\in\boldsymbol{Q}(T,\boldsymbol{D})}|\mathbf{n}(Q)|\Big).$$

Example 2.12. Suppose $t := 6$. For the vertex $T := (-1,1,1,1,1,-1) \in \{1,-1\}^t$ of the hypercube graph $\boldsymbol{H}(t,2)$ with its symmetric cycle $\boldsymbol{D}$, depicted in Figure 2.7, we have

$$\begin{aligned}\sum_{0\leq i<j\leq \mathfrak{q}(T)-1} d(Q^i,Q^j) &= 36\\ &= \Big(1-\frac{1}{t}|\mathbf{n}(T)| + \frac{1}{t}\sum_{Q\in\boldsymbol{Q}(T,\boldsymbol{D})}|\mathbf{n}(Q)|\Big)\\ &\quad\times\Big(-|\mathbf{n}(T)| + \sum_{Q\in\boldsymbol{Q}(T,\boldsymbol{D})}|\mathbf{n}(Q)|\Big)\\ &= \Big(1-\frac{1}{6}\cdot 2+\frac{1}{6}\cdot 14\Big)\cdot\Big(-2+14\Big).\end{aligned}$$

For convenience, we now reformulate relations (2.12)–(2.14) and Remark 2.11 via the scalar products of vertices:

Remark 2.13. Let $\boldsymbol{D}$ be a symmetric cycle in a hypercube graph $\boldsymbol{H}(t,2)$. Let $T \in \{1,-1\}^t$ be a vertex of the graph $\boldsymbol{H}(t,2)$, such that $T \not\in \mathrm{V}(\boldsymbol{D})$. We have:

(i)

$$\sum_{0\leq i<j\leq \mathfrak{q}(T)-1} \langle Q^i,Q^j\rangle = \frac{1}{2}(1-\mathfrak{q}(T))t,$$

$$\mathfrak{q}(T) = 1-\frac{2}{t}\sum_{0\leq i<j\leq \mathfrak{q}(T)-1}\langle Q^i,Q^j\rangle.$$

(ii)

$$\sum_{0\le i<j\le \mathfrak{q}(T)-1} \langle Q^i, Q^j\rangle = \binom{\mathfrak{q}(T)}{2}t - 2\Big(1 - \frac{1}{t}|\mathfrak{n}(T)| + \frac{1}{t}\sum_{Q\in \boldsymbol{Q}(T,\boldsymbol{D})} |\mathfrak{n}(Q)|\Big) \times\Big(-|\mathfrak{n}(T)| + \sum_{Q\in \boldsymbol{Q}(T,\boldsymbol{D})} |\mathfrak{n}(Q)|\Big)\,.$$

Example 2.14. Suppose $t := 6$. For the vertex $T := (-1, 1, 1, 1, 1, -1) \in \{1, -1\}^t$ of the hypercube graph $\boldsymbol{H}(t, 2)$ with its symmetric cycle $\boldsymbol{D}$, depicted in Figure 2.7, we have

$$\sum_{0\le i<j\le \mathfrak{q}(T)-1} \langle Q^i, Q^j\rangle = \underbrace{\langle Q^0, Q^1\rangle}_{2} + \underbrace{\langle Q^0, Q^2\rangle}_{-2} + \underbrace{\langle Q^0, Q^3\rangle}_{-4} + \underbrace{\langle Q^0, Q^4\rangle}_{0} + \underbrace{\langle Q^1, Q^2\rangle}_{2} + \underbrace{\langle Q^1, Q^3\rangle}_{-4} + \underbrace{\langle Q^1, Q^4\rangle}_{-4} + \underbrace{\langle Q^2, Q^3\rangle}_{0} + \underbrace{\langle Q^2, Q^4\rangle}_{-4} + \underbrace{\langle Q^3, Q^4\rangle}_{2} = -12$$

$$= \frac{1}{2}(1 - \mathfrak{q}(T))t = \frac{1}{2}(1-5)6\,,$$

and

$$\mathfrak{q}(T) = 5 = 1 - \frac{2}{t}\sum_{0\le i<j\le \mathfrak{q}(T)-1} \langle Q^i, Q^j\rangle = 1 - \frac{2}{6}(-12)\,.$$

We also have

$$\sum_{0\le i<j\le \mathfrak{q}(T)-1} \langle Q^i, Q^j\rangle = -12$$

$$= \binom{\mathfrak{q}(T)}{2}t - 2\Big(1 - \frac{1}{t}|\mathfrak{n}(T)| + \frac{1}{t}\sum_{Q\in \boldsymbol{Q}(T,\boldsymbol{D})} |\mathfrak{n}(Q)|\Big) \times\Big(-|\mathfrak{n}(T)| + \sum_{Q\in \boldsymbol{Q}(T,\boldsymbol{D})} |\mathfrak{n}(Q)|\Big)$$

$$= \binom{5}{2}6 - 2(1 - \frac{1}{6}\cdot 2 + \frac{1}{6}\cdot 14)\cdot(-2+14)\,.$$

Now, if T' and T'' are two vertices of a hypercube graph $\boldsymbol{H}(t, 2)$, then we have

$$\begin{aligned} t - 2d(T', T'') &= \langle T', T''\rangle = \Big\langle \sum_{Q' \in \boldsymbol{Q}(T',\boldsymbol{D})}, \sum_{Q'' \in \boldsymbol{Q}(T'',\boldsymbol{D})} \Big\rangle \\ &= \sum_{\substack{Q' \in \boldsymbol{Q}(T',\boldsymbol{D}),\\ Q'' \in \boldsymbol{Q}(T'',\boldsymbol{D})}} \langle Q', Q''\rangle = \sum_{\substack{Q' \in \boldsymbol{Q}(T',\boldsymbol{D}),\\ Q'' \in \boldsymbol{Q}(T'',\boldsymbol{D})}} (t - 2d(Q', Q'')) \\ &= \mathfrak{q}(T')\mathfrak{q}(T'')t - 2 \sum_{\substack{Q' \in \boldsymbol{Q}(T',\boldsymbol{D}),\\ Q'' \in \boldsymbol{Q}(T'',\boldsymbol{D})}} d(Q', Q'') \, . \end{aligned}$$

Remark 2.15. If $T', T'' \in \{1, -1\}^t$ are vertices of the hypercube graph $\boldsymbol{H}(t, 2)$ with its symmetric cycle $\boldsymbol{D}$, then we have

$$-d(T', T'') + \sum_{\substack{Q' \in \boldsymbol{Q}(T',\boldsymbol{D}),\\ Q'' \in \boldsymbol{Q}(T'',\boldsymbol{D})}} d(Q', Q'') = \frac{1}{2}\big(\mathfrak{q}(T')\mathfrak{q}(T'') - 1\big)t \, .$$

Example 2.16. Suppose $t := 6$. Pick the vertices $T' := (-1, 1, 1, 1, 1, -1) \in \{1, -1\}^t$ and $T'' := (1, -1, -1, 1, 1, 1)$ of the hypercube graph $\boldsymbol{H}(t, 2)$ with its symmetric cycle $\boldsymbol{D} := (D^0, D^1, \ldots, D^{2t-1}, D^0)$, depicted in Figure 2.8. We have

$$\begin{aligned} &-d(T', T'') + \sum_{\substack{Q' \in \boldsymbol{Q}(T',\boldsymbol{D}),\\ Q'' \in \boldsymbol{Q}(T'',\boldsymbol{D})}} d(Q', Q'') \\ &= -4 + \underbrace{d(Q'^0, Q''^0)}_{1} + \underbrace{d(Q'^0, Q''^1)}_{3} + \underbrace{d(Q'^0, Q''^2)}_{6} \\ &+ \underbrace{d(Q'^1, Q''^0)}_{3} + \underbrace{d(Q'^1, Q''^1)}_{1} + \underbrace{d(Q'^1, Q''^2)}_{4} \\ &+ \underbrace{d(Q'^2, Q''^0)}_{5} + \underbrace{d(Q'^2, Q''^1)}_{1} + \underbrace{d(Q'^2, Q''^2)}_{2} \\ &+ \underbrace{d(Q'^3, Q''^0)}_{4} + \underbrace{d(Q'^3, Q''^1)}_{4} + \underbrace{d(Q'^3, Q''^2)}_{1} \\ &+ \underbrace{d(Q'^4, Q''^0)}_{2} + \underbrace{d(Q'^4, Q''^1)}_{6} + \underbrace{d(Q'^4, Q''^2)}_{3} \\ &= 42 = \frac{1}{2}\big(\mathfrak{q}(T')\mathfrak{q}(T'') - 1\big)t = \frac{1}{2}(5 \cdot 3 - 1)6 \, . \end{aligned}$$

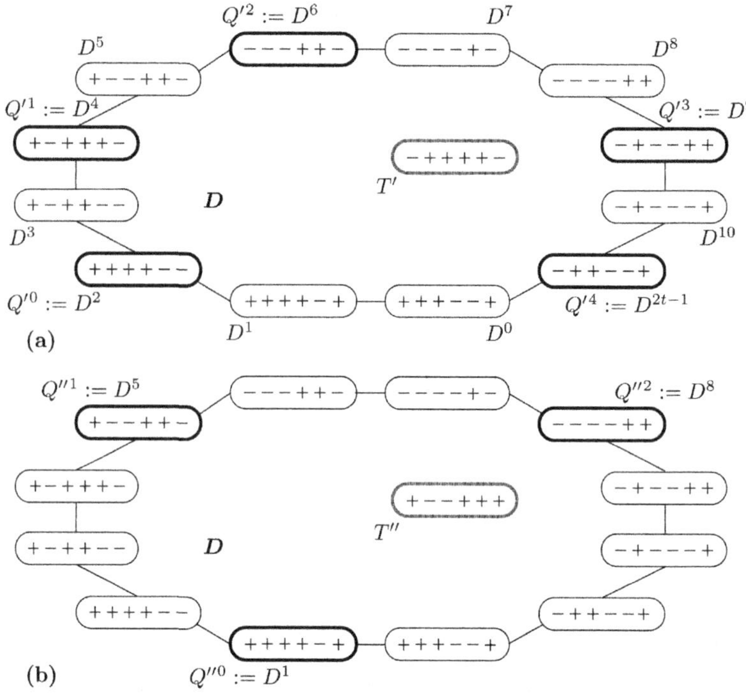

Figure 2.8 Two instances of the same symmetric cycle $\boldsymbol{D} := (D^0, D^1, \ldots, D^{2t-1}, D^0)$ in the hypercube graph $\boldsymbol{H}((t := 6), 2)$, and two vertices T' and T'' of the graph.
(a) The set $\boldsymbol{Q}(T', \boldsymbol{D}) := \{Q'^0, Q'^1, Q'^2, Q'^3, Q'^4\}$ is the unique inclusion-minimal subset of the set $\mathrm{V}(\boldsymbol{D})$ such that $T' = \sum_{Q' \in \boldsymbol{Q}(T', \boldsymbol{D})} Q'$.
(b) The set $\boldsymbol{Q}(T'', \boldsymbol{D}) := \{Q''^0, Q''^1, Q''^2\}$ is the unique inclusion-minimal subset of the set $\mathrm{V}(\boldsymbol{D})$ such that $T'' = \sum_{Q'' \in \boldsymbol{Q}(T'', \boldsymbol{D})} Q''$.

2.1.10 A Basic Statistic on Vertex Decompositions in a Hypercube Graph with Respect to a Symmetric Cycle

For an arbitrary symmetric cycle $\boldsymbol{D}$ in a hypercube graph $\boldsymbol{H}(t, 2)$, in view of Eq. (13.7) of Ref. [5], we have

$$\sum_{T \in \{1,-1\}^t} \underbrace{|\boldsymbol{Q}(T, \boldsymbol{D})|}_{\mathfrak{q}(T)} = 2^{t-1} t \,. \tag{2.17}$$

Recall that this quantity coincides with the number of edges of the graph $\boldsymbol{H}(t, 2)$.

By Theorem 13.6 of Ref. [5], for any *odd* integer ℓ, $1 \le \ell \le t$, we have

$$c_\ell(t) := |\{T \in \{1,-1\}^t \colon \mathfrak{q}(T) = \ell\}| = 2\tbinom{t}{\ell} . \tag{2.18}$$

Note that if t is *even*, then for any *odd* integer ℓ, $1 \le \ell < t$, we have

$$c_\ell(t) = c_{t-\ell}(t) .$$

If t is *odd*, then for any *odd* integer ℓ, $1 \le \ell \le t$, we have

$$\ell c_\ell(t) = (1 + t - \ell) c_{1+t-\ell}(t) .$$

Example 2.17. Suppose $t := 3$. For a symmetric cycle $\boldsymbol{D} := (D^0, D^1, \ldots, D^{2t-1}, D^0)$ in the hypercube graph $\boldsymbol{H}(t, 2)$, depicted in Figure 2.4(a), we have

$$\sum_{T\in\{1,-1\}^t} \mathfrak{q}(T) = \underbrace{\mathfrak{q}(T)}_{3} + \underbrace{\mathfrak{q}(-T)}_{3} + \sum_{D\in\mathrm{V}(\boldsymbol{D})} \underbrace{\mathfrak{q}(D)}_{1} = 12 = 2^{t-1}t = 2^{3-1}3 .$$

We also have

$$|\{T \in \{1,-1\}^t \colon \mathfrak{q}(T) = 1\}| = 2\tbinom{t}{1} := 2\tbinom{3}{1} = 6 ,$$

and

$$|\{T \in \{1,-1\}^t \colon \mathfrak{q}(T) = 3\}| = 2\tbinom{t}{3} := 2\tbinom{3}{3} = 2 .$$

2.1.11 Equinumerous Decompositions of a Vertex in a Hypercube Graph with Respect to Its Symmetric Cycles

In this subsection, we will count the number of symmetric cycles $\boldsymbol{D}$ in a hypercube graph $\boldsymbol{H}(t, 2)$ that provide equinumerous decompositions $\boldsymbol{Q}(T, \boldsymbol{D})$ of an arbitrary vertex $T \in \{1, -1\}^t$.

It is easy to see that each of the 2^t vertices of the graph $\boldsymbol{H}(t, 2)$ belongs to $\frac{1}{2}t!$ symmetric cycles. Indeed, the graph $\boldsymbol{H}(t, 2)$ regarded as the Hasse diagram of the *tope poset* of the oriented matroid $\mathcal{H} := (E_t, \{1, -1\}^t)$ can be *based* at any vertex $B \in \{1, -1\}^t$, and such a poset, with its least element B contained in $t!$ *maximal chains*, is a principal order ideal of a *binomial poset* whose *factorial function* is $n!$; see Example 3.18.3b in Ref. [7]. It now suffices to note that any symmetric cycle of $\boldsymbol{H}(t, 2)$ containing B as its vertex is a union of two maximal chains of the tope poset based at B. Thus, there are $\frac{1}{2t} \cdot 2^t \cdot \frac{1}{2}t!$ symmetric cycles in $\boldsymbol{H}(t, 2)$:

$$\#\{\boldsymbol{D} : \boldsymbol{D} \text{ symmetric cycle of } \boldsymbol{H}(t, 2)\} = 2^{t-2}(t-1)!\,.$$

See Sequence A002866 in online Ref. [8] on the corresponding integer sequence.

Recall that for any symmetric cycle $\boldsymbol{D}$ of the hypercube graph $\boldsymbol{H}(t, 2)$, by (2.17) and (2.18) we have

$$\sum_{T \in \{1,-1\}^t} \mathfrak{q}(T) = 2^{t-1}t = 2 \sum_{\substack{1 \le \ell \le t: \\ j \text{ odd}}} \ell\binom{t}{\ell}\,.$$

As a consequence, we have

$$\sum_{\substack{\boldsymbol{D}: \\ \boldsymbol{D} \text{ symmetric cycle of } \boldsymbol{H}(t,2)}} \sum_{T \in \{1,-1\}^t} \mathfrak{q}(T) = 2^{t-1}(t-1)! \sum_{\substack{1 \le \ell \le t: \\ \ell \text{ odd}}} \ell\binom{t}{\ell}\,.$$

Since

$$\sum_{\substack{\boldsymbol{D}: \\ \boldsymbol{D} \text{ symmetric cycle of } \boldsymbol{H}(t,2)}} \mathfrak{q}(T) = \frac{1}{2^t} 2^{t-1}(t-1)! \sum_{\substack{1 \le \ell \le t: \\ \ell \text{ odd}}} \ell\binom{t}{\ell}$$
$$= \frac{(t-1)!}{2} \sum_{\substack{1 \le \ell \le t: \\ \ell \text{ odd}}} \ell\binom{t}{\ell}\,,$$

for any vertex $T \in \{1, -1\}^t$ of the graph $\boldsymbol{H}(t, 2)$, we come to the following result:

Proposition 2.18. *For any vertex $T \in \{1, -1\}^t$ of the hypercube graph $\boldsymbol{H}(t, 2)$, and for an* odd *integer ℓ, $1 \le \ell \le t$, we have*

$$\#\{\boldsymbol{D} : \boldsymbol{D} \text{ symmetric cycle of } \boldsymbol{H}(t, 2),\ \mathfrak{q}(T, \boldsymbol{D}) = \ell\} = \frac{(t-1)!}{2}\binom{t}{\ell}\,.$$

Example 2.19. Suppose $t := 3$, and consider the vertex $T := (-1, 1, 1)$ of the hypercube graph $\boldsymbol{H}(t, 2)$, depicted in Figure 2.9(a).

There are $2^{t-2}(t-1)! = 2^{3-2}(3-1)! = 4$ symmetric cycles in the graph $\boldsymbol{H}(t, 2)$. We have

$$\#\{\boldsymbol{D} : \boldsymbol{D} \text{ symm. cycle of } \boldsymbol{H}(t, 2),\ \mathfrak{q}(T, \boldsymbol{D}) = 1\}$$
$$= \frac{(t-1)!}{2}\binom{t}{1} = \frac{(3-1)!}{2}\binom{3}{1} = 3\,;$$

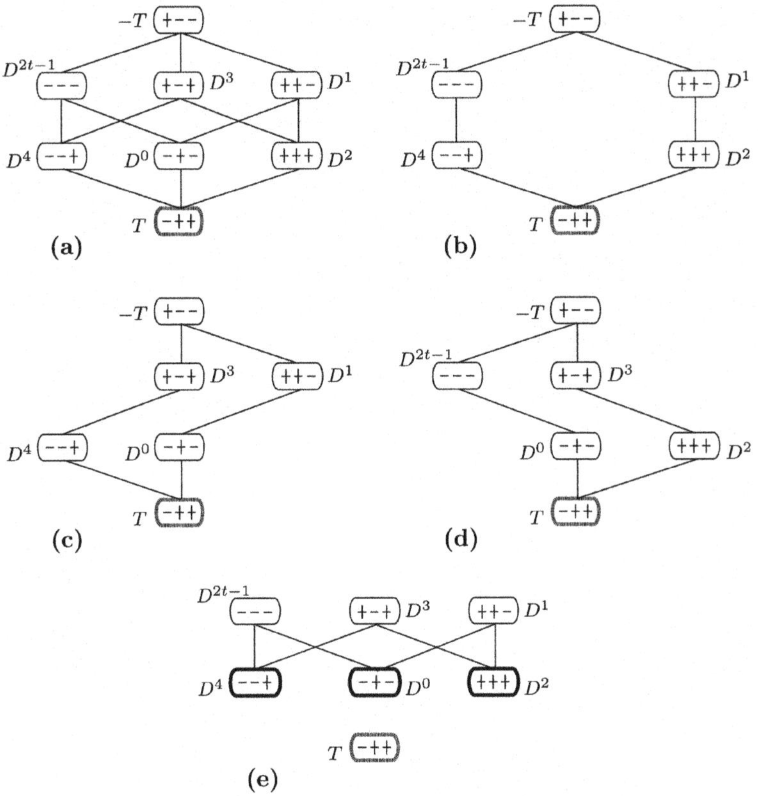

Figure 2.9 **(a)** The hypercube graph $\boldsymbol{H}((t := 3), 2)$, and its vertex T.
(b)(c)(d) The three symmetric cycles $\boldsymbol{D}$ of the graph $\boldsymbol{H}(t, 2)$ such that $\mathrm{q}(T, \boldsymbol{D}) = 1$.
(e) The unique symmetric cycle $\boldsymbol{D}$ of the graph $\boldsymbol{H}(t, 2)$ such that $\mathrm{q}(T, \boldsymbol{D}) = 3$. Here $\boldsymbol{Q}(T, \boldsymbol{D}) = \{D^0, D^2, D^4\}$.

see Figures 2.9(b), (c) and (d); and we have

$$\#\{\boldsymbol{D} : \boldsymbol{D} \text{ symm. cycle of } \boldsymbol{H}(t, 2),\ \mathrm{q}(T, \boldsymbol{D}) = 3\}$$
$$= \frac{(t-1)!}{2}\binom{t}{3} = \frac{(3-1)!}{2}\binom{3}{3} = 1\ ;$$

see Figure 2.9(e).

2.1.12 Change of Cycle, Change of Basis

If we are interested in the decompositions of a vertex T of a hypercube graph $\boldsymbol{H}(t, 2)$ with respect to its symmetric cycles $\boldsymbol{D}'$ and $\boldsymbol{D}''$, then changing cycles is essentially the same as changing from one (maximal positive) basis of the space $\mathbb{R}^t$ to another. Therefore, we have

$$\mathbf{x}(T, \boldsymbol{D}'') = \mathbf{x}(T, \boldsymbol{D}') \cdot \mathbf{M}(\boldsymbol{D}')\mathbf{M}(\boldsymbol{D}'')^{-1} .$$

Example 2.20. Suppose $t := 6$, and pick the vertex $T := (1, -1, -1, 1, 1, 1)$ of the hypercube graph $\boldsymbol{H}(t, 2)$.

We associate with a symmetric cycle $\boldsymbol{R} := (R^0, R^1, \ldots, R^{2t-1}, R^0)$ of the graph $\boldsymbol{H}(t, 2)$, depicted in Figure 2.10(a), the matrix

$$\mathbf{M}(\boldsymbol{R}) := \begin{pmatrix} 1 & 1 & 1 & 1 & 1 & 1 \\ -1 & 1 & 1 & 1 & 1 & 1 \\ -1 & -1 & 1 & 1 & 1 & 1 \\ -1 & -1 & -1 & 1 & 1 & 1 \\ -1 & -1 & -1 & -1 & 1 & 1 \\ -1 & -1 & -1 & -1 & -1 & 1 \end{pmatrix},$$

and we associate with a symmetric cycle $\boldsymbol{D} := (D^0, D^1, \ldots, D^{2t-1}, D^0)$, depicted in Figure 2.10(b), the matrices

$$\mathbf{M}(\boldsymbol{D}) := \begin{pmatrix} 1 & 1 & 1 & -1 & -1 & 1 \\ 1 & 1 & 1 & 1 & -1 & 1 \\ 1 & 1 & 1 & 1 & -1 & -1 \\ 1 & -1 & 1 & 1 & -1 & -1 \\ 1 & -1 & 1 & 1 & 1 & -1 \\ 1 & -1 & -1 & 1 & 1 & -1 \end{pmatrix},$$

and

$$\mathbf{M}(\boldsymbol{D})^{-1} = \frac{1}{2} \begin{pmatrix} 1 & 0 & 0 & 0 & 0 & 1 \\ 0 & 0 & 1 & -1 & 0 & 0 \\ 0 & 0 & 0 & 0 & 1 & -1 \\ -1 & 1 & 0 & 0 & 0 & 0 \\ 0 & 0 & 0 & -1 & 1 & 0 \\ 0 & 1 & -1 & 0 & 0 & 0 \end{pmatrix}.$$

The unique inclusion-minimal subset $\boldsymbol{Q}(T, \boldsymbol{R})$ of the vertex set $\mathrm{V}(\boldsymbol{R})$, such that $T = \sum_{Q \in \boldsymbol{Q}(T,\boldsymbol{R})} Q$, is the set $\{R^0, R^3, R^7\}$, that is,

$$\mathbf{x}(T, \boldsymbol{R}) = (1, -1, \;\; 0, \;\; 1, \;\; 0, \;\; 0) .$$

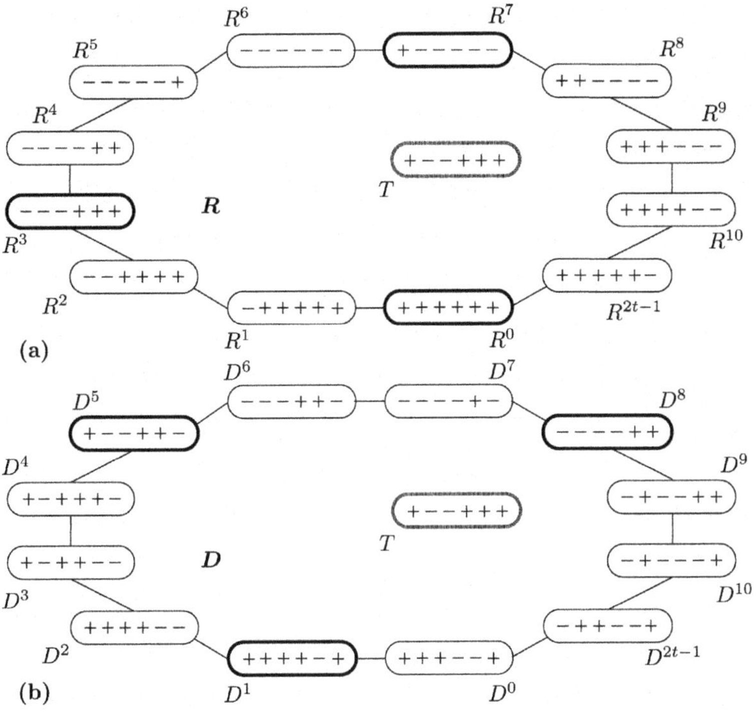

Figure 2.10 A vertex T of the hypercube graph $\boldsymbol{H}((t := 6), 2)$, and two symmetric cycles $\boldsymbol{R}$ and $\boldsymbol{D}$ in the graph.
(a) The set $\boldsymbol{Q}(T, \boldsymbol{R}) = \{R^0, R^3, R^7\}$ is the unique inclusion-minimal subset of the set $\mathrm{V}(\boldsymbol{R})$ such that $T = \sum_{Q \in \boldsymbol{Q}(T,\boldsymbol{R})} Q$.
(b) The set $\boldsymbol{Q}(T, \boldsymbol{D}) := \{D^1, D^5, D^8\}$ is the unique inclusion-minimal subset of the set $\mathrm{V}(\boldsymbol{D})$ such that $T = \sum_{Q \in \boldsymbol{Q}(T,\boldsymbol{D})} Q$.

In order to find the decomposition of the vertex T with respect to the symmetric cycle $\boldsymbol{D}$, we use the formula

$$\mathbf{x}(T, \boldsymbol{D}) = \mathbf{x}(T, \boldsymbol{R}) \cdot \mathbf{M}(\boldsymbol{R})\mathbf{M}(\boldsymbol{D})^{-1},$$

and we see that

$$\mathbf{x}(T, \boldsymbol{D}) = (0, \quad 1, -1, \quad 0, \quad 0, \quad 1),$$

that is, $\boldsymbol{Q}(T, \boldsymbol{D}) = \{D^1, D^5, D^8\}$.

2.1.13 Circular Translations of Vertex Decompositions

Remark 2.21. Let $T \in \{1, -1\}^t$ be a vertex of the hypercube graph $\boldsymbol{H}(t, 2)$, and let $\boldsymbol{D} := (D^0, D^1, \ldots, D^{2t-1}, D^0)$ be a symmetric cycle in $\boldsymbol{H}(t, 2)$. Assume that

$$(D^0, D^1, \ldots, D^{2t-1}) =: \vec{\mathrm{V}}(\boldsymbol{D}) \supset \boldsymbol{Q}(T, \boldsymbol{D}) \\ = (D^{i_0}, D^{i_1}, \ldots, D^{i_{\mathfrak{q}(T)-1}}),$$

for some indices $i_0 < i_1 < \cdots < \mathfrak{q}(T) - 1$. For any $s \in \mathbb{Z}$, we have

$$\sum_{0 \leq j \leq \mathfrak{q}(T)-1} D^{(i_j+s) \bmod 2t} \in \{1, -1\}^t .$$

Example 2.22. Suppose $t := 6$, and pick the vertex $T := (1, -1, -1, 1, 1, 1)$ of the hypercube graph $\boldsymbol{H}(t, 2)$. For a symmetric cycle $\boldsymbol{D} := (D^0, D^1, \ldots, D^{2t-1}, D^0)$ in the graph $\boldsymbol{H}(t, 2)$, depicted in Figure 2.11, we have $\boldsymbol{Q}(T, \boldsymbol{D}) = \{D^1, D^5, D^8\}$.

Now suppose $s := 2$, and note that the sum $\sum_{D^k \in \boldsymbol{Q}(T,\boldsymbol{D})} D^{k+s} := D^{1+2} + D^{5+2} + D^{8+2} = D^3 + D^7 + D^{10} = \mathrm{T}^{(-)}$ is also a vertex of the graph $\boldsymbol{H}(t, 2)$.

2.2 Coherent Vertex Decompositions in Hypercube Graphs $\boldsymbol{H}(t, 2)$ and $\widetilde{\boldsymbol{H}}(t, 2)$

Let $\widetilde{\boldsymbol{H}}(t, 2)$ denote the *hypercube graph* on its vertex set $\{0, 1\}^t$. For vertices $\widetilde{T}'$ and $\widetilde{T}''$, the pair $\{\widetilde{T}', \widetilde{T}''\}$ by definition is an *edge* of $\widetilde{\boldsymbol{H}}(t, 2)$ if and only if the *Hamming distance* $|\{e \in E_t \colon \widetilde{T}'(e) \neq \widetilde{T}''(e)\}|$ between the tuples $\widetilde{T}'$ and $\widetilde{T}''$ is 1.

Following Section 1 of Ref. [9], we recall that a family of interesting graphs related to the hypercube graph $\widetilde{\boldsymbol{H}}(t, 2)$ includes, in particular, *Fibonacci cubes* (see Refs. [10, 11, 12]), *Lucas cubes* (see Ref. [13]), *generalized Fibonacci cubes* (see Ref. [14]), *k-Fibonacci cubes* (see Ref. [15]), *Fibonacci-run graphs* (see Refs. [9, 16]) and *daisy cubes* (see Ref. [17]).

We define a *symmetric cycle* $\widetilde{\boldsymbol{D}}$ in the hypercube graph $\widetilde{\boldsymbol{H}}(t, 2)$ to be a $2t$-cycle, with its *vertex sequence*

$$\vec{\mathrm{V}}(\widetilde{\boldsymbol{D}}) := (\widetilde{D}^0, \widetilde{D}^1, \ldots, \widetilde{D}^{2t-1}),$$

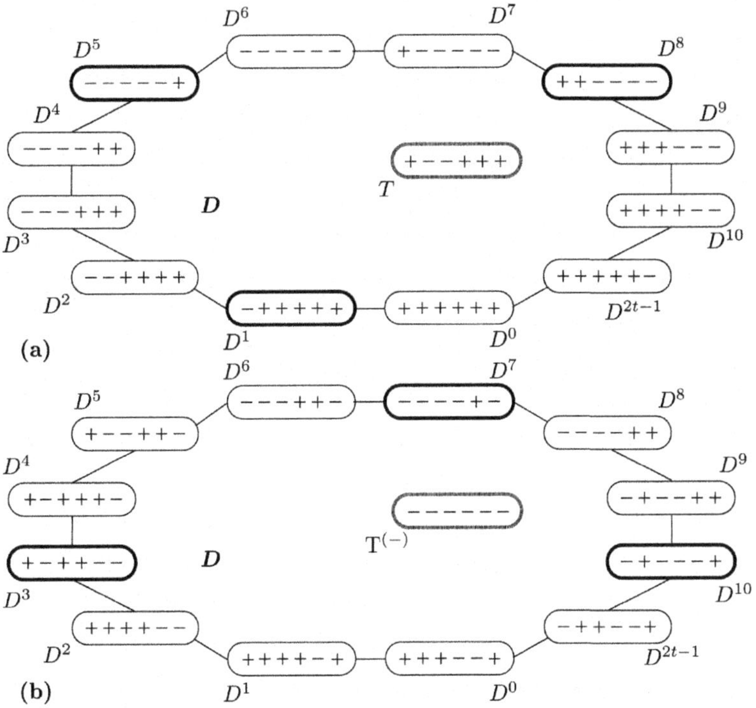

Figure 2.11 Two instances of the same symmetric cycle $\boldsymbol{D} := (D^0, D^1, \ldots, D^{2t-1}, D^0)$ in the hypercube graph $\boldsymbol{H}((t := 6), 2)$.
(a) Pick the vertex $T := (1, -1, -1, 1, 1, 1) \in \{1, -1\}^t$ of the graph $\boldsymbol{H}(t, 2)$. The set $\boldsymbol{Q}(T, \boldsymbol{D}) = \{D^1, D^5, D^8\}$ is the unique inclusion-minimal subset of the set $\mathrm{V}(\boldsymbol{D})$ such that $T = \sum_{Q \in \boldsymbol{Q}(T,\boldsymbol{R})} Q$.
(b) Suppose $s := 2$. The sum $\sum_{D^k \in \boldsymbol{Q}(T,\boldsymbol{R})} D^{k+s} = D^{1+2} + D^{5+2} + D^{8+2} = D^3 + D^7 + D^{10}$ is the vertex $\mathrm{T}^{(-)}$ of the graph $\boldsymbol{H}(t, 2)$.

and with its *vertex set*

$$\mathrm{V}(\widetilde{\boldsymbol{D}}) := \{\widetilde{D}^0, \widetilde{D}^1, \ldots, \widetilde{D}^{2t-1}\},$$

such that

$$\widetilde{D}^{k+t} := \mathrm{T}^{(+)} - \widetilde{D}^k, \quad 0 \le k \le t-1;$$

cf. definition (2.1)–(2.3).

Coherent decompositions of vertices in hypercube graphs $\widetilde{\boldsymbol{H}}(t, 2)$ and $\boldsymbol{H}(t, 2)$ are implemented by means of the standard conversions

$$\{0, 1\} \to \{1, -1\}: \quad x \mapsto 1 - 2x, \qquad 0 \mapsto 1, \qquad 1 \mapsto -1,$$

and

$$\{1,-1\} \to \{0,1\}: \qquad z \mapsto \frac{1}{2}(1-z)\,, \quad 1 \mapsto 0\,, \quad -1 \mapsto 1\,,$$

whose derived bijective maps are

$$\{0,1\}^t \to \{1,-1\}^t: \qquad \widetilde{T} \mapsto \mathrm{T}^{(+)} - 2\widetilde{T}\,, \tag{2.19}$$

and

$$\{1,-1\}^t \to \{0,1\}^t: \qquad T \mapsto \frac{1}{2}(\mathrm{T}^{(+)} - T)\,. \tag{2.20}$$

Given a symmetric cycle $\boldsymbol{D}$ in the hypercube graph $\boldsymbol{H}(t,2)$ on its vertex set $\{1,-1\}^t$, let us regard bijection (2.20) as the (de)composition

$$\{1,-1\}^t \xrightarrow{(2.20)} \{0,1\}^t:$$

$$\begin{aligned}
T &= \sum_{Q\in\boldsymbol{Q}(T,\boldsymbol{D})} Q \\
&\mapsto \frac{1}{2}\Big(\mathrm{T}^{(+)} - \sum_{Q\in\boldsymbol{Q}(T,\boldsymbol{D})} Q\Big) \\
&= \frac{1}{2}\Big(\mathrm{q}(T,\boldsymbol{D})\mathrm{T}^{(+)} - (\mathrm{q}(T,\boldsymbol{D})-1)\mathrm{T}^{(+)} - \sum_{Q\in\boldsymbol{Q}(T,\boldsymbol{D})} Q\Big) \\
&= -\frac{1}{2}(\mathrm{q}(T,\boldsymbol{D})-1)\mathrm{T}^{(+)} + \sum_{Q\in\boldsymbol{Q}(T,\boldsymbol{D})} \frac{1}{2}(\mathrm{T}^{(+)} - Q)\,.
\end{aligned}$$

Remark 2.23. For a symmetric cycle $\widetilde{\boldsymbol{D}}$ in the hypercube graph $\widetilde{\boldsymbol{H}}(t, 2)$ on its vertex set $\{0,1\}^t$, and for any vertex $\widetilde{T}$ of $\widetilde{\boldsymbol{H}}(t,2)$, there exists a *unique inclusion-minimal* subset $\widetilde{\boldsymbol{Q}}(\widetilde{T},\widetilde{\boldsymbol{D}}) \subset \mathrm{V}(\widetilde{\boldsymbol{D}})$, of *odd* cardinality, such that

$$\widetilde{T} = -\frac{1}{2}(|\widetilde{\boldsymbol{Q}}(\widetilde{T},\widetilde{\boldsymbol{D}})| - 1)\mathrm{T}^{(+)} + \sum_{\substack{\widetilde{Q}\in\widetilde{\boldsymbol{Q}}(\widetilde{T},\widetilde{\boldsymbol{D}}):\\ \widetilde{Q}\neq(0,\ldots,0)}} \widetilde{Q}\,.$$

Example 2.24. Suppose $t := 6$. Figure 2.12 depicts two coherent symmetric cycles $\boldsymbol{D}$ and $\widetilde{\boldsymbol{D}}$ in the hypercube graphs $\boldsymbol{H}(t,2)$ and $\widetilde{\boldsymbol{H}}(t,2)$, respectively. The vertex sets $\mathrm{V}(\boldsymbol{D})$ and $\mathrm{V}(\widetilde{\boldsymbol{D}})$ of the cycles are in one-to-one correspondences (2.20) and (2.19) with each other.

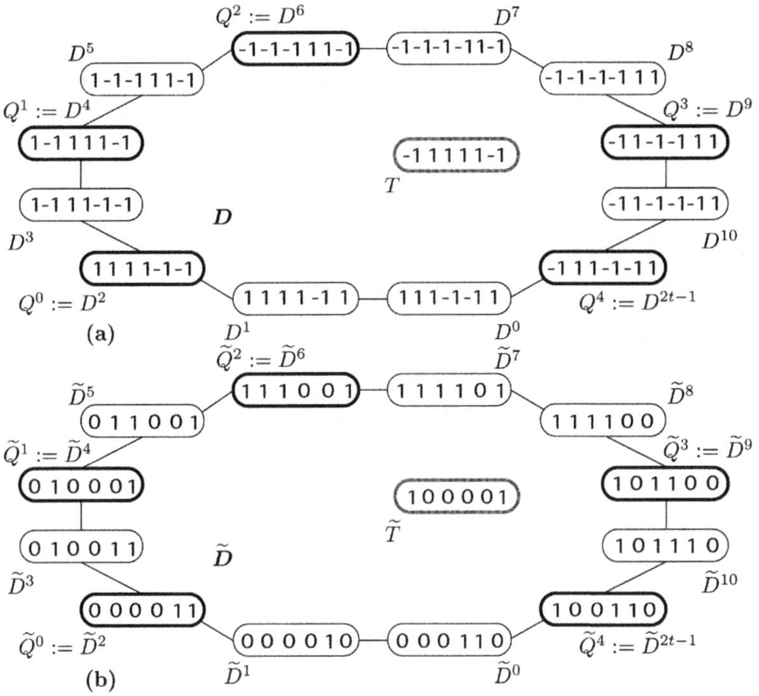

Figure 2.12 Coherent symmetric cycles $\boldsymbol{D}$ and $\widetilde{\boldsymbol{D}}$ in the hypercube graphs $\boldsymbol{H}((t := 6), 2)$ and $\widetilde{\boldsymbol{H}}(t, 2)$. The bijections between the vertex sets $\mathrm{V}(\boldsymbol{D})$ and $\mathrm{V}(\widetilde{\boldsymbol{D}})$ of the cycles are given in (2.20) and (2.19).
(a) The set $\boldsymbol{Q}(T, \boldsymbol{D}) := \{Q^0, Q^1, \ldots, Q^4\}$ is the unique inclusion-minimal subset of the vertex set $\mathrm{V}(\boldsymbol{D})$ of the cycle $\boldsymbol{D}$, such that $T = \sum_{Q \in \boldsymbol{Q}(T, \boldsymbol{D})} Q$.
(b) The set $\widetilde{\boldsymbol{Q}}(\widetilde{T}, \widetilde{\boldsymbol{D}}) := \{\widetilde{Q}^0, \widetilde{Q}^1, \ldots, \widetilde{Q}^4\}$ is the unique inclusion-minimal subset of the vertex set $\mathrm{V}(\widetilde{\boldsymbol{D}})$ of the cycle $\widetilde{\boldsymbol{D}}$, such that $\widetilde{T} = -\frac{1}{2}(|\widetilde{\boldsymbol{Q}}(\widetilde{T}, \widetilde{\boldsymbol{D}})| - 1)\mathrm{T}^{(+)} + \sum_{\widetilde{Q} \in \widetilde{\boldsymbol{Q}}(\widetilde{T}, \widetilde{\boldsymbol{D}})} \widetilde{Q}$.

For the vertex $\widetilde{T} := (1, 0, 0, 0, 0, 1) \in \{0, 1\}^t$ of the graph $\widetilde{\boldsymbol{H}}(t, 2)$, the set

$$\begin{aligned}\widetilde{\boldsymbol{Q}}(\widetilde{T}, \widetilde{\boldsymbol{D}}) :&= \left\{ \underbrace{\widetilde{Q}^0}_{\widetilde{D}^2}, \underbrace{\widetilde{Q}^1}_{\widetilde{D}^4}, \underbrace{\widetilde{Q}^2}_{\widetilde{D}^6}, \underbrace{\widetilde{Q}^3}_{\widetilde{D}^9}, \underbrace{\widetilde{Q}^4}_{\widetilde{D}^{2t-1}} \right\} \\ &= \{(0, 0, 0, 0, 1, 1), (0, 1, 0, 0, 0, 1), (1, 1, 1, 0, 0, 1), \\ &\quad (1, 0, 1, 1, 0, 0), (1, 0, 0, 1, 1, 0)\}\end{aligned}$$

is the unique inclusion-minimal subset, of odd cardinality, of the vertex set $\mathrm{V}(\widetilde{\boldsymbol{D}})$ such that

$$\widetilde{T} = -\frac{1}{2}(|\widetilde{\boldsymbol{Q}}(\widetilde{T}, \widetilde{\boldsymbol{D}})| - 1)\mathrm{T}^{(+)} + \sum_{\widetilde{Q} \in \widetilde{\boldsymbol{Q}}(\widetilde{T}, \widetilde{\boldsymbol{D}})} \widetilde{Q} \, .$$

Indeed, we have

$$\widetilde{T} := (1, 0, 0, 0, 0, 1) = -\frac{1}{2}(5-1)\cdot(1, 1, 1, 1, 1, 1)+(3, 2, 2, 2, 2, 3) \, .$$

In practice, it is convenient to translate decomposition problems, for the hypercube graph $\widetilde{\boldsymbol{H}}(t, 2)$ and its symmetric cycles, to the hypercube graph $\boldsymbol{H}(t, 2)$ on its vertex set $\{1, -1\}^t$, and then to send their solutions (found more or less easily) back to the graph $\widetilde{\boldsymbol{H}}(t, 2)$.

Let $X, Y \in \{1, -1\}^t$ be two vertices of the hypercube graph $\boldsymbol{H}(t, 2)$, where t is *even*, and let $\widetilde{X}, \widetilde{Y} \in \{0, 1\}^t$ be the corresponding vertices of the hypercube graph $\widetilde{\boldsymbol{H}}(t, 2)$, namely, $\widetilde{X} := \frac{1}{2}(\mathrm{T}^{(+)} - X)$ and $\widetilde{Y} := \frac{1}{2}(\mathrm{T}^{(+)} - Y)$. Let $\mathtt{hwt}(\widetilde{T}) := \langle \widetilde{T}, \mathrm{T}^{(+)} \rangle$ denote the *Hamming weight* of a vertex $\widetilde{T} \in \{0, 1\}^t$, that is, the number of 1's in $\widetilde{T}$. We have

$$\langle X, Y \rangle = 0 \iff \langle \widetilde{X}, \widetilde{Y} \rangle = \frac{2(\mathtt{hwt}(\widetilde{X}) + \mathtt{hwt}(\widetilde{Y})) - t}{4} \, ,$$

and if $4|t$ (i.e., t is divisible by 4), then we have

$$|\mathfrak{n}(X)| = |\mathfrak{n}(Y)| =: s \, , \ \ \langle X, Y \rangle = 0$$
$$\iff \mathtt{hwt}(\widetilde{X}) = \mathtt{hwt}(\widetilde{Y}) =: s \, , \ \ \langle \widetilde{X}, \widetilde{Y} \rangle = s - \frac{t}{4} \, .$$

Chapter 3

Vertex Decompositions in Hypercube Graphs, and Dehn–Sommerville Type Relations

An arrangement of distinct oriented lines crossing the origin (0, 0) of the two-dimensional Euclidean space $\mathbb{R}^2$ on a piece of paper represents implicitly a rank 2 system of homogeneous strict linear inequalities. If there is no point lying on the positive sides of all lines of the arrangement, then the inclusion-maximal positive parts of topes of the oriented matroid realized by the arrangement are the multi-indices of maximal feasible subsystems of an infeasible system. If the system has at least five maximal feasible subsystems, then the numbers of feasible subsystems satisfy relations derived from the Dehn–Sommerville equations for the face numbers of the boundary complexes of simplicial convex polytopes. In such a situation, the tope set of our oriented matroid, regarded as the vertex set of a symmetric cycle in a hypercube graph, leads us to Dehn–Sommerville type relations for a certain abstract simplicial complex associated with the decomposition set for the positive vertex of the graph. In this chapter, we present Dehn–Sommerville type relations that are valid for arbitrary vertices of hypercube graphs with sufficiently large decomposition sets.

Symmetric Cycles
Andrey O. Matveev

ISBN 978-981-4968-81-2 (Hardcover), 978-1-003-43832-8 (eBook)
www.jennystanford.com

In Section 3.1 we recall some enumerative results on rank 2 infeasible systems of linear inequalities related to arrangements of oriented lines in the plane. In Section 3.2 we present Dehn–Sommerville type relations for the numbers of faces of abstract simplicial complexes associated with large-size decomposition sets for vertices of hypercube graphs.

3.1 Dehn–Sommerville Equations for the Feasible Subsystems of Rank 2 Infeasible Systems of Homogeneous Strict Linear Inequalities

Let $\boldsymbol{D}$ be a symmetric cycle in the hypercube graph $\boldsymbol{H}(t,2)$ on its vertex set $\{1,-1\}^t$, such that $\mathrm{V}(\boldsymbol{D}) \not\ni \mathrm{T}^{(+)}$, that is, the positive vertex $\mathrm{T}^{(+)}$ is not a vertex of the cycle $\boldsymbol{D}$.

As mentioned in Section 2.1.1, the vertex set $\mathrm{V}(\boldsymbol{D})$ of the cycle $\boldsymbol{D}$ is the set of topes of a *rank* 2 *oriented matroid* $\mathcal{H}_{\boldsymbol{D}} := (E_t, \mathrm{V}(\boldsymbol{D}))$. Thus, we can consider the representation of $\mathcal{H}_{\boldsymbol{D}}$ by a certain central line arrangement

$$\{\mathbf{x} \in \mathbb{R}^2 \colon \langle \boldsymbol{a}_e, \mathbf{x}\rangle = 0,\ \ e \in E_t\} \tag{3.1}$$

in the plane $\mathbb{R}^2$, with the set of corresponding normal vectors

$$\boldsymbol{A} := \{\boldsymbol{a}_e \colon e \in E_t\}\ . \tag{3.2}$$

If the cardinality of the unique inclusion-minimal subset $\boldsymbol{Q}(\mathrm{T}^{(+)}, \boldsymbol{D}) \subset \mathrm{V}(\boldsymbol{D})$, with the property $\sum_{Q \in \boldsymbol{Q}(\mathrm{T}^{(+)}, \boldsymbol{D})} Q = \mathrm{T}^{(+)}$, is not too small, namely

$$\mathfrak{q}(\mathrm{T}^{(+)}) := \mathfrak{q}(\mathrm{T}^{(+)}, \boldsymbol{D}) := |\boldsymbol{Q}(\mathrm{T}^{(+)}, \boldsymbol{D})| \ge 5\ ,$$

then for the set of normal vectors (3.2) we have

$$|\{\boldsymbol{a} \in \boldsymbol{A} \colon \boldsymbol{a} \in \mathbf{C}_{>}\}| \ge 2\ , \tag{3.3}$$

for any *open half-plane* $\mathbf{C}_{>} \subset \mathbb{R}^2$ bounded by a one-dimensional subspace of $\mathbb{R}^2$; see, e.g., the proof of Proposition 2.33 in Ref. [18]. As noted several times in Section 2.1, we have

$$\boldsymbol{Q}(\mathrm{T}^{(+)}, \boldsymbol{D}) = \mathbf{max}^{+}(\mathrm{V}(\boldsymbol{D}))\ ,$$

that is, the decomposition set $\boldsymbol{Q}(\mathrm{T}^{(+)}, \boldsymbol{D})$ for the positive vertex $\mathrm{T}^{(+)}$ with respect to the cycle $\boldsymbol{D}$ is precisely the subset $\mathbf{max}^{+}(\mathrm{V}(\boldsymbol{D})) \subset \mathrm{V}(\boldsymbol{D})$ of vertices with inclusion-maximal positive parts.

Example 3.1. Suppose $t := 6$, and consider a symmetric cycle $\boldsymbol{D}$ in the hypercube graph $\boldsymbol{H}(t, 2)$, depicted in Figure 3.1(a). The positive vertex $\mathrm{T}^{(+)}$ is not a vertex of the cycle $\boldsymbol{D}$.

The vertex set $\mathrm{V}(\boldsymbol{D})$ of the cycle $\boldsymbol{D}$ is the set of topes of a rank 2 oriented matroid $\mathcal{H}_{\boldsymbol{D}}$. A central arrangement of oriented lines in the plane $\mathbb{R}^2$ that realizes $\mathcal{H}_{\boldsymbol{D}}$ is depicted in Figure 3.1(b). We denote by $\boldsymbol{A} := \{\boldsymbol{a}_1, \ldots, \boldsymbol{a}_t\}$ the set of normal vectors defining the lines of the arrangement.

Since for the unique inclusion-minimal subset $\boldsymbol{Q}(\mathrm{T}^{(+)}, \boldsymbol{D}) \subset \mathrm{V}(\boldsymbol{D})$, such that $\sum_{Q \in \boldsymbol{Q}(\mathrm{T}^{(+)}, \boldsymbol{D})} Q = \mathrm{T}^{(+)}$, we have $\mathfrak{q}(\mathrm{T}^{(+)}) \geq 5$, any open half-plane $\mathbf{C}_{>}$ bounded by a one-dimensional subspace of the space $\mathbb{R}^2$ contains at least two vectors of the set $\boldsymbol{A}$; see Figure 3.1(c).

Denote by ν_j the number of *feasible* subsystems, of cardinality j, of the *infeasible system* of *homogeneous strict linear inequalities*

$$\{\langle \boldsymbol{a}_e, \mathbf{x}\rangle > 0 : \mathbf{x} \in \mathbb{R}^2,\ \boldsymbol{a}_e \in \boldsymbol{A}\} \tag{3.4}$$

associated with the arrangement (3.1). Since the condition (3.3) is fulfilled, the quantities ν_j satisfy the relations (where 'x' is a formal variable)

$$\begin{cases} \nu_j = \binom{t}{j}, \ \text{if}\ 0 \leq j \leq 2\,, \\ \nu_{t-1} = \nu_t = 0\,, \\ \sum_{j=3}^{t} \left(\binom{t}{j} - \nu_j\right) (\mathrm{x}-1)^{t-j} = -\sum_{j=3}^{t} (-1)^j \left(\binom{t}{j} - \nu_j\right) \mathrm{x}^{t-j} \end{cases} \tag{3.5}$$

called the *Dehn–Sommerville equations for the feasible subsystems of the system* (3.4); see, e.g., Proposition 3.53 in Ref. [18].

In particular,

- if $t := 5$, then

$$\nu_3 = \binom{t}{2} - t\,;$$

- if $t := 6$, then

$$\nu_3 = \binom{t}{3} - 2t + 4\,,$$
$$\nu_4 = \binom{t}{2} - 3t + 6\,;$$

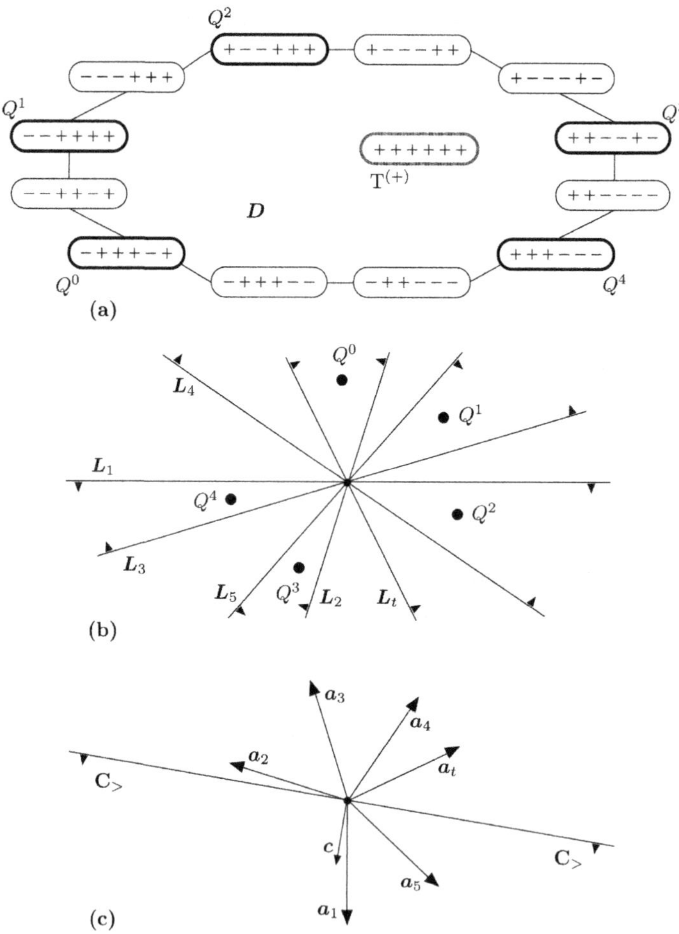

Figure 3.1 **(a)** A symmetric cycle $\boldsymbol{D}$ in the hypercube graph $\boldsymbol{H}((t := 6), 2)$. The positive vertex does not belong to the cycle: $\mathrm{T}^{(+)} \not\in \mathrm{V}(\boldsymbol{D})$. The subset $\boldsymbol{Q}(\mathrm{T}^{(+)}, \boldsymbol{D}) = \{Q^0, Q^1, \ldots, Q^4\}$ of the vertex set $\mathrm{V}(\boldsymbol{D})$ of the cycle $\boldsymbol{D}$ consists of the vertices with *inclusion-maximal positive parts*.
(b) The vertex set $\mathrm{V}(\boldsymbol{D})$ of the cycle $\boldsymbol{D}$ is the set of topes of the rank 2 oriented matroid $\mathcal{H}_{\boldsymbol{D}}$ represented by a central line arrangement $\{\boldsymbol{L}_1, \ldots, \boldsymbol{L}_t\}$ in the plane $\mathbb{R}^2$. The regions of the arrangement marked by disks correspond to the vertices of the graph $\boldsymbol{H}(t, 2)$ that have *inclusion-maximal positive parts*.
(c) The set of normal vectors $\boldsymbol{A} := \{\boldsymbol{a}_e \colon e \in E_t\}$ defining the oriented lines $\boldsymbol{L}_e := \{\mathbf{x} \in \mathbb{R}^2 \colon \langle \boldsymbol{a}_e, \mathbf{x}\rangle = 0\}$ of the arrangement depicted in Figure **(b)**. Since $\mathrm{q}(\mathrm{T}^{(+)}) \geq 5$, we have $|\boldsymbol{A} \cap \mathbf{C}_>| \geq 2$, for any open half-plane $\mathbf{C}_> := \{\mathbf{x} \in \mathbb{R}^2 \colon \langle \boldsymbol{c}, \mathbf{x}\rangle > 0\}$ bounded by a one-dimensional subspace $\{\mathbf{x} \in \mathbb{R}^2 \colon \langle \boldsymbol{c}, \mathbf{x}\rangle = 0\} \subset \mathbb{R}^2$, where $\boldsymbol{c} \neq \mathbf{0}$.

- if $t := 7$, then

$$\nu_4 = 2\nu_3 - 2\tbinom{t}{4} + \tbinom{t}{3} ,$$
$$\nu_5 = \nu_3 - \tbinom{t}{4} + \tbinom{t}{2} - t .$$

For any j, where $3 \le j \le t-2$, by Corollary 3.55(i) of Ref. [18] we have

$$\tbinom{t}{j} - \nu_j = -\sum_{i=3}^{j} (-1)^i \tbinom{t-i}{j-i} \Big(\tbinom{t}{i} - \nu_i \Big) . \qquad (3.6)$$

Recall also that

$$\sum_{j=1}^{t-2} (-1)^j \nu_j = 0 .$$

3.2 Dehn–Sommerville Type Relations for Vertex Decompositions in Hypercube Graphs. I

Given a vertex $T \in \{1,-1\}^t$ of the hypercube graph $\boldsymbol{H}(t,2)$ with its symmetric cycle $\boldsymbol{D}$, consider the corresponding decomposition

$$T = \sum_{Q \in \boldsymbol{Q}(T,\boldsymbol{D})} Q$$

of T with respect to the cycle $\boldsymbol{D}$, for a unique inclusion-minimal subset $\boldsymbol{Q}(T,\boldsymbol{D}) \subset \mathrm{V}(\boldsymbol{D})$. Suppose that

$$\mathfrak{q}(T) := \mathfrak{q}(T,\boldsymbol{D}) := |\boldsymbol{Q}(T,\boldsymbol{D})| \ge 5 ,$$

and consider the oriented matroid

$$\mathcal{Y}_{T,\boldsymbol{D}} := {}_{-\mathfrak{n}(T)}(\mathcal{H}_{\boldsymbol{D}})$$

obtained from the rank 2 oriented matroid $\mathcal{H}_{\boldsymbol{D}} := (E_t, \mathrm{V}(\boldsymbol{D}))$, with its set of topes $\mathrm{V}(\boldsymbol{D})$, by *reorientation* of $\mathcal{H}_{\boldsymbol{D}}$ on the *negative part* $\mathfrak{n}(T)$ of the tope T.

Associate with the abstract simplicial *complex*

$$\begin{aligned}\Delta &:= \Delta_{\text{acyclic}}(\mathcal{Y}_{T,\boldsymbol{D}}) \\ &:= \big\{ A \subset E_t \colon \exists F \in {}_{-\mathfrak{n}(T)}\mathrm{V}(\boldsymbol{D}) \text{ with } \mathfrak{p}(F) \supseteq A \big\} \qquad (3.7)\end{aligned}$$

of *acyclic subsets* of the ground set of the oriented matroid $\mathcal{Y}_{T,\boldsymbol{D}}$ its *long f-vector*

$$\boldsymbol{f}(\Delta;t) := \big(f_0(\Delta;t),\ f_1(\Delta;t), \ldots,\ f_t(\Delta;t)\big) \in \mathbb{N}^{t+1} \tag{3.8}$$

defined by

$$f_j(\Delta;t) := \#\{F \in \Delta\colon |F| = j\}\,, \quad 0 \le j \le t\,. \tag{3.9}$$

In view of (3.5), we have

$$\begin{cases} f_j(\Delta;t) = \binom{t}{j}, \ \text{if}\ 0 \le j \le 2\,, \\ f_{t-1}(\Delta;t) = f_t(\Delta;t) = 0\,, \\ \sum_{j=3}^{t}\Big(\binom{t}{j} - f_j(\Delta;t)\Big)\,(\mathrm{x}-1)^{t-j} \\ \qquad = -\sum_{j=3}^{t}(-1)^j\Big(\binom{t}{j} - f_j(\Delta;t)\Big)\,\mathrm{x}^{t-j}\,. \end{cases}$$

Long f-vectors of face systems similar to those defined by (3.8) and (3.9) are discussed in Chapters 2 and 3 of Ref. [5].

Let us return to the symmetric cycle $\boldsymbol{D}$, and associate with the decomposition set $\boldsymbol{Q}(T,\,\boldsymbol{D})$ for the vertex T the abstract simplicial complex $\boldsymbol{\Lambda} := \boldsymbol{\Lambda}(T,\,\boldsymbol{D})$ whose facet family is defined to be the family

$$\big\{E_t - \mathbf{S}(T,\,Q)\colon\ Q \in \boldsymbol{Q}(T,\,\boldsymbol{D})\big\}\,, \tag{3.10}$$

where $\mathbf{S}(T,\,Q)$ is the *separation set* of the vertices T and Q. By construction, the complex $\boldsymbol{\Lambda}$ and the complex Δ defined by (3.7) coincide, and we come to the following result:

Proposition 3.2. *Let $\boldsymbol{D}$ be a symmetric cycle in a hypercube graph $\boldsymbol{H}(t,\,2)$. Let $T \in \{1,-1\}^t$ be a vertex of the graph $\boldsymbol{H}(t,\,2)$, such that for the unique inclusion-minimal subset of vertices $\boldsymbol{Q}(T,\,\boldsymbol{D}) \subset \mathrm{V}(\boldsymbol{D})$ with the property*

$$\sum_{Q \in \boldsymbol{Q}(T,\boldsymbol{D})} Q = T\,,$$

we have

$$\mathfrak{q}(T,\,\boldsymbol{D}) \ge 5\,.$$

(i) *The components of the long f-vector $\boldsymbol{f}(\boldsymbol{\Lambda};t)$ of the complex*

$$\begin{aligned} \boldsymbol{\Lambda} &:= \boldsymbol{\Lambda}(T,\,\boldsymbol{D}) \\ &= \boldsymbol{\Lambda}(-T,\,\boldsymbol{D}) \end{aligned}$$

whose family of facets is defined by (3.10) *satisfy the* Dehn–Sommerville *type* relations

$$\begin{cases} f_j(\mathbf{\Lambda};t) = \binom{t}{j}, \text{ if } 0 \le j \le 2\,, \\ f_{t-1}(\mathbf{\Lambda};t) = f_t(\mathbf{\Lambda};t) = 0\,, \\ \sum_{j=3}^{t} \left(\binom{t}{j} - f_j(\mathbf{\Lambda};t)\right)(\mathrm{x}-1)^{t-j} \\ \qquad = -\sum_{j=3}^{t}(-1)^j\left(\binom{t}{j} - f_j(\mathbf{\Lambda};t)\right)\mathrm{x}^{t-j}\,. \end{cases}$$

In particular,

- *if* $t := 5$, *then*

$$f_3(\mathbf{\Lambda};t) = \binom{t}{2} - t = 5\,;$$

- *if* $t := 6$, *then*

$$f_3(\mathbf{\Lambda};t) = \binom{t}{3} - 2t + 4 = 12\,,$$
$$f_4(\mathbf{\Lambda};t) = \binom{t}{2} - 3t + 6 = 3\,;$$

- *if* $t := 7$, *then*

$$\begin{aligned} f_3(\mathbf{\Lambda};t) &= \tfrac{1}{2} f_4(\mathbf{\Lambda};t) + \binom{t}{4} - \tfrac{1}{2}\binom{t}{3} &&= \tfrac{1}{2}\left(f_4(\mathbf{\Lambda};t) + 35\right) \\ &= f_5(\mathbf{\Lambda};t) + \binom{t}{4} - \binom{t}{2} + t &&= f_5(\mathbf{\Lambda};t) + 21\,, \\ f_4(\mathbf{\Lambda};t) &= 2 f_3(\mathbf{\Lambda};t) - 2\binom{t}{4} + \binom{t}{3} &&= 2 f_3(\mathbf{\Lambda};t) - 35 \\ &= 2 f_5(\mathbf{\Lambda};t) + \binom{t}{3} - 2\binom{t}{2} + 2t &&= 2 f_5(\mathbf{\Lambda};t) + 7\,, \\ f_5(\mathbf{\Lambda};t) &= f_3(\mathbf{\Lambda};t) - \binom{t}{4} + \binom{t}{2} - t &&= f_3(\mathbf{\Lambda};t) - 21 \\ &= \tfrac{1}{2} f_4(\mathbf{\Lambda};t) - \tfrac{1}{2}\binom{t}{3} + \binom{t}{2} - t &&= \tfrac{1}{2}\left(f_4(\mathbf{\Lambda};t) - 35\right) \\ & &&\quad +14\,; \end{aligned}$$

note that the quantity $f_4(\mathbf{\Lambda};7)$ *is always* odd.

(ii) *For any* j, $3 \le j \le t-2$, *we have*

$$\binom{t}{j} - f_j(\mathbf{\Lambda};t) = -\sum_{i=3}^{j}(-1)^i\binom{t-i}{j-i}\left(\binom{t}{i} - f_i(\mathbf{\Lambda};t)\right).$$

(iii) *We have*

$$\sum_{j=1}^{t-2}(-1)^j f_j(\mathbf{\Lambda};t) = 0\,.$$

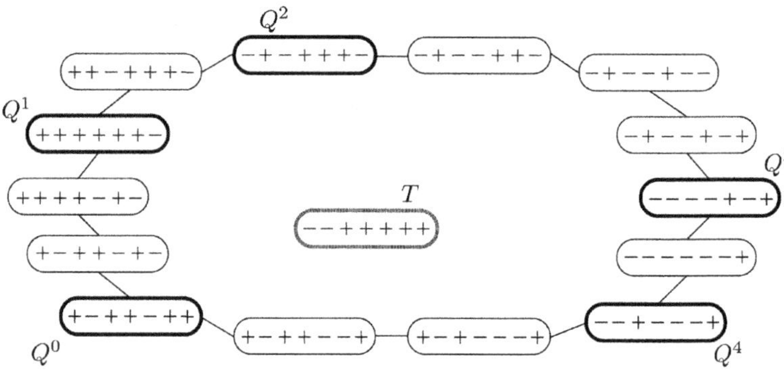

Figure 3.2 A symmetric cycle $\boldsymbol{D}$ of the hypercube graph $\boldsymbol{H}(7, 2)$. For the vertex $T := (-1, -1, 1, 1, 1, 1, 1)$ of the graph $\boldsymbol{H}(7, 2)$ we have $T = Q^0 + \cdots + Q^4$, for a unique inclusion-minimal subset $\boldsymbol{Q}(T, \boldsymbol{D}) := \{Q^0, \ldots, Q^4\} \subset \mathrm{V}(\boldsymbol{D})$.

Example 3.3. Pick the vertex

$$T := (-1, -1, \; 1, \; 1, \; 1, \; 1, \; 1)$$

of the hypercube graph $\boldsymbol{H}(7, 2)$. Consider a symmetric cycle $\boldsymbol{D}$ of the graph $\boldsymbol{H}(7, 2)$, depicted in Figure 3.2. We have $T = \sum_{Q \in \boldsymbol{Q}(T, \boldsymbol{D})} Q$, for a unique inclusion-minimal subset $\boldsymbol{Q}(T, \boldsymbol{D}) := \{Q^0, \ldots, Q^4\} \subset \mathrm{V}(\boldsymbol{D})$, where

$$\begin{aligned}
Q^0 &:= (\;\;1, -1, \;\;1, \;\;1, -1, \;\;1, \;\;1)\,,\\
Q^1 &:= (\;\;1, \;\;1, \;\;1, \;\;1, \;\;1, \;\;1, -1)\,,\\
Q^2 &:= (-1, \;\;1, -1, \;\;1, \;\;1, \;\;1, -1)\,,\\
Q^3 &:= (-1, -1, -1, -1, \;\;1, -1, \;\;1)\,,\\
Q^4 &:= (-1, -1, \;\;1, -1, -1, -1, \;\;1)\,.
\end{aligned}$$

The separation sets $\mathbf{S}(T, Q^i)$ are as follows:

$$\mathbf{S}(T, Q^0) = \{1, 5\}\,, \quad \mathbf{S}(T, Q^1) = \{1, 2, 7\}\,, \quad \mathbf{S}(T, Q^2) = \{2, 3, 7\}\,,$$
$$\mathbf{S}(T, Q^3) = \{3, 4, 6\}\,, \quad \mathbf{S}(T, Q^4) = \{4, 5, 6\}\,.$$

The facet family of the abstract simplicial complex $\boldsymbol{\Lambda} := \boldsymbol{\Lambda}(T, \boldsymbol{D})$ defined by (3.10) is

$$\Big\{ \underbrace{\{2, 3, 4, 6, 7\}}_{E_7 - \mathbf{S}(T, Q^0)}, \underbrace{\{3, 4, 5, 6\}}_{E_7 - \mathbf{S}(T, Q^1)}, \underbrace{\{1, 4, 5, 6\}}_{E_7 - \mathbf{S}(T, Q^2)}, \underbrace{\{1, 2, 5, 7\}}_{E_7 - \mathbf{S}(T, Q^3)}, \underbrace{\{1, 2, 3, 7\}}_{E_7 - \mathbf{S}(T, Q^4)} \Big\}\,.$$

It is easy to check that the components of the long f-vector

$$\boldsymbol{f}(\boldsymbol{\Lambda};7)=\left(\underset{\overset{\uparrow}{0}}{1},\ \overset{\overset{1}{\downarrow}}{7},\ \underset{\overset{\uparrow}{2}}{21},\ \overset{\overset{3}{\downarrow}}{22},\ \underset{\overset{\uparrow}{4}}{9},\ \overset{\overset{5}{\downarrow}}{1},\ \underset{\overset{\uparrow}{6}}{0},\ \overset{\overset{7}{\downarrow}}{0}\right)$$

of the complex $\boldsymbol{\Lambda}$ satisfy the relations given in Proposition 3.2.

Chapter 4

Vertex Decompositions in Hypercube Graphs, and Orthogonality Relations

If a vertex of a hypercube graph admits its large-size decomposition with respect to a symmetric cycle of the graph, then the face numbers of an abstract simplicial complex associated in a natural way with the decomposition set satisfy Dehn–Sommerville type relations; see the previous Chapter 3. In the present chapter, we consider large-size decomposition sets for vertices of hypercube graphs $\boldsymbol{H}(s, 2)$ and $\boldsymbol{H}(t, 2)$, where the parameters s and t have different parity. We show that in addition to the individual Dehn–Sommmerville type relations, enumerative descriptions of the decomposition sets for the vertices satisfy a certain common orthogonality relation.

In Section 4.1 we recall Dehn–Sommerville type relations that are valid for large-size decomposition sets for vertices of a hypercube graph. In Section 4.2 we present a common orthogonality relation that establishes a connection between enumerative properties of large-size decomposition sets in the above graphs $\boldsymbol{H}(s, 2)$ and $\boldsymbol{H}(t, 2)$.

Symmetric Cycles
Andrey O. Matveev

ISBN 978-981-4968-81-2 (Hardcover), 978-1-003-43832-8 (eBook)
www.jennystanford.com

4.1 Dehn–Sommerville Type Relations for Vertex Decompositions in Hypercube Graphs. II

Let $\boldsymbol{H}(s, 2)$ and $\boldsymbol{H}(t, 2)$ be the hypercube graphs on their vertex sets $\{1, -1\}^s$ and $\{1, -1\}^t$, respectively. Throughout this chapter, we assume that

$$s < t\,, \quad \text{and} \quad s \not\equiv t \pmod 2\,.$$

Let $\boldsymbol{D}'$ be a *symmetric* $2s$*-cycle* in the graph $\boldsymbol{H}(s, 2)$, and let $\boldsymbol{D}''$ be a *symmetric* $2t$*-cycle* in the graph $\boldsymbol{H}(t, 2)$.

Let $T' \in \{1, -1\}^s$ and $T'' \in \{1, -1\}^t$ be vertices of the graphs $\boldsymbol{H}(s, 2)$ and $\boldsymbol{H}(t, 2)$, such that for the *unique inclusion-minimal subsets* $\boldsymbol{Q}(T', \boldsymbol{D}') \subset \mathrm{V}(\boldsymbol{D}')$ and $\boldsymbol{Q}(T'', \boldsymbol{D}'') \subset \mathrm{V}(\boldsymbol{D}'')$, with the properties

$$\sum_{Q' \in \boldsymbol{Q}(T', \boldsymbol{D}')} Q' = T'\,, \quad \text{and} \quad \sum_{Q'' \in \boldsymbol{Q}(T'', \boldsymbol{D}'')} Q'' = T''\,,$$

we have

$$\mathfrak{q}(T', \boldsymbol{D}') \geq 5\,, \quad \text{and} \quad \mathfrak{q}(T'', \boldsymbol{D}'') \geq 5\,.$$

Let

$$\boldsymbol{\Lambda}'' := \boldsymbol{\Lambda}(T'', \boldsymbol{D}'')$$

be an abstract simplicial complex on the vertex set E_t, with its facet family

$$\big\{E_t - \mathbf{S}(T'', Q'') \colon Q'' \in \boldsymbol{Q}(T'', \boldsymbol{D}'')\big\}\,, \tag{4.1}$$

where $\mathbf{S}(T'', Q'')$ is the *separation set* of the vertices T'' and Q''. See the previous Chapter 3 on such complexes.

Associate with the complex $\boldsymbol{\Lambda}''$ its *long* f*-vector*

$$\boldsymbol{f}(\boldsymbol{\Lambda}''; t) := \big(f_0(\boldsymbol{\Lambda}''; t), f_1(\boldsymbol{\Lambda}''; t), \ldots, f_t(\boldsymbol{\Lambda}''; t)\big) \in \mathbb{N}^{t+1}$$

defined by

$$f_j(\boldsymbol{\Lambda}''; t) := \#\{F \in \boldsymbol{\Lambda}'' \colon |F| = j\}\,, \quad 0 \leq j \leq t\,.$$

Throughout the chapter, the components of all vectors, as well as the rows and columns of matrices, are indexed starting with zero.

Define vectors $\boldsymbol{\beta}(t;t) \in \mathbb{P}^{t+1}$ and $\boldsymbol{\beta}(s;t) \in \mathbb{N}^{t+1}$ by

$$\boldsymbol{\beta}(t;t) := \left(\binom{t}{0}, \binom{t}{1}, \ldots, \binom{t}{t}\right),$$
$$\boldsymbol{\beta}(s;t) := \left(\binom{s}{0}, \binom{s}{1}, \ldots, \binom{s}{t}\right).$$

Let $\mathbf{U}(t)$ denote the square *backward identity matrix* of order $t+1$, whose (i, j)th entry is the Kronecker delta $\delta_{i+j,t}$. We denote by $\mathbf{T}(t)$ the square *forward shift matrix* of order $t+1$, whose (i, j)th entry is $\delta_{j-i,1}$.

Note that the vector

$$\boldsymbol{f}(\boldsymbol{\Omega}'';t) := \big(\boldsymbol{\beta}(t;t) - \boldsymbol{f}(\boldsymbol{\Lambda}'';t)\big)\mathbf{U}(t) \in \mathbb{N}^{t+1} \tag{4.2}$$

is the *long f-vector* of the *boundary complex* $\boldsymbol{\Omega}''$ of a $(t-3)$-dimensional *simplicial* convex *polytope* with t vertices, according to the argument given in Chapter 3, and by Proposition 3.51(a) of Ref. [18].

Now associate with the abstract simplicial complex

$$\boldsymbol{\Lambda}' := \boldsymbol{\Lambda}(T', \boldsymbol{D}')$$

on the vertex set E_s, and with the facet family

$$\left\{E_s - \mathbf{S}(T', Q') \colon Q' \in \boldsymbol{Q}(T', \boldsymbol{D}')\right\}, \tag{4.3}$$

its *long f-vector*

$$\boldsymbol{f}(\boldsymbol{\Lambda}';t) \in \mathbb{N}^{t+1},$$

defined by $f_j(\boldsymbol{\Lambda}';t) := \#\{F \in \boldsymbol{\Lambda}' \colon |F| = j\}, 0 \le j \le t$.

The vector

$$\boldsymbol{f}(\boldsymbol{\Omega}';t) := \big(\boldsymbol{\beta}(s;t) - \boldsymbol{f}(\boldsymbol{\Lambda}';t)\big)\mathbf{T}(t)^{t-s}\mathbf{U}(t) \in \mathbb{N}^{t+1} \tag{4.4}$$

is the *long f-vector* of the *boundary complex* $\boldsymbol{\Omega}'$ of an $(s-3)$-dimensional *simplicial* convex *polytope* with s vertices; again, see Chapter 3, and Proposition 3.51(a) in Ref. [18].

Define the (i, j)th entry of a square matrix

$$\mathbf{S}(t)$$

of order $t+1$ to be $(-1)^{j-i}\binom{t-i}{j-i}$.

Recall that the standard *h-vectors* $\boldsymbol{h}(\boldsymbol{\Omega}'') \in \mathbb{N}^{t-2}$ and $\boldsymbol{h}(\boldsymbol{\Omega}') \in \mathbb{N}^{s-2}$ of the *boundary complexes* $\boldsymbol{\Omega}''$ and $\boldsymbol{\Omega}'$ of *simplicial polytopes* both satisfy the *Dehn–Sommerville relations*:

$$h_k(\boldsymbol{\Omega}'') = h_{t-k-3}(\boldsymbol{\Omega}''), \qquad 0 \le k \le t-3; \tag{4.5}$$
$$h_k(\boldsymbol{\Omega}') = h_{s-k-3}(\boldsymbol{\Omega}'), \qquad 0 \le k \le s-3. \tag{4.6}$$

The *Dehn–Sommerville relations* for the standard *h-vectors* of the *boundary complexes* of *simplicial polytopes*, whose particular variants are given in (4.5) and (4.6), are discussed, e.g., in Section 8.3 of Ref. [19].

As a consequence, our *long h-vectors*

$$\boldsymbol{h}(\boldsymbol{\Omega}'';t) := \big(\boldsymbol{\beta}(t;t) - \boldsymbol{f}(\boldsymbol{\Lambda}'';t)\big)\mathbf{U}(t)\mathbf{S}(t) \in \mathbb{Z}^{t+1} \tag{4.7}$$

and

$$\boldsymbol{h}(\boldsymbol{\Omega}';t) := \big(\boldsymbol{\beta}(s;t) - \boldsymbol{f}(\boldsymbol{\Lambda}';t)\big)\mathbf{T}(t)^{t-s}\mathbf{U}(t)\mathbf{S}(t) \in \mathbb{Z}^{t+1} \tag{4.8}$$

of the complexes $\boldsymbol{\Omega}''$ and $\boldsymbol{\Omega}'$ satisfy the *Dehn–Sommerville* type *relations*

$$h_k(\boldsymbol{\Omega}'';t) = -h_{t-k}(\boldsymbol{\Omega}'';t)\,, \qquad 0 \le k \le t\,; \tag{4.9}$$

$$h_k(\boldsymbol{\Omega}';t) = h_{t-k}(\boldsymbol{\Omega}';t)\,, \qquad 0 \le k \le t\,. \tag{4.10}$$

On the *Dehn–Sommerville* type *relations* for *long h-vectors* illustrated by (4.9) and (4.10); see Section 2.3 of Ref. [5].

4.2 Orthogonality Relations for Vertex Decompositions in Hypercube Graphs

Since the maximal face $[t]$ of the simplex $\mathbf{2}^{[t]}$ does not belong to the complexes $\boldsymbol{\Omega}'$ and $\boldsymbol{\Omega}''$, we have

$$\sum_{k=0}^{t} h_k(\boldsymbol{\Omega}';t) =: \big\langle \boldsymbol{h}(\boldsymbol{\Omega}';t), \boldsymbol{\iota}(t)\big\rangle$$

$$= \big\langle \boldsymbol{h}(\boldsymbol{\Omega}'';t), \boldsymbol{\iota}(t)\big\rangle := \sum_{k=0}^{t} h_k(\boldsymbol{\Omega}'';t) = 0\,, \tag{4.11}$$

for the vector of all one's

$$\boldsymbol{\iota}(t) := (1, 1, \ldots, 1) \in \mathbb{N}^{t+1}\,.$$

Relations (4.11) follow immediately from Eq. (2.3) of Proposition 2.1 in Ref. [5].

For positive integers k, define abstract simplicial complexes $\overline{\mathbf{2}^{[k]}}$, with their *long h-vectors* $\boldsymbol{h}(\overline{\mathbf{2}^{[k]}};t) := \boldsymbol{f}(\overline{\mathbf{2}^{[k]}};t) \cdot \mathbf{S}(t)$, to be the *boundary complexes* of the *simplices* $\mathbf{2}^{[k]}$ by

$$\overline{\mathbf{2}^{[k]}} := \mathbf{2}^{[k]} - \{[k]\}\,.$$

In view of (4.10), the long h-vector $\boldsymbol{h}(\boldsymbol{\Omega}')$ lies either in the linear span

$$\operatorname{span}\left(\boldsymbol{h}(\overline{\mathbf{2}^{[1]}};t),\ \boldsymbol{h}(\overline{\mathbf{2}^{[3]}};t),\ldots,\ \boldsymbol{h}(\overline{\mathbf{2}^{[s-2]}};t)\right), \tag{4.12}$$

when t is *even*, or in the linear span

$$\operatorname{span}\left(\boldsymbol{h}(\overline{\mathbf{2}^{[2]}};t),\ \boldsymbol{h}(\overline{\mathbf{2}^{[4]}};t),\ldots,\ \boldsymbol{h}(\overline{\mathbf{2}^{[s-2]}};t)\right), \tag{4.13}$$

when t is *odd*. The *eigenspaces* of the *backward identity matrix* $\mathbf{U}(t)$ that correspond to its *eigenvalue* 1, and whose subspaces are represented as the linear spans (4.12) and (4.13), are described in Section 3.1 of Ref. [5].

The Dehn–Sommerville type relations (4.9) and (4.10) imply that $\boldsymbol{h}(\boldsymbol{\Omega}';t)$ is a *left eigenvector* of the backward identity matrix $\mathbf{U}(t)$ that corresponds to its eigenvalue 1, while $\boldsymbol{h}(\boldsymbol{\Omega}'';t)$ is a *right eigenvector* of $\mathbf{U}(t)$ that corresponds to the other eigenvalue -1. By the *principle of biorthogonality* (see, e.g., Theorem 1.4.7(a) in Ref. [20]) we have

$$\left\langle \boldsymbol{h}(\boldsymbol{\Omega}';t),\ \boldsymbol{h}(\boldsymbol{\Omega}'';t)\right\rangle = 0\,. \tag{4.14}$$

In other words, together with relation (4.14), definitions (4.7) and (4.8) yield

$$\begin{aligned}\big(\boldsymbol{\beta}(s;t) - \boldsymbol{f}(\boldsymbol{\Lambda}';t)\big)\mathbf{T}(t)^{t-s}\mathbf{U}(t)\mathbf{S}(t)\mathbf{S}(t)^{\top}\mathbf{U}(t)&\\ \times\big(\boldsymbol{\beta}(t;t)^{\top} - \boldsymbol{f}(\boldsymbol{\Lambda}'';t)^{\top}\big) &= 0\,.\end{aligned}$$

Note that the $(i,\ j)$th entry of the square matrix

$$\mathbf{M}(t) := \mathbf{U}(t)\mathbf{S}(t)\mathbf{S}(t)^{\top}\mathbf{U}(t)$$

of order $t+1$ is $(-1)^{i+j}\binom{i+j}{i}$.

Let us sum up our conclusions:

Proposition 4.1. *Let $\boldsymbol{H}(s,2)$ and $\boldsymbol{H}(t,2)$ be the hypercube graphs on their vertex sets $\{1,-1\}^s$ and $\{1,-1\}^t$, respectively, where*

$$s < t\,, \quad \text{and} \quad s \not\equiv t \pmod 2\,.$$

Let $\mathbf{D}'$ be a symmetric cycle in the graph $\boldsymbol{H}(s,2)$, and let $\mathbf{D}''$ be a symmetric cycle in the graph $\boldsymbol{H}(t,2)$.

Suppose that $T' \in \{1,-1\}^s$ *and* $T'' \in \{1,-1\}^t$ *are vertices of the graphs* $\boldsymbol{H}(s,2)$ *and* $\boldsymbol{H}(t,2)$*, such that for the unique inclusion-minimal subsets* $\boldsymbol{Q}(T',\boldsymbol{D}') \subset \mathrm{V}(\boldsymbol{D}')$ *and* $\boldsymbol{Q}(T'',\boldsymbol{D}'') \subset \mathrm{V}(\boldsymbol{D}'')$ *with the properties*

$$\sum_{Q' \in \boldsymbol{Q}(T',\boldsymbol{D}')} Q' = T' , \quad \textit{and} \quad \sum_{Q'' \in \boldsymbol{Q}(T'',\boldsymbol{D}'')} Q'' = T'' ,$$

we have

$$\mathfrak{q}(T',\boldsymbol{D}') \geq 5 , \quad \textit{and} \quad \mathfrak{q}(T'',\boldsymbol{D}'') \geq 5 .$$

The long f*-vectors* $\boldsymbol{f}(\boldsymbol{\Lambda}'';t)$ *and* $\boldsymbol{f}(\boldsymbol{\Lambda}';t)$ *of the complexes* $\boldsymbol{\Lambda}''$ *and* $\boldsymbol{\Lambda}'$ *whose families of facets are defined by* (4.1) *and* (4.3)*, respectively, satisfy the* orthogonality relation

$$\big(\boldsymbol{\beta}(s;t) - \boldsymbol{f}(\boldsymbol{\Lambda}';t)\big)\mathbf{T}(t)^{t-s} \cdot \mathbf{M}(t) \cdot \big(\boldsymbol{\beta}(t;t)^{\top} - \boldsymbol{f}(\boldsymbol{\Lambda}'';t)^{\top}\big) = 0 .$$

Example 4.2. Consider a symmetric cycle $\boldsymbol{D}'$ of the hypercube graph $\boldsymbol{H}(6,2)$, depicted in Figure 4.1.

For the vertex

$$T' := \mathrm{T}^{(+)} := (\ 1,\ 1,\ 1,\ 1,\ 1,\ 1)$$

of the graph $\boldsymbol{H}(6,2)$ we have $T' = \sum_{Q' \in \boldsymbol{Q}(T',\boldsymbol{D}')} Q'$, for the unique inclusion-minimal subset

$$\begin{aligned}\boldsymbol{Q}(T',\boldsymbol{D}') = \{\ & Q'^{0} := (-1,\ 1,\ 1,\ 1,-1,\ 1) ,\\ & Q'^{1} := (-1,-1,\ 1,\ 1,\ 1,\ 1) ,\\ & Q'^{2} := (\ 1,-1,-1,\ 1,\ 1,\ 1) ,\\ & Q'^{3} := (\ 1,\ 1,-1,-1,\ 1,-1) ,\\ & Q'^{4} := (\ 1,\ 1,\ 1,-1,-1,-1)\ \}\end{aligned}$$

of the vertex set $\mathrm{V}(\boldsymbol{D}')$ of the cycle $\boldsymbol{D}'$.

The long f-vector $\boldsymbol{f}'(\boldsymbol{\Lambda}';7)$ of the corresponding abstract simplicial complex $\boldsymbol{\Lambda}'$, with its facet family

$$\Big\{ \underbrace{\{2,3,4,6\}}_{E_6 - \mathbf{S}(T,Q'^0)} , \underbrace{\{3,4,5,6\}}_{E_6 - \mathbf{S}(T,Q'^1)} , \underbrace{\{1,4,5,6\}}_{E_6 - \mathbf{S}(T,Q'^2)} , \underbrace{\{1,2,5\}}_{E_6 - \mathbf{S}(T,Q'^3)} , \underbrace{\{1,2,3\}}_{E_6 - \mathbf{S}(T,Q'^4)} \Big\} ,$$

is

$$\boldsymbol{f}(\boldsymbol{\Lambda}';7) = \left(\begin{array}{c} \overset{1}{\downarrow}\ \ \overset{3}{\downarrow}\ \ \overset{5}{\downarrow}\ \ \overset{7}{\downarrow} \\ 1, 6, 15, 12, 3, 0, 0, 0 \\ \underset{0}{\uparrow}\ \ \underset{2}{\uparrow}\ \ \underset{4}{\uparrow}\ \ \underset{6}{\uparrow} \end{array} \right) .$$

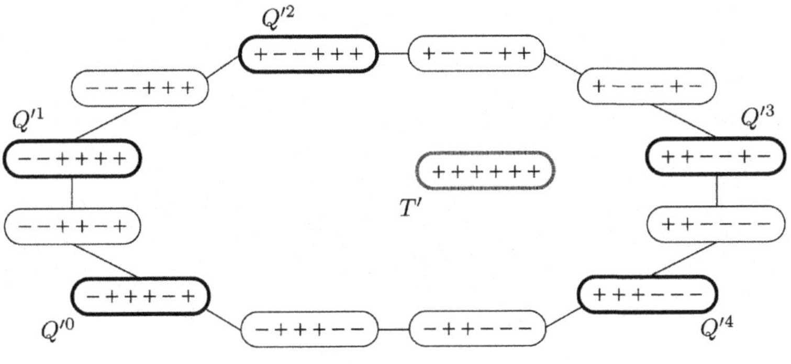

Figure 4.1 A symmetric cycle $\boldsymbol{D}'$ in the hypercube graph $\boldsymbol{H}(6, 2)$. For the positive vertex $T' := \mathrm{T}^{(+)} := (1, 1, 1, 1, 1, 1)$ of the graph $\boldsymbol{H}(6, 2)$ we have $T' = Q'^0 + \cdots + Q'^4$, for a unique inclusion-minimal subset $\boldsymbol{Q}(T', \boldsymbol{D}') =: \{Q'^0, \ldots, Q'^4\} \subset \mathrm{V}(\boldsymbol{D}')$.

Considering also the vector

$$\boldsymbol{\beta}(6;7) := \left(\underset{\underset{0}{\uparrow}}{1}, \overset{\overset{1}{\downarrow}}{6}, \underset{\underset{2}{\uparrow}}{15}, \overset{\overset{3}{\downarrow}}{20}, \underset{\underset{4}{\uparrow}}{15}, \overset{\overset{5}{\downarrow}}{6}, \underset{\underset{6}{\uparrow}}{1}, \overset{\overset{7}{\downarrow}}{0} \right) ,$$

we see that

$$\big(\boldsymbol{\beta}(6;7) - \boldsymbol{f}(\boldsymbol{\Lambda}';7)\big)\mathbf{T}(7)^{7-6} = \left(\underset{\underset{0}{\uparrow}}{0}, \overset{\overset{1}{\downarrow}}{0}, \underset{\underset{2}{\uparrow}}{0}, \overset{\overset{3}{\downarrow}}{0}, \underset{\underset{4}{\uparrow}}{8}, \overset{\overset{5}{\downarrow}}{12}, \underset{\underset{6}{\uparrow}}{6}, \overset{\overset{7}{\downarrow}}{1} \right)$$

and

$$\big(\boldsymbol{\beta}(6;7) - \boldsymbol{f}(\boldsymbol{\Lambda}';7)\big)\mathbf{T}(7)^{7-6}\mathbf{U}(7)\mathbf{S}(7) = (1, -1, -3, 3, 3, -3, -1, 1) .$$

Let us now turn to Example 3.3, change the notation used there, and consider the long f-vector

$$\boldsymbol{f}(\boldsymbol{\Lambda}'';7) = \left(\underset{\underset{0}{\uparrow}}{1}, \overset{\overset{1}{\downarrow}}{7}, \underset{\underset{2}{\uparrow}}{21}, \overset{\overset{3}{\downarrow}}{22}, \underset{\underset{4}{\uparrow}}{9}, \overset{\overset{5}{\downarrow}}{1}, \underset{\underset{6}{\uparrow}}{0}, \overset{\overset{7}{\downarrow}}{0} \right)$$

of the complex $\boldsymbol{\Lambda}''$, with 5 facets, associated with the decomposition $T'' = \sum_{Q'' \in \boldsymbol{Q}(T'', \boldsymbol{D}'')} Q''$ of a vertex T'' with respect to a symmetric cycle $\boldsymbol{D}''$ in the hypercube graph $\boldsymbol{H}(7, 2)$.

Since

$$\boldsymbol{\beta}(7;7) := \left(\underset{\underset{0}{\uparrow}}{1}, \overset{\overset{1}{\downarrow}}{7}, \underset{\underset{2}{\uparrow}}{21}, \overset{\overset{3}{\downarrow}}{35}, \underset{\underset{4}{\uparrow}}{35}, \overset{\overset{5}{\downarrow}}{21}, \underset{\underset{6}{\uparrow}}{7}, \overset{\overset{7}{\downarrow}}{1} \right) ,$$

we obtain

$$\big(\boldsymbol{\beta}(7;7) - \boldsymbol{f}(\boldsymbol{\Lambda}'';7)\big)\mathbf{U}(7)\mathbf{S}(7) = (1, 0, -1, -4, 4, 1, 0, -1)\,.$$

According to Proposition 4.1, we have

$$\begin{aligned}&\big(\boldsymbol{\beta}(6;7) - \boldsymbol{f}(\boldsymbol{\Lambda}';7)\big)\mathbf{T}(7)^{7-6}\cdot\underbrace{\mathbf{U}(7)\mathbf{S}(7)\mathbf{S}(7)^{\top}\mathbf{U}(7)}_{=:\mathbf{M}(7)}\\&\qquad\qquad\times\big(\boldsymbol{\beta}(7;7)^{\top} - \boldsymbol{f}(\boldsymbol{\Lambda}'';7)^{\top}\big)\\&= (1, -1, -3, 3, 3, -3, -1, 1)\cdot(1, 0, -1, -4, 4, 1, 0, -1)^{\top} = 0\,.\end{aligned}$$

Chapter 5

Distinguished Symmetric Cycles in Hypercube Graphs and Computation-free Vertex Decompositions

A *symmetric cycle* $\boldsymbol{D} := (D^0, D^1, \ldots, D^{2t-1}, D^0)$ in the hypercube graph $\boldsymbol{H}(t, 2)$ on the vertex set $\{1, -1\}^t$ is defined to be a $2t$-cycle such that

$$D^{k+t} = -D^k , \quad 0 \le k \le t-1 . \tag{5.1}$$

In this chapter, we describe explicitly the decompositions of vertices $T \in \{1, -1\}^t$ of the graph $\boldsymbol{H}(t, 2)$ with respect to a *distinguished* symmetric cycle $\boldsymbol{R} := (R^0, R^1, \ldots, R^{2t-1}, R^0)$, with its *vertex sequence*

$$\vec{\mathrm{V}}(\boldsymbol{R}) := (R^0, R^1, \ldots, R^{2t-1}) ,$$

and with its *vertex set*

$$\mathrm{V}(\boldsymbol{R}) := \{R^0, R^1, \ldots, R^{2t-1}\} ,$$

described as follows:

$$\begin{aligned} R^0 &:= \mathrm{T}^{(+)} , \\ R^s &:= {}_{-[s]}R^0 , \quad 1 \le s \le t-1 , \end{aligned} \tag{5.2}$$

Symmetric Cycles
Andrey O. Matveev

ISBN 978-981-4968-81-2 (Hardcover), 978-1-003-43832-8 (eBook)
www.jennystanford.com

and

$$R^{k+t} := -R^k \,, \quad 0 \le k \le t-1 \,. \tag{5.3}$$

Note that the *negative part* $\mathfrak{n}(R^i)$ of a vertex R^i of the cycle $\boldsymbol{R}$ constitutes one (possibly empty) *interval* of the ground set E_t. In practice this simple feature leads to quite impressive consequences: No matter how large the dimension t of the discrete hypercube $\{1, -1\}^t$ is, the four assertions of Proposition 5.9 allow us to find the *linear algebraic* decompositions

$$T = \sum_{Q \in \boldsymbol{Q}(T,\boldsymbol{R})} Q \tag{5.4}$$

of any vertices T of the graph $\boldsymbol{H}(t, 2)$ into *inclusion-minimal* and *linearly independent* subsets $\boldsymbol{Q}(T, \boldsymbol{R}) \subset \mathrm{V}(\boldsymbol{R})$ of the vertex set of the cycle $\boldsymbol{R}$ in an *explicit* and *computation-free* way.

Dealing with the distinguished symmetric cycle $\boldsymbol{R}$, in this chapter we develop a *WYSIWYG*-approach[1] to decomposing. Indeed, given any vertex $T \in \{1, -1\}^t$ out of the monstrous total number 2^t of vertices of the graph $\boldsymbol{H}(t, 2)$, we merely look at the negative part $\mathfrak{n}(T) \subseteq E_t$ of T, catch the endpoints of the intervals composing the set $\mathfrak{n}(T)$, and at the same time we build up the vector $\boldsymbol{x}(T, \boldsymbol{R})$ that describes the decomposition set $\boldsymbol{Q}(T, \boldsymbol{R})$ for the vertex T with respect to the cycle $\boldsymbol{R}$.

In Section 5.1 we give a few comments on our distinguished symmetric cycles in hypercube graphs. Proposition 5.9 and Theorem 5.15 are the main results of Section 5.2. They concern the interval structure of the negative parts of vertices of hypercube graphs, the corresponding computation-free decompositions with respect to the distinguished symmetric cycles, and statistics on decompositions. In Sections 5.3 and 5.4, we discuss equinumerous decompositions of vertices. In Section 5.5 we mention that vector descriptions of vertex decompositions with respect to arbitrary symmetric cycles are valuations on the Boolean lattices of subsets of the vertex sets of hypercube graphs.

---

[1]For software developers, *WYSIWYG* may sound as *What You See Is What You Get*.

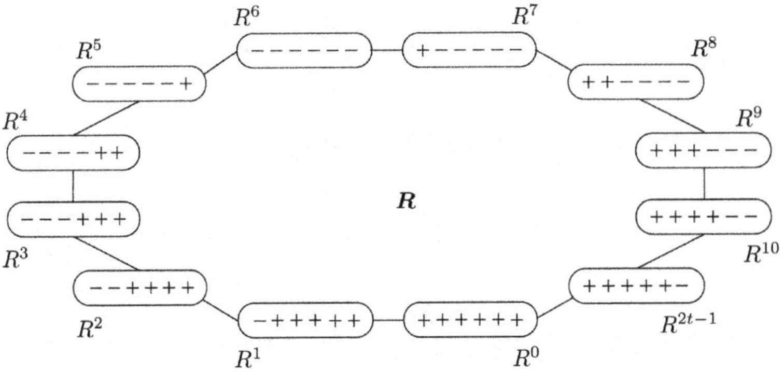

Figure 5.1 The distinguished symmetric cycle $\boldsymbol{R} := (R^0, R^1, \ldots, R^{2t-1}, R^0)$ in the hypercube graph $\boldsymbol{H}((t := 6), 2)$, defined by (5.2)(5.3).

5.1 Distinguished Symmetric Cycles in Hypercube Graphs

Let us begin with an illustrative example.

Example 5.1. Figure 5.1 depicts the distinguished symmetric cycle $\boldsymbol{R}$ in the hypercube graph $\boldsymbol{H}((t := 6), 2)$, defined by (5.2)(5.3).

Following the approach we have taken in Section 2.1.4, let us consider the nonsingular matrix

$$\mathbf{M} := \mathbf{M}(\boldsymbol{R}) := \begin{pmatrix} R^0 \\ R^1 \\ \vdots \\ R^{t-1} \end{pmatrix} \in \mathbb{R}^{t\times t} \tag{5.5}$$

whose rows are the vertices of the cycle $\boldsymbol{R}$, given in (5.2). The ith row $(\mathbf{M}^{-1})_i$, $1 \le i \le t$, of the inverse matrix $\mathbf{M}^{-1}$ of $\mathbf{M}$ is

$$(\mathbf{M}^{-1})_i = \begin{cases} \frac{1}{2} \cdot \big(\sigma(i) - \sigma(i+1)\big), & \text{if } i \neq t, \\ \frac{1}{2} \cdot \big(\sigma(1) + \sigma(t)\big), & \text{if } i = t, \end{cases} \tag{5.6}$$

where $\sigma(s) := (0, \ldots, \underset{\underset{s}{\uparrow}}{1}, \ldots, 0)$ are unit vectors of the standard basis of the space $\mathbb{R}^t$.

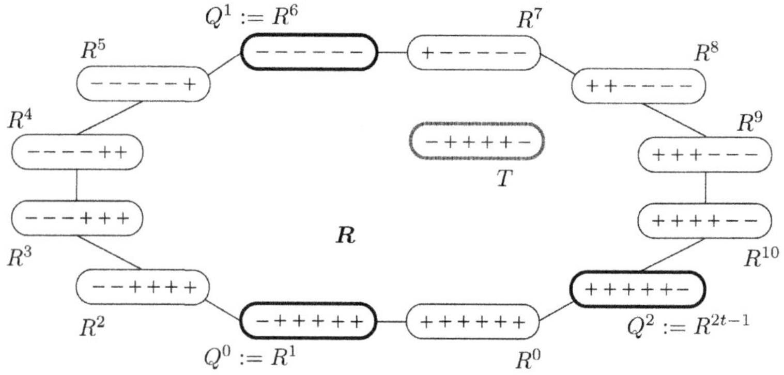

Figure 5.2 The distinguished symmetric cycle $\boldsymbol{R} := (R^0, R^1, \ldots, R^{2t-1}, R^0)$ in the hypercube graph $\boldsymbol{H}((t := 6), 2)$, defined by (5.2)(5.3). The set $\boldsymbol{Q}(T, \boldsymbol{R}) := \{Q^0, Q^1, Q^2\}$ is the unique inclusion-minimal subset of the vertex set $\mathrm{V}(\boldsymbol{R})$ of the cycle $\boldsymbol{R}$ such that $T = \sum_{Q \in \boldsymbol{Q}(T,\boldsymbol{R})} Q$.

As shown in Section 2.1.4, for any vertex $T \in \{1, -1\}^t$ of the hypercube graph $\boldsymbol{H}(t, 2)$ there exists a unique *inclusion-minimal* subset $\boldsymbol{Q}(T, \boldsymbol{R}) \subset \mathrm{V}(\boldsymbol{R})$ of the vertex set of the cycle $\boldsymbol{R}$, with the property (5.4). The set $\boldsymbol{Q}(T, \boldsymbol{R}) \subset \mathrm{V}(\boldsymbol{R})$ is of *odd* cardinality.

If we define a row vector $\boldsymbol{x} := \boldsymbol{x}(T) := \boldsymbol{x}(T, \boldsymbol{R}) \in \{-1, 0, 1\}^t$, as earlier in Subsection 2.1.4, by

$$\boldsymbol{x} := T \cdot \mathbf{M}^{-1} ,$$

then

$$\boldsymbol{Q}(T, \boldsymbol{R}) = \{x_i \cdot R^{i-1} : x_i \neq 0\} ,$$

and

$$\mathrm{q}(T) := \mathrm{q}(T, \boldsymbol{R}) := |\boldsymbol{Q}(T, \boldsymbol{R})| = |\{i \in E_t : x_i \neq 0\}| = \|\boldsymbol{x}(T)\|^2 .$$

We have $\|\boldsymbol{x}(T) \cdot \mathbf{M}\|^2 = \|T\|^2 = t$, and $\langle \boldsymbol{x}(T), \mathrm{T}^{(+)} \rangle = \sum_{e \in E_t} x_e(T) = T(t)$. We will see in Section 5.2 that if $x_e(T) \neq 0$, then $T(e) = x_e(T)$.

Example 5.2. We associate two matrices with our distinguished symmetric cycle $\boldsymbol{R}$ of the hypercube graph $\boldsymbol{H}((t := 6), 2)$, depicted in Figures 5.1 and 5.2:

$$\mathbf{M} := \mathbf{M}(\boldsymbol{R}) = \begin{pmatrix} 1 & 1 & 1 & 1 & 1 & 1 \\ -1 & 1 & 1 & 1 & 1 & 1 \\ -1 & -1 & 1 & 1 & 1 & 1 \\ -1 & -1 & -1 & 1 & 1 & 1 \\ -1 & -1 & -1 & -1 & 1 & 1 \\ -1 & -1 & -1 & -1 & -1 & 1 \end{pmatrix}, \tag{5.7}$$

$$\mathbf{M}^{-1} = \frac{1}{2}\begin{pmatrix} 1 & -1 & 0 & 0 & 0 & 0 \\ 0 & 1 & -1 & 0 & 0 & 0 \\ 0 & 0 & 1 & -1 & 0 & 0 \\ 0 & 0 & 0 & 1 & -1 & 0 \\ 0 & 0 & 0 & 0 & 1 & -1 \\ 1 & 0 & 0 & 0 & 0 & 1 \end{pmatrix};$$

cf. (5.5) and (5.6).

For the vertex $T := (-1, 1, 1, 1, 1, -1) \in \{1, -1\}^t$ of the graph $\boldsymbol{H}(t, 2)$ we have

$$\boldsymbol{x} := \boldsymbol{x}(T) = T \cdot \mathbf{M}^{-1} = (-1, 1, 0, 0, 0, -1)\,. \tag{5.8}$$

Thus,

$$\begin{aligned}\boldsymbol{Q}(T, \boldsymbol{R}) = \{x_i \cdot R^{i-1} \colon x_i \neq 0\} &= \{\underbrace{-R^0}_{R^6}, R^1, \underbrace{-R^5}_{R^{2t-1}}\} \\ &= \{R^1, R^6, R^{2t-1}\}\,;\end{aligned}$$

see Figure 5.2.

In Proposition 5.9 of this chapter, we describe explicitly the vectors $\boldsymbol{x}(T, \boldsymbol{R})$ associated with vertices T of the hypercube graph $\boldsymbol{H}(t, 2)$ and with its distinguished symmetric cycle $\boldsymbol{R}$ defined by (5.2)(5.3).

5.2 Separation Sets, Negative Parts, and Decompositions of Vertices of Hypercube Graphs

If $T', T'' \in \{1, -1\}^t$ are two vertices of the hypercube graph $\boldsymbol{H}(t, 2)$, then

$$T'' = T' - 2 \sum_{s \in \mathbf{S}(T', T'')} T'(s)\sigma(s)\,, \tag{5.9}$$

where $\mathbf{S}(T', T'')$ is the separation set of the vertices T' and T''. For the vertices in the ordered collection (5.2) we have $\boldsymbol{x}(R^{s-1}) = \boldsymbol{\sigma}(s)$, $s \in E_t$.

Relation (5.9) implies that

$$\boldsymbol{x}(T'') = \boldsymbol{x}(T') - 2\Big(\sum_{s \in \mathbf{S}(T',T'')} T'(s)\boldsymbol{\sigma}(s)\Big) \cdot \mathbf{M}^{-1} . \tag{5.10}$$

In particular, for any vertex $T \in \{1, -1\}^t$ of the graph $\boldsymbol{H}(t, 2)$ we have

$$\begin{aligned}\boldsymbol{x}(T) &= \boldsymbol{x}(\mathrm{T}^{(+)}) - 2\Big(\sum_{s \in \mathfrak{n}(T)} \boldsymbol{\sigma}(s)\Big) \cdot \mathbf{M}^{-1} \\ &= \boldsymbol{\sigma}(1) - 2\Big(\sum_{s \in \mathfrak{n}(T)} \boldsymbol{\sigma}(s)\Big) \cdot \mathbf{M}^{-1} ,\end{aligned} \tag{5.11}$$

where, as earlier, $\mathfrak{n}(T)$ denotes the negative part of the vertex T.

Since

$$|\,\mathfrak{n}(T)| = \frac{1}{2}\big(t - \langle T, \mathrm{T}^{(+)}\rangle\big) = \frac{1}{2}\big(t - \sum_{e \in E_t} T(e)\big) ,$$

we have

$$|\,\mathfrak{n}(T)| = j$$

for some integer j if and only if

$$\begin{aligned}\langle T, \mathrm{T}^{(+)}\rangle &= \boldsymbol{x}(T) \cdot \mathbf{M} \cdot (\mathrm{T}^{(+)})^\top \\ &= \boldsymbol{x}(T) \cdot (t, t-2, t-4, \ldots, -(t-2))^\top = t - 2j .\end{aligned}$$

Note also that for any two vertices T' and T'' of the graph $\boldsymbol{H}(t, 2)$ we have

$$\begin{aligned}|\,\mathfrak{n}(T') \cap \mathfrak{n}(T'')| &= \frac{1}{4}\langle T' - \mathrm{T}^{(+)}, T'' - \mathrm{T}^{(+)}\rangle \\ &= \frac{1}{4}\big(t + \langle T', T''\rangle - \langle T' + T'', \mathrm{T}^{(+)}\rangle\big)\end{aligned}$$

and

$$|\,\mathfrak{n}(T') \cup \mathfrak{n}(T'')| = \frac{1}{4}\big(3t - \langle T', T''\rangle - \langle T' + T'', \mathrm{T}^{(+)}\rangle\big) .$$

Example 5.3. Consider the vertices

$$\begin{aligned}T &:= (-1,\ \ 1,\ \ 1,\ \ 1,\ \ 1, -1) \in \{1, -1\}^t ,\\ T' &:= (-1, -1,\ \ 1,\ \ 1, -1,\ \ 1) ,\end{aligned}$$

and

$$T'' := (\ \ 1, -1, -1, \ \ 1, \ \ 1, \ \ 1)$$

of the hypercube graph $\boldsymbol{H}(t, 2)$, where $t := 6$. We have

$$\begin{aligned}
|\,\mathfrak{n}(T)| &= |\{1, 6\}| = \frac{1}{2}\big(t - \langle T, \mathrm{T}^{(+)}\rangle\big) \\
&= \frac{1}{2}(6 - 2) = 2\,, \\
|\,\mathfrak{n}(T') \cap \mathfrak{n}(T'')| &= |\{2\}| = \frac{1}{4}\big(t + \langle T', T''\rangle - \langle T' + T'', \mathrm{T}^{(+)}\rangle\big) \\
&= \frac{1}{4}(6 + 0 - 2) = 1\,,
\end{aligned}$$

and

$$\begin{aligned}
|\,\mathfrak{n}(T') \cup \mathfrak{n}(T'')| &= |\{1, 2, 3, 5\}| \\
&= \frac{1}{4}\big(3t - \langle T', T''\rangle - \langle T' + T'', \mathrm{T}^{(+)}\rangle\big) \\
&= \frac{1}{4}(3 \cdot 6 - 0 - 2) = 4\,.
\end{aligned}$$

Remark 5.4. For the symmetric cycle $\boldsymbol{R}$ in the hypercube graph $\boldsymbol{H}(t, 2)$, defined by (5.2)(5.3), let $\mathbf{x}, \mathbf{x}', \mathbf{x}'' \in \{-1, 0, 1\}^t$ be row vectors such that the tuples $T := \mathbf{x} \cdot \mathbf{M}(\boldsymbol{R})$, $T' := \mathbf{x}' \cdot \mathbf{M}(\boldsymbol{R})$, and $T'' := \mathbf{x}'' \cdot \mathbf{M}(\boldsymbol{R})$ are all vertices of $\boldsymbol{H}(t, 2)$.

(i) We have

$$\begin{aligned}
|\,\mathfrak{n}(T)| &= \frac{1}{2}\Big(t - \sum_{i \in E_t} x_i \cdot (t - 2(i-1))\Big) \\
&= \frac{t}{2}\Big(1 - \sum_{e \in E_t} x_e\Big) + \sum_{i \in [2,t]} x_i \cdot (i-1) \\
&= \begin{cases} t + \sum_{i \in [2,t]} x_i \cdot (i-1)\,, & \text{if } \langle \mathbf{x}, \mathrm{T}^{(+)}\rangle = -1\,, \\ \sum_{i \in [2,t]} x_i \cdot (i-1)\,, & \text{if } \langle \mathbf{x}, \mathrm{T}^{(+)}\rangle = \ \ 1 \end{cases} \\
&= \begin{cases} t + 1 + \sum_{i \in E_t} x_i \cdot i\,, & \text{if } \langle \mathbf{x}, \mathrm{T}^{(+)}\rangle = -1\,, \\ -1 + \sum_{i \in E_t} x_i \cdot i\,, & \text{if } \langle \mathbf{x}, \mathrm{T}^{(+)}\rangle = \ \ 1\,. \end{cases}
\end{aligned}$$

(ii) We have

$$
\begin{aligned}
&|\mathfrak{n}(T')\cap\mathfrak{n}(T'')|\\
&=\frac{1}{4}\big(t+\langle \mathbf{x}'\cdot\mathbf{M},\mathbf{x}''\cdot\mathbf{M}\rangle-\langle(\mathbf{x}'+\mathbf{x}'')\cdot\mathbf{M},\mathrm{T}^{(+)}\rangle\big)\\
&=\begin{cases}
\frac{3t}{4}+1+\frac{1}{4}\langle \mathbf{x}'\cdot\mathbf{M},\mathbf{x}''\cdot\mathbf{M}\rangle+\frac{1}{2}\sum_{i\in E_t}(x_i'+x_i'')\cdot i\,, \\
\qquad\text{if } \langle\mathbf{x}'',\mathrm{T}^{(+)}\rangle=\langle\mathbf{x}',\mathrm{T}^{(+)}\rangle=-1\,,\\
\frac{t}{4}+\frac{1}{4}\langle \mathbf{x}'\cdot\mathbf{M},\mathbf{x}''\cdot\mathbf{M}\rangle+\frac{1}{2}\sum_{i\in E_t}(x_i'+x_i'')\cdot i\,, \\
\qquad\text{if } \langle\mathbf{x}'',\mathrm{T}^{(+)}\rangle=-\langle\mathbf{x}',\mathrm{T}^{(+)}\rangle\,,\\
-\frac{t}{4}-1+\frac{1}{4}\langle \mathbf{x}'\cdot\mathbf{M},\mathbf{x}''\cdot\mathbf{M}\rangle+\frac{1}{2}\sum_{i\in E_t}(x_i'+x_i'')\cdot i\,, \\
\qquad\text{if } \langle\mathbf{x}'',\mathrm{T}^{(+)}\rangle=\langle\mathbf{x}',\mathrm{T}^{(+)}\rangle=1\,,
\end{cases}
\end{aligned}
$$

and

$$
\begin{aligned}
&|\mathfrak{n}(T')\cup\mathfrak{n}(T'')|\\
&=\frac{1}{4}\big(3t-\langle \mathbf{x}'\cdot\mathbf{M},\mathbf{x}''\cdot\mathbf{M}\rangle-\langle(\mathbf{x}'+\mathbf{x}'')\cdot\mathbf{M},\mathrm{T}^{(+)}\rangle\big)\\
&=\begin{cases}
\frac{5t}{4}+1-\frac{1}{4}\langle \mathbf{x}'\cdot\mathbf{M},\mathbf{x}''\cdot\mathbf{M}\rangle+\frac{1}{2}\sum_{i\in E_t}(x_i'+x_i'')\cdot i\,, \\
\qquad\text{if } \langle\mathbf{x}'',\mathrm{T}^{(+)}\rangle=\langle\mathbf{x}',\mathrm{T}^{(+)}\rangle=-1\,,\\
\frac{3t}{4}-\frac{1}{4}\langle \mathbf{x}'\cdot\mathbf{M},\mathbf{x}''\cdot\mathbf{M}\rangle+\frac{1}{2}\sum_{i\in E_t}(x_i'+x_i'')\cdot i\,, \\
\qquad\text{if } \langle\mathbf{x}'',\mathrm{T}^{(+)}\rangle=-\langle\mathbf{x}',\mathrm{T}^{(+)}\rangle\,,\\
\frac{t}{4}-1-\frac{1}{4}\langle \mathbf{x}'\cdot\mathbf{M},\mathbf{x}''\cdot\mathbf{M}\rangle+\frac{1}{2}\sum_{i\in E_t}(x_i'+x_i'')\cdot i\,, \\
\qquad\text{if } \langle\mathbf{x}'',\mathrm{T}^{(+)}\rangle=\langle\mathbf{x}',\mathrm{T}^{(+)}\rangle=1\,.
\end{cases}
\end{aligned}
$$

Example 5.5. Suppose $t := 6$. Let $\mathbf{x}, \mathbf{x}', \mathbf{x}'' \in \{-1, 0, 1\}^t$ be the following row vectors:

$$
\begin{aligned}
\mathbf{x} &:= (-1,\ \ 1,\ \ 0,\ \ 0,\ \ 0,-1)\,,\\
\mathbf{x}' &:= (\ \ 0,\ \ 0,\ \ 1,\ \ 0,-1,\ \ 1)\,,\\
\mathbf{x}'' &:= (\ \ 1,-1,\ \ 0,\ \ 1,\ \ 0,\ \ 0)\,.
\end{aligned}
$$

Let $\mathbf{M} := \mathbf{M}(\boldsymbol{R})$ be the matrix given in (5.7) and associated with the symmetric cycle (5.2)(5.3) of the hypercube graph $\boldsymbol{H}(t, 2)$. The row vectors

$$
\begin{aligned}
T &:= \mathbf{x}\cdot\mathbf{M} = (-1,\ \ 1,\ \ 1,\ \ 1,\ \ 1,-1)\in\{1,-1\}^t\,,\\
T' &:= \mathbf{x}'\cdot\mathbf{M} = (-1,-1,\ \ 1,\ \ 1,-1,\ \ 1)\,,
\end{aligned}
$$

and

$$T'' := \mathbf{x}'' \cdot \mathbf{M} = (\ 1, -1, -1, \ 1, \ 1, \ 1)\,,$$

that appeared earlier in Example 5.3, are vertices of the graph $\boldsymbol{H}(t, 2)$.

(i) Since $\langle \mathbf{x}, \mathrm{T}^{(+)} \rangle = -1$, by Remark 5.4(i) we have

$$|\mathfrak{n}(T)| = t + 1 + \sum_{i \in E_t} x_i \cdot i = 6 + 1 + (-1 + 2 - 6) = 2\,.$$

(ii) Since $\langle \mathbf{x}'', \mathrm{T}^{(+)} \rangle = \langle \mathbf{x}', \mathrm{T}^{(+)} \rangle = 1$, by Remark 5.4(ii) we have

$$|\mathfrak{n}(T') \cap \mathfrak{n}(T'')| = -\frac{t}{4} - 1 + \frac{1}{4}\langle \underbrace{\mathbf{x}' \cdot \mathbf{M}}_{T'}, \underbrace{\mathbf{x}'' \cdot \mathbf{M}}_{T''} \rangle + \frac{1}{2}\sum_{i \in E_t} (x'_i + x''_i) \cdot i$$

$$= -\frac{6}{4} - 1 + \frac{1}{4} \cdot 0 + \frac{1}{2}(1 \cdot 1 - 1 \cdot 2 + 1 \cdot 3 + 1 \cdot 4 - 1 \cdot 5 + 1 \cdot 6) = 1\,,$$

and

$$|\mathfrak{n}(T') \cup \mathfrak{n}(T'')| = \frac{t}{4} - 1 - \frac{1}{4}\langle \mathbf{x}' \cdot \mathbf{M}, \mathbf{x}'' \cdot \mathbf{M} \rangle + \frac{1}{2}\sum_{i \in E_t} (x'_i + x''_i) \cdot i$$

$$= \frac{6}{4} - 1 - \frac{1}{4} \cdot 0 + \frac{1}{2}(1 \cdot 1 - 1 \cdot 2 + 1 \cdot 3 + 1 \cdot 4 - 1 \cdot 5 + 1 \cdot 6) = 4\,.$$

Since

$$\{1, -1\}^t = \Big\{\mathrm{T}^{(+)} - 2\sum_{s \in A} \sigma(s) \colon A \subseteq E_t\Big\}\,,$$

we have

$$\big\{\mathbf{x}(T) \colon T \in \{1, -1\}^t\big\} = \Big\{\sigma(1) - 2\Big(\sum_{s \in A} \sigma(s)\Big) \cdot \mathbf{M}^{-1} \colon A \subseteq E_t\Big\}\,.$$

If $s \in E_t$, then we define a row vector $\boldsymbol{y}(s) := \boldsymbol{y}(s; t) \in \{-1, 0, 1\}^t$ by

$$\boldsymbol{y}(s) := \mathbf{x}(\{s\}) := \mathbf{x}({}_{-s}\mathrm{T}^{(+)}) = \sigma(1) - 2\sigma(s) \cdot \mathbf{M}^{-1}\,,$$

that is,

$$\boldsymbol{y}(s) := \begin{cases} \sigma(2)\,, & \text{if } s = 1\,, \\ \sigma(1) - \sigma(s) + \sigma(s+1)\,, & \text{if } 1 < s < t\,, \\ -\sigma(t)\,, & \text{if } s = t\,. \end{cases} \qquad (5.12)$$

Remark 5.6. (See also Section 5.5.) If $A \subseteq E_t$, then

$$\begin{aligned}\boldsymbol{x}(A) &= -\boldsymbol{x}(E_t - A)\\ &= (1 - |A|) \cdot \boldsymbol{\sigma}(1) + \sum_{s \in A} \boldsymbol{y}(s)\end{aligned}$$

and, as a consequence, we have

$$\left\{\boldsymbol{x}(T) \colon T \in \{1, -1\}^t\right\} = \left\{(1 - |A|) \cdot \boldsymbol{\sigma}(1) + \sum_{s \in A} \boldsymbol{y}(s) \colon A \subseteq E_t\right\} .$$

Example 5.7. For $t := 6$, the row vectors $\boldsymbol{y}(s) := \boldsymbol{y}(s; t) \in \{-1, 0, 1\}^t$ given in (5.12) are as follows:

$$\begin{aligned}\boldsymbol{y}(1) &= (0,\ \ 1,\ \ 0,\ \ 0,\ \ 0,\ \ 0)\ ,\\ \boldsymbol{y}(2) &= (1, -1,\ \ 1,\ \ 0,\ \ 0,\ \ 0)\ ,\\ \boldsymbol{y}(3) &= (1,\ \ 0, -1,\ \ 1,\ \ 0,\ \ 0)\ ,\\ \boldsymbol{y}(4) &= (1,\ \ 0,\ \ 0, -1,\ \ 1,\ \ 0)\ ,\\ \boldsymbol{y}(5) &= (1,\ \ 0,\ \ 0,\ \ 0, -1,\ \ 1)\ ,\\ \boldsymbol{y}(t) &= (0,\ \ 0,\ \ 0,\ \ 0,\ \ 0, -1)\ .\end{aligned}$$

Considering the vertex $T := (-1, 1, 1, 1, 1, -1) = {}_{-\{1,t\}}\mathrm{T}^{(+)} \in \{1, -1\}^t$ of the hypercube graph $\boldsymbol{H}(t, 2)$ with its distinguished symmetric cycle $\boldsymbol{R}$ defined in (5.2)(5.3), by Remark 5.6 we have

$$\begin{aligned}\boldsymbol{x}(T) &= (1 - |\{1, t\}|) \cdot \boldsymbol{\sigma}(1) + \big(\boldsymbol{y}(1) + \boldsymbol{y}(t)\big)\\ &= -(1, 0, 0, 0, 0, 0) + \big((0, 1, 0, 0, 0, 0) + (0, 0, 0, 0, 0, -1)\big)\\ &= (-1, 1, 0, 0, 0, -1)\ ;\end{aligned}$$

cf. Eq. (5.8).

For subsets $A \subseteq E_t$, relations (5.11) imply the following:

$$\{1, t\} \cap A = \{1\} \implies \boldsymbol{x}(A) = \boldsymbol{\sigma}(1) - (\underbrace{\boldsymbol{\sigma}(1) - \boldsymbol{\sigma}(2)}_{2\cdot(\mathbf{M}^{-1})_1}) - \sum_{i \in A - \{1\}} (\underbrace{\boldsymbol{\sigma}(i) - \boldsymbol{\sigma}(i+1)}_{2\cdot(\mathbf{M}^{-1})_i})\ ;$$

$$\{1, t\} \cap A = \{1, t\} \implies \boldsymbol{x}(A) = \boldsymbol{\sigma}(1) - (\underbrace{\boldsymbol{\sigma}(1) - \boldsymbol{\sigma}(2)}_{2\cdot(\mathbf{M}^{-1})_1}) - (\underbrace{\boldsymbol{\sigma}(1) + \boldsymbol{\sigma}(t)}_{2\cdot(\mathbf{M}^{-1})_t}) - \sum_{i \in A - \{1,t\}} (\underbrace{\boldsymbol{\sigma}(i) - \boldsymbol{\sigma}(i+1)}_{2\cdot(\mathbf{M}^{-1})_i})\ ;$$

$$|\{1,t\}\cap A|=0 \implies \boldsymbol{x}(A)=\sigma(1)-\sum_{i\in A}(\underbrace{\sigma(i)-\sigma(i+1)}_{2\cdot(\mathbf{M}^{-1})_i});$$

$$\{1,t\}\cap A=\{t\} \implies \boldsymbol{x}(A)=\sigma(1)-(\underbrace{\sigma(1)+\sigma(t)}_{2\cdot(\mathbf{M}^{-1})_t}) - \sum_{i\in A-\{t\}}(\underbrace{\sigma(i)-\sigma(i+1)}_{2\cdot(\mathbf{M}^{-1})_i}).$$

We arrive at the following conclusion:

Remark 5.8. If $A\subseteq E_t$, then

$$\boldsymbol{x}(A)=-\boldsymbol{x}(E_t-A)=$$
$$\begin{cases}\sigma(2)-\sum_{i\in A-\{1\}}(\sigma(i)-\sigma(i+1)), & \text{if } \{1,t\}\cap A=\{1\},\\ -\sigma(1)+\sigma(2)-\sigma(t) & \\ \quad -\sum_{i\in A-\{1,t\}}(\sigma(i)-\sigma(i+1)), & \text{if } \{1,t\}\cap A=\{1,t\},\\ \sigma(1)-\sum_{i\in A}(\sigma(i)-\sigma(i+1)), & \text{if } |\{1,t\}\cap A|=0,\\ -\sigma(t)-\sum_{i\in A-\{t\}}(\sigma(i)-\sigma(i+1)), & \text{if } \{1,t\}\cap A=\{t\}.\end{cases}$$

Since the sums appearing in Remark 5.8 depend only on the endpoints of intervals that compose the sets A, we obtain the following explicit descriptions of the decompositions of vertices of the hypercube graph $\boldsymbol{H}(t,2)$:

Proposition 5.9. *Let* $\boldsymbol{R}$ *be the symmetric cycle in the hypercube graph* $\boldsymbol{H}(t,2)$*, defined by* (5.2)(5.3).

Let A *be a nonempty subset of the set* E_t*, and let*

$$A=[i_1,j_1]\,\dot\cup\,[i_2,j_2]\,\dot\cup\,\cdots\,\dot\cup\,[i_\varrho,j_\varrho] \tag{5.13}$$

be its partition into inclusion-maximal intervals such that

$$j_1+2\le i_2,\quad j_2+2\le i_3,\quad \ldots,\quad j_{\varrho-1}+2\le i_\varrho, \tag{5.14}$$

for some $\varrho:=\varrho(A)$.

(i) *If* $\{1,t\}\cap A=\{1\}$*, then*

$$|\boldsymbol{Q}({}_{-A}\mathrm{T}^{(+)},\boldsymbol{R})|=2\varrho-1,$$
$$\boldsymbol{x}({}_{-A}\mathrm{T}^{(+)},\boldsymbol{R})=\sum_{1\le k\le\varrho}\sigma(j_k+1)-\sum_{2\le\ell\le\varrho}\sigma(i_\ell).$$

(ii) *If* $\{1, t\} \cap A = \{1, t\}$, *then*

$$|\boldsymbol{Q}({}_{-A}\mathrm{T}^{(+)}, \boldsymbol{R})| = 2\varrho - 1 ,$$

$$\boldsymbol{x}({}_{-A}\mathrm{T}^{(+)}, \boldsymbol{R}) = -\boldsymbol{\sigma}(1) + \sum_{1 \le k \le \varrho - 1} \boldsymbol{\sigma}(j_k + 1) - \sum_{2 \le \ell \le \varrho} \boldsymbol{\sigma}(i_\ell) .$$

(iii) *If* $|\{1, t\} \cap A| = 0$, *then*

$$|\boldsymbol{Q}({}_{-A}\mathrm{T}^{(+)}, \boldsymbol{R})| = 2\varrho + 1 ,$$

$$\boldsymbol{x}({}_{-A}\mathrm{T}^{(+)}, \boldsymbol{R}) = \boldsymbol{\sigma}(1) + \sum_{1 \le k \le \varrho} \boldsymbol{\sigma}(j_k + 1) - \sum_{1 \le \ell \le \varrho} \boldsymbol{\sigma}(i_\ell) .$$

(iv) *If* $\{1, t\} \cap A = \{t\}$, *then*

$$|\boldsymbol{Q}({}_{-A}\mathrm{T}^{(+)}, \boldsymbol{R})| = 2\varrho - 1 ,$$

$$\boldsymbol{x}({}_{-A}\mathrm{T}^{(+)}, \boldsymbol{R}) = \sum_{1 \le k \le \varrho - 1} \boldsymbol{\sigma}(j_k + 1) - \sum_{1 \le \ell \le \varrho} \boldsymbol{\sigma}(i_\ell) .$$

In particular, we have

$$1 \le j < t \implies \boldsymbol{x}({}_{-[j]}\mathrm{T}^{(+)}) = \boldsymbol{\sigma}(j + 1) ;$$

$$\boldsymbol{x}(\mathrm{T}^{(-)}) = -\boldsymbol{\sigma}(1) ;$$

$$1 < i < j < t \implies \boldsymbol{x}({}_{-[i,j]}\mathrm{T}^{(+)}) = \boldsymbol{\sigma}(1) - \boldsymbol{\sigma}(i) + \boldsymbol{\sigma}(j+1) ; \quad (5.15)$$

$$1 < i \le t \implies \boldsymbol{x}({}_{-[i,t]}\mathrm{T}^{(+)}) = -\boldsymbol{\sigma}(i) .$$

Example 5.10. Suppose $t := 6$, and turn to the distinguished symmetric cycle $\boldsymbol{R}$ in the hypercube graph $\boldsymbol{H}(t, 2)$, defined by (5.2)(5.3).

(i) Let us pick the subset $A := \{1, 2, 4, 5\}$ of the set E_t. Following the description (5.13)(5.14), we regard the set A as the disjoint union of inclusion-maximal intervals

$$A = [i_1, j_1] \,\dot{\cup}\, [i_\varrho, j_\varrho] ,$$

where $\varrho = 2$, and

$$i_1 = 1 , \quad j_1 = 2 , \quad i_\varrho = 4 , \quad \text{and} \quad j_\varrho = 5 .$$

Since $\{1, t\} \cap A = \{1\}$, by Proposition 5.9(i) we have

$$\mathfrak{q}(A) := |\boldsymbol{Q}({}_{-A}\mathrm{T}^{(+)}, \boldsymbol{R})| = 2\varrho - 1 = 2 \cdot 2 - 1 = 3 ,$$

$$\boldsymbol{x}(A) := \boldsymbol{x}({}_{-A}\mathrm{T}^{(+)}, \boldsymbol{R}) = \sum_{1 \le k \le \varrho} \boldsymbol{\sigma}(j_k + 1) - \sum_{2 \le \ell \le \varrho} \boldsymbol{\sigma}(i_\ell)$$

$$= \big(\boldsymbol{\sigma}(3) + \boldsymbol{\sigma}(t)\big) - \boldsymbol{\sigma}(4) = (0, 0, 1, -1, 0, 1) .$$

Thus, we have $\boldsymbol{Q}({}_{-A}\mathrm{T}^{(+)}, \boldsymbol{R}) = \{R^2, \underbrace{-R^3}_{R^9}, R^5\} = \{R^2, R^5, R^9\}$;

see Figure 5.3.

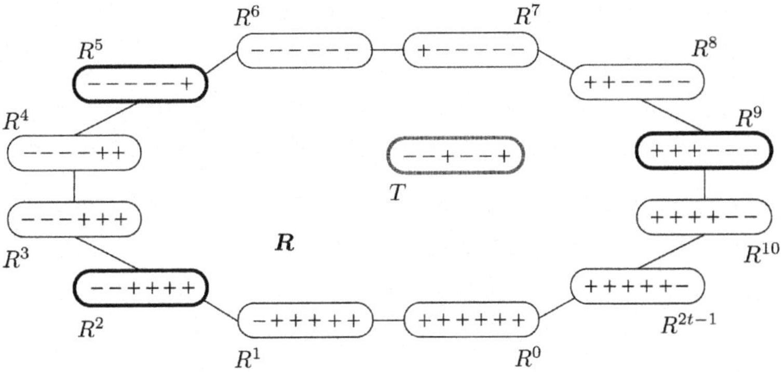

Figure 5.3 The distinguished symmetric cycle $\boldsymbol{R} := (R^0, R^1, \ldots, R^{2t-1}, R^0)$ in the hypercube graph $\boldsymbol{H}((t := 6), 2)$, defined by (5.2)(5.3).
For the vertex $T := {}_{-A}\mathrm{T}^{(+)} \in \{1, -1\}^t$ of the graph $\boldsymbol{H}(t, 2)$, where $A := \{1, 2, 4, 5\}$, we have $\{1, t\} \cap A = \{1\}$.
The set $\boldsymbol{Q}(T, \boldsymbol{R}) = \{R^2, R^5, R^9\}$ is the unique inclusion-minimal subset of the vertex set $\mathrm{V}(\boldsymbol{R})$ of the cycle $\boldsymbol{R}$ such that $T = \sum_{Q \in \boldsymbol{Q}(T, \boldsymbol{R})} Q$.

(ii) The subset $A := \{1, 3, t\} \subset E_t$ can be seen as the disjoint union of inclusion-maximal intervals

$$A = [i_1, j_1] \,\dot{\cup}\, [i_2, j_2] \,\dot{\cup}\, [i_\varrho, j_\varrho] \,,$$

where $\varrho = 3$, and

$$i_1 = j_1 = 1 \,, \quad i_2 = j_2 = 3 \,, \quad \text{and} \quad i_\varrho = j_\varrho = t \,.$$

Since $\{1, t\} \cap A = \{1, t\}$, by Proposition 5.9(ii) we have

$$\mathfrak{q}(A) = 2\varrho - 1 = 2 \cdot 3 - 1 = 5 \,,$$

$$\begin{aligned} \boldsymbol{x}(A) &= -\boldsymbol{\sigma}(1) + \sum_{1 \le k \le \varrho - 1} \boldsymbol{\sigma}(j_k + 1) - \sum_{2 \le \ell \le \varrho} \boldsymbol{\sigma}(i_\ell) \\ &= -\boldsymbol{\sigma}(1) + \big(\boldsymbol{\sigma}(2) + \boldsymbol{\sigma}(4)\big) - \big(\boldsymbol{\sigma}(3) + \boldsymbol{\sigma}(6)\big) \\ &= (-1, 1, -1, 1, 0, -1) \,, \end{aligned}$$

and indeed, we have $\boldsymbol{Q}({}_{-A}\mathrm{T}^{(+)}, \boldsymbol{R}) = \{\underbrace{-R^0}_{R^6}, R^1, \underbrace{-R^2}_{R^8}, R^3, \underbrace{-R^5}_{R^{2t-1}}\} = \{R^1, R^3, R^6, R^8, R^{2t-1}\}$; see Figure 5.4.

(iii) The subset $A := \{2, 3, 5\} \subset E_t$ is the disjoint union of inclusion-maximal intervals

$$A = [i_1, j_1] \,\dot{\cup}\, [i_\varrho, j_\varrho] \,,$$

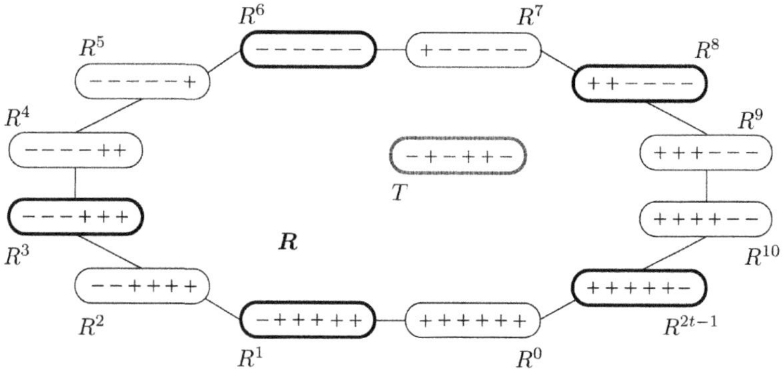

Figure 5.4 The distinguished symmetric cycle $\boldsymbol{R} := (R^0, R^1, \ldots, R^{2t-1}, R^0)$ in the hypercube graph $\boldsymbol{H}((t := 6), 2)$, defined by (5.2)(5.3).
For the vertex $T := {}_{-A}\mathrm{T}^{(+)} \in \{1, -1\}^t$ of the graph $\boldsymbol{H}(t, 2)$, where $A := \{1, 3, t\}$, we have $\{1, t\} \cap A = \{1, t\}$.
The set $\boldsymbol{Q}(T, \boldsymbol{R}) = \{R^1, R^3, R^6, R^8, R^{2t-1}\}$ is the unique inclusion-minimal subset of the vertex set $\mathrm{V}(\boldsymbol{R})$ of the cycle $\boldsymbol{R}$ such that $T = \sum_{Q \in \boldsymbol{Q}(T,\boldsymbol{R})} Q$.

where $\varrho = 2$, and

$$i_1 = 2\,, \quad j_1 = 3\,, \quad \text{and} \quad i_\varrho = j_\varrho = 5\,.$$

Since $|\{1, t\} \cap A| = 0$, by Proposition 5.9(iii) we have

$$\begin{aligned}
\mathfrak{q}(A) &= 2\varrho + 1 = 2 \cdot 2 + 1 = 5\,, \\
\boldsymbol{x}(A) &= \boldsymbol{\sigma}(1) + \sum_{1 \le k \le \varrho} \boldsymbol{\sigma}(j_k + 1) - \sum_{1 \le \ell \le \varrho} \boldsymbol{\sigma}(i_\ell) \\
&= \boldsymbol{\sigma}(1) + \big(\boldsymbol{\sigma}(4) + \boldsymbol{\sigma}(6)\big) - \big(\boldsymbol{\sigma}(2) + \boldsymbol{\sigma}(5)\big) \\
&= (1, -1, 0, 1, -1, 1)\,.
\end{aligned}$$

We have $\boldsymbol{Q}({}_{-A}\mathrm{T}^{(+)}, \boldsymbol{R}) = \{R^0, \underbrace{-R^1}_{R^7}, R^3, \underbrace{-R^4}_{R^{10}}, R^5\} = \{R^0, R^3, R^5, R^7, R^{10}\}$; see Figure 5.5.

(iv) The subset $A := \{3, 5, t\}$ of the set E_t is the disjoint union of inclusion-maximal intervals

$$A = [i_1, j_1] \,\dot{\cup}\, [i_\varrho, j_\varrho]\,,$$

where $\varrho = 2$, and

$$i_1 = j_1 = 3\,, \quad i_\varrho = 5\,, \quad \text{and} \quad j_\varrho = t\,.$$

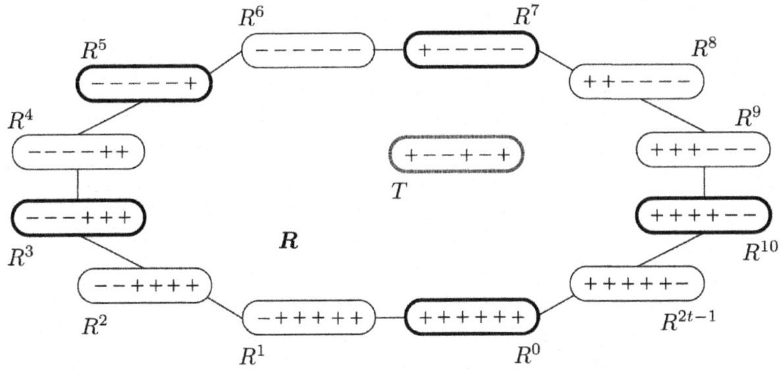

Figure 5.5 The distinguished symmetric cycle $\boldsymbol{R} := (R^0, R^1, \ldots, R^{2t-1}, R^0)$ in the hypercube graph $\boldsymbol{H}((t := 6), 2)$, defined by (5.2)(5.3).
For the vertex $T := {}_{-A}\mathrm{T}^{(+)} \in \{1, -1\}^t$ of the graph $\boldsymbol{H}(t, 2)$, where $A := \{2, 3, 5\}$, we have $|\{1, t\} \cap A| = 0$.
The set $\boldsymbol{Q}(T, \boldsymbol{R}) = \{R^0, R^3, R^5, R^7, R^{10}\}$ is the unique inclusion-minimal subset of the vertex set $\mathrm{V}(\boldsymbol{R})$ of the cycle $\boldsymbol{R}$ such that $T = \sum_{Q \in \boldsymbol{Q}(T,\boldsymbol{R})} Q$.

Since $\{1, t\} \cap A = \{t\}$, by Proposition 5.9(iv) we have

$$\mathfrak{q}(A) = 2\varrho - 1 = 2 \cdot 2 - 1 = 3\,,$$

$$\boldsymbol{x}(A) = \sum_{1 \le k \le \varrho - 1} \boldsymbol{\sigma}(j_k + 1) - \sum_{1 \le \ell \le \varrho} \boldsymbol{\sigma}(i_\ell)$$

$$= \boldsymbol{\sigma}(4) - \big(\boldsymbol{\sigma}(3) + \boldsymbol{\sigma}(5)\big) = (0, 0, -1, 1, -1, 0)\,.$$

Indeed, we have $\boldsymbol{Q}({}_{-A}\mathrm{T}^{(+)}, \boldsymbol{R}) = \{\underbrace{-R^2}_{R^8}, R^3, \underbrace{-R^4}_{R^{10}}\} = \{R^3, R^8, R^{10}\}$; see Figure 5.6.

Note that for any vertex $T \in \{1, -1\}^t$ of the hypercube graph $\boldsymbol{H}(t, 2)$, and for elements $e \in E_t$, we have

$$x_e(T) \neq 0 \implies x_e(T) = T(e)\,.$$

Note also that if $x_e(T') \neq 0$ and $x_e(T'') \neq 0$, for vertices T' and T'', and for some $e \in E_t$, then

$$e \notin \mathbf{S}(T', T'') \implies x_e(T'') = x_e(T')\,;$$

$$e \in \mathbf{S}(T', T'') \implies x_e(T'') = -x_e(T')\,.$$

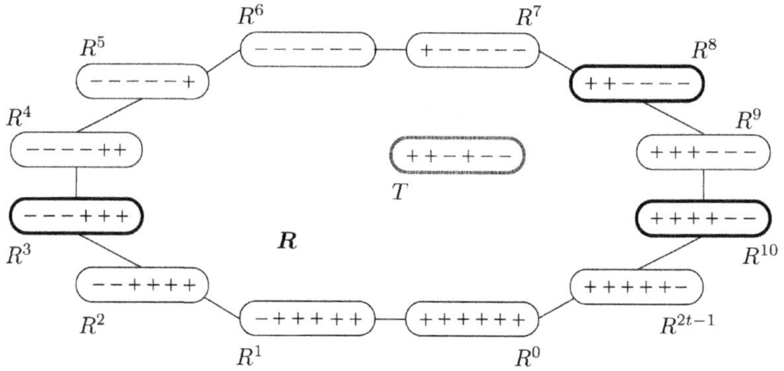

Figure 5.6 The distinguished symmetric cycle $\boldsymbol{R} := (R^0, R^1, \ldots, R^{2t-1}, R^0)$ in the hypercube graph $\boldsymbol{H}((t := 6), 2)$, defined by (5.2)(5.3).
For the vertex $T := {}_{-A}\mathrm{T}^{(+)} \in \{1, -1\}^t$ of the graph $\boldsymbol{H}(t, 2)$, where $A := \{3, 5, t\}$, we have $\{1, t\} \cap A = \{t\}$.
The set $\boldsymbol{Q}(T, \boldsymbol{R}) = \{R^3, R^8, R^{10}\}$ is the unique inclusion-minimal subset of the vertex set $\mathrm{V}(\boldsymbol{R})$ of the cycle $\boldsymbol{R}$ such that $T = \sum_{Q \in \boldsymbol{Q}(T,\boldsymbol{R})} Q$.

Recall that a *composition* of a positive integer n with m positive parts by definition is any tuple $(c_1, \ldots, c_m) \in \mathbb{P}^m$, such that $\sum_{i=1}^{m} c_i = n$.

We let $\mathrm{c}(m; n) := \binom{n-1}{m-1}$ denote the total number of *compositions* of a positive integer n with m positive parts.

Remark 5.11.

(i) It follows directly from the basic enumerative result on *compositions of integers* (see, e.g., page 17 of Ref. [21], Theorem 1.3 in Ref. [22], and page 18 of Ref. [7]) that there are precisely $2\mathrm{c}(2\varrho; t) = 2\binom{t-1}{2\varrho-1}$ subsets $A \subset E_t$, with their partitions (5.13)(5.14) into inclusion-maximal intervals, such that $|\{1, t\} \cap A| = 1$.

(ii) There are $\mathrm{c}(2\varrho - 1; t) = \binom{t-1}{2(\varrho-1)}$ subsets $A \subseteq E_t$, with their partitions (5.13)(5.14), such that $|\{1, t\} \cap A| = 2$.

(iii) There are $\mathrm{c}(2\varrho + 1; t) = \binom{t-1}{2\varrho}$ subsets $A \subset E_t$, with their partitions (5.13)(5.14), such that $|\{1, t\} \cap A| = 0$.

Example 5.12. Suppose $t := 5$.

(i) According to Remark 5.11(i), there are $2\binom{t-1}{2\varrho-1} := 2\binom{5-1}{2\cdot 2-1} = 8$ subsets $A \subset E_t$, with their partitions (5.13)(5.14) into $\varrho := 2$ intervals, such that $|\{1, t\} \cap A| = 1$:

$$\{1,\ 2,\ \ ,\ 4,\ \ \} \subset E_t\,, \tag{5.16}$$

$$\{1,\ \ ,\ 3,\ \ ,\ \ \}\,,$$

$$\{1,\ \ ,\ 3,\ 4,\ \ \}\,, \tag{5.17}$$

$$\{1,\ \ ,\ \ ,\ 4,\ \ \}\,,$$

and

$$\{\ \ ,\ 2,\ \ ,\ \ ,\ t\} \subset E_t\,,$$

$$\{\ \ ,\ 2,\ 3,\ \ ,\ t\}\,, \tag{5.18}$$

$$\{\ \ ,\ 2,\ \ ,\ 4,\ t\}\,, \tag{5.19}$$

$$\{\ \ ,\ \ ,\ 3,\ \ ,\ t\}\,.$$

(ii) By Remark 5.11(ii), there are $\binom{t-1}{2(\varrho-1)} := \binom{5-1}{2(2-1)} = 6$ subsets $A \subset E_t$, with their partitions (5.13)(5.14) into $\varrho := 2$ intervals, such that $|\{1, t\} \cap A| = 2$:

$$\{1,\ 2,\ 3,\ \ ,\ t\} \subset E_t\,,$$

$$\{1,\ 2,\ \ ,\ 4,\ t\}\,,$$

$$\{1,\ 2,\ \ ,\ \ ,\ t\}\,, \tag{5.20}$$

$$\{1,\ \ ,\ 3,\ 4,\ t\}\,,$$

$$\{1,\ \ ,\ \ ,\ 4,\ t\}\,, \tag{5.21}$$

$$\{1,\ \ ,\ \ ,\ \ ,\ t\}\,.$$

(iii) By Remark 5.11(iii), since $\binom{t-1}{2\varrho} := \binom{5-1}{2\cdot 2} = 1$, there is only one subset $A \subset E_t$, with its partition (5.13)(5.14) into $\varrho := 2$ intervals, such that $|\{1, t\} \cap A| = 0$:

$$\{\ \ ,\ 2,\ \ ,\ 4,\ \ \} \subset E_t\,.$$

As recalled in Subsection 2.1.10, for an odd integer $\ell \in E_t$, there are $2\binom{t}{\ell}$ vertices T of the hypercube graph $\boldsymbol{H}(t, 2)$ such that $\mathfrak{q}(T) := \mathfrak{q}(T, \boldsymbol{R}) := |\boldsymbol{Q}(T, \boldsymbol{R})| = \ell$; see Theorem 13.6 of Ref. [5].

Lemma 5.13. *Let ℓ be an odd integer such that $3 \leq \ell \leq t$. Consider the symmetric cycle $\boldsymbol{R}$ in the hypercube graph $\boldsymbol{H}(t, 2)$,* *defined by* (5.2)(5.3), *and the subset of vertices*

$$\{T \in \{1, -1\}^t \colon \mathfrak{q}(T) = \ell\} . \tag{5.22}$$

(i) (a) *In the set* (5.22) *there are* $\binom{t-1}{\ell}$ *vertices* T *whose negative parts* $\mathfrak{n}(T)$ *are disjoint unions*

$$\begin{aligned} \mathfrak{n}(T) = [i_1, j_1] \,\dot{\cup}\, [i_2, j_2] \,\dot{\cup}\, \cdots \,\dot{\cup}\, [i_{(\ell+1)/2}, j_{(\ell+1)/2}] \colon \\ j_1 + 2 \le i_2, \quad j_2 + 2 \le i_3, \quad \ldots, \quad j_{(\ell-1)/2} + 2 \le i_{(\ell+1)/2} , \end{aligned} \tag{5.23}$$

of $\frac{\ell+1}{2}$ *inclusion-maximal intervals of* E_t, *and*

$$\{1, t\} \cap \mathfrak{n}(T) = \{i_1\} = \{1\} .$$

More precisely, if

$$\frac{\ell+1}{2} \le j \le t - \frac{\ell+1}{2} ,$$

then in the set (5.22) *there are*

$$\mathrm{c}\left(\frac{\ell+1}{2}; j\right) \cdot \mathrm{c}\left(\frac{\ell+1}{2}; t-j\right) = \binom{j-1}{(\ell-1)/2}\binom{t-j-1}{(\ell-1)/2}$$

vertices T *whose negative parts* $\mathfrak{n}(T)$, *of cardinality* j, *are disjoint unions* (5.23) *of* $\frac{\ell+1}{2}$ *intervals of* E_t, *such that* $\{1, t\} \cap \mathfrak{n}(T) = \{i_1\} = \{1\}$.

(b) *In the set* (5.22) *there are* $\binom{t-1}{\ell}$ *vertices* T *whose negative parts* $\mathfrak{n}(T)$ *are disjoint unions* (5.23) *of* $\frac{\ell+1}{2}$ *inclusion-maximal intervals of* E_t, *and*

$$\{1, t\} \cap \mathfrak{n}(T) = \{j_{(\ell+1)/2}\} = \{t\} .$$

More precisely, if $\frac{\ell+1}{2} \le j \le t - \frac{\ell+1}{2}$, *then in the set* (5.22) *there are* $\binom{j-1}{(\ell-1)/2}\binom{t-j-1}{(\ell-1)/2}$ *vertices* T *whose negative parts* $\mathfrak{n}(T)$, *with* $|\,\mathfrak{n}(T)| = j$, *are disjoint unions* (5.23) *of* $\frac{\ell+1}{2}$ *intervals of* E_t, *such that* $\{1, t\} \cap \mathfrak{n}(T) = \{j_{(\ell+1)/2}\} = \{t\}$.

(ii) *In the set* (5.22) *there are* $\binom{t-1}{\ell-1}$ *vertices* T *whose negative parts* $\mathfrak{n}(T)$ *are disjoint unions* (5.23) *of* $\frac{\ell+1}{2}$ *inclusion-maximal intervals of* E_t, *and*

$$\{1, t\} \cap \mathfrak{n}(T) = \{i_1, j_{(\ell+1)/2}\} = \{1, t\} .$$

More precisely, if

$$\frac{\ell+1}{2} \le j \le t - \frac{\ell-1}{2} ,$$

then in the set (5.22) *there are*

$$\mathrm{c}\left(\frac{\ell+1}{2};j\right)\cdot\mathrm{c}\left(\frac{\ell-1}{2};t-j\right)=\binom{j-1}{(\ell-1)/2}\binom{t-j-1}{(\ell-3)/2}$$

vertices T *whose negative parts* $\mathfrak{n}(T)$, *of cardinality* j, *are disjoint unions* (5.23) *of* $\frac{\ell+1}{2}$ *intervals of* E_t, *such that* $\{1, t\} \cap \mathfrak{n}(T) = \{i_1, j_{(\ell+1)/2}\} = \{1, t\}$.

(iii) *In the set* (5.22) *there are* $\binom{t-1}{\ell-1}$ *vertices* T *whose negative parts* $\mathfrak{n}(T)$ *are disjoint unions*

$$\begin{aligned}\mathfrak{n}(T) = [i_1, j_1] \,\dot{\cup}\, [i_2, j_2] \,\dot{\cup}\, \cdots \,\dot{\cup}\, [i_{(\ell-1)/2}, j_{(\ell-1)/2}]:\\ j_1+2 \le i_2,\ \ j_2+2 \le i_3,\ \ \ldots,\ \ j_{(\ell-3)/2}+2 \le i_{(\ell-1)/2}\,,\end{aligned} \tag{5.24}$$

of $\frac{\ell-1}{2}$ *inclusion-maximal intervals of* E_t, *and*

$$|\{1, t\} \cap \mathfrak{n}(T)| = 0\,.$$

More precisely, if

$$\frac{\ell-1}{2} \le j \le t - \frac{\ell+1}{2}\,,$$

then in the set (5.22) *there are*

$$\mathrm{c}\left(\frac{\ell-1}{2};j\right)\cdot\mathrm{c}\left(\frac{\ell+1}{2};t-j\right)=\binom{j-1}{(\ell-3)/2}\binom{t-j-1}{(\ell-1)/2}$$

vertices T *whose negative parts* $\mathfrak{n}(T)$, *of cardinality* j, *are disjoint unions* (5.24) *of* $\frac{\ell-1}{2}$ *intervals of* E_t, *such that* $|\{1, t\} \cap \mathfrak{n}(T)| = 0$.

Example 5.14. Suppose $t := 5$, $j := 3$, and $\ell := 3$. Consider the symmetric cycle $\boldsymbol{R}$ in the hypercube graph $\boldsymbol{H}(t, 2)$, defined by (5.2)(5.3), and the subset of vertices

$$\{T \in \{1, -1\}^t \colon \mathfrak{q}(T) = \ell\}\,. \tag{5.25}$$

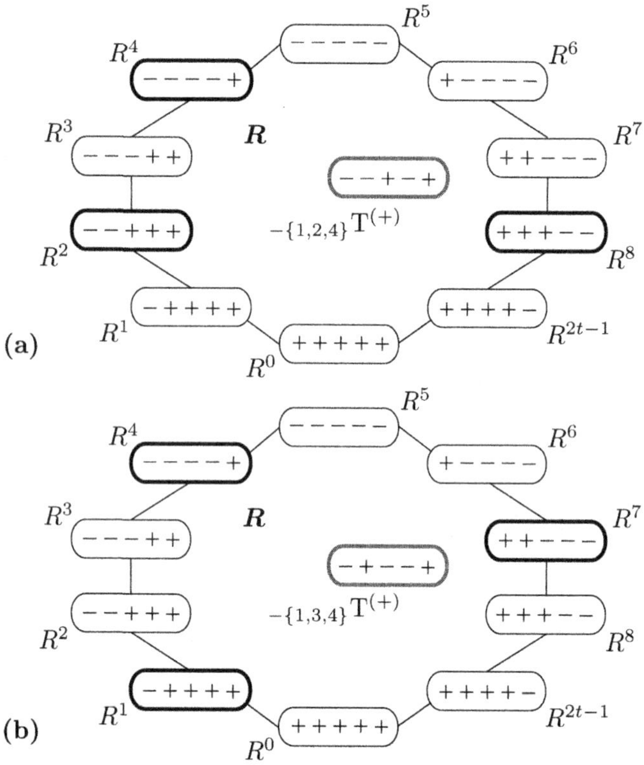

Figure 5.7 Two instances of the same distinguished symmetric cycle $\boldsymbol{R} := (R^0, R^1, \ldots, R^{2t-1}, R^0)$ in the hypercube graph $\boldsymbol{H}((t := 5), 2)$, defined by (5.2)(5.3), and the two vertices T (that admit decompositions of size $\ell := 3$) of the graph, with the negative parts of cardinality $j := 3$, such that $\{1, t\} \cap \mathfrak{n}(T) = \{1\}$.
(a) The vertex $T := {}_{-\{1,2,4\}}\mathrm{T}^{(+)}$, with the corresponding decomposition set $\boldsymbol{Q}(T, \boldsymbol{R}) = \{R^2, R^4, R^8\}$.
(b) The vertex $T := {}_{-\{1,3,4\}}\mathrm{T}^{(+)}$, with the corresponding decomposition set $\boldsymbol{Q}(T, \boldsymbol{R}) = \{R^1, R^4, R^7\}$.

(i)(a) By Lemma 5.13(i)(a), in the set (5.25) there are

$$\binom{j-1}{(\ell-1)/2}\binom{t-j-1}{(\ell-1)/2} := \binom{3-1}{(3-1)/2}\binom{5-3-1}{(3-1)/2} = 2$$

vertices T whose negative parts $\mathfrak{n}(T)$, of cardinality $j := 3$, are disjoint unions (5.23) of $\frac{\ell+1}{2} := \frac{3+1}{2} = 2$ inclusion-maximal intervals of E_t, such that $\{1, t\} \cap \mathfrak{n}(T) = \{i_1\} = \{1\}$.

These negative parts $\mathfrak{n}(T)$ are the sets (5.16) and (5.17):

$$\{1,\ 2,\ \ ,\ 4,\ \ \} \subset E_t\,,$$
$$\{1,\ \ ,\ 3,\ 4,\ \ \}\,,$$

and the two corresponding vertices T are

$$_{-\{1,2,4\}}\mathrm{T}^{(+)} := (-1,\ -1,\quad 1,\ -1,\quad 1) \in \{1,\ -1\}^t\,,$$

and

$$_{-\{1,3,4\}}\mathrm{T}^{(+)} := (-1,\quad 1,\ -1,\ -1,\quad 1)\,,$$

respectively.

Proposition 5.9(i) implies that

$$\boldsymbol{x}(\{1,\ 2,\ 4\}) := \boldsymbol{x}({}_{-\{1,2,4\}}\mathrm{T}^{(+)},\ \boldsymbol{R}) = (0,\quad 0,\quad 1,\ -1,\quad 1)\,,$$

and

$$\boldsymbol{x}(\{1,\ 3,\ 4\}) = (0,\quad 1,\ -1,\quad 0,\quad 1)\,;$$

see Figure 5.7.

(b) By Lemma 5.13(i)(b), in the set (5.25) there are two vertices T whose negative parts $\mathfrak{n}(T)$, with $|\,\mathfrak{n}(T)| = 3$, are disjoint unions (5.23) of two inclusion-maximal intervals of E_t, such that $\{1,\ t\} \cap \mathfrak{n}(T) = \{j_2\} = \{t\}$.

These are the vertices

$$_{-\{2,3,t\}}\mathrm{T}^{(+)} := (1,\ -1,\ -1,\quad 1,\ -1) \in \{1,\ -1\}^t\,,$$

and

$$_{-\{2,4,t\}}\mathrm{T}^{(+)} := (1,\ -1,\quad 1,\ -1,\ -1)\,,$$

with their negative parts (5.18) and (5.19):

$$\{\ \ ,\ 2,\ 3,\ \ ,\ t\} \subset E_t$$
$$\{\ \ ,\ 2,\ \ ,\ 4,\ t\}\,.$$

By Proposition 5.9(iv) we have

$$\boldsymbol{x}(\{2,\ 3,\ t\}) = (0,\ -1,\quad 0,\quad 1,\ -1)\,,$$

and

$$\boldsymbol{x}(\{2,\ 4,\ t\}) = (0,\ -1,\quad 1,\ -1,\quad 0)\,;$$

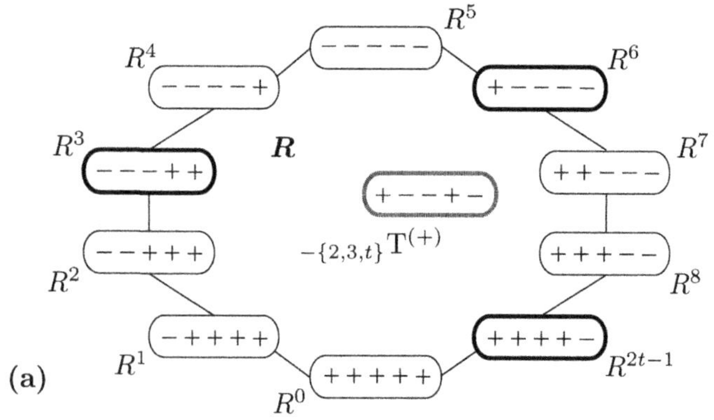

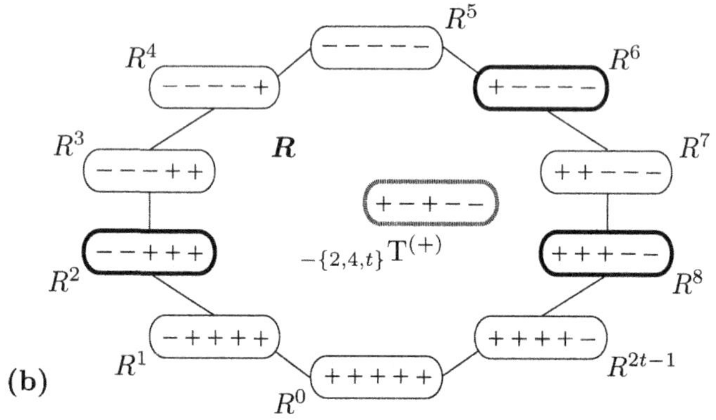

Figure 5.8 Two instances of the same distinguished symmetric cycle $\boldsymbol{R} := (R^0, R^1, \ldots, R^{2t-1}, R^0)$ in the hypercube graph $\boldsymbol{H}((t := 5), 2)$, defined by (5.2)(5.3), and the two vertices T (that admit decompositions of size $\ell := 3$) of the graph, with the negative parts of cardinality $j := 3$, such that $\{1, t\} \cap \mathfrak{n}(T) = \{t\}$.
(a) The vertex $T := {}_{-\{2,3,t\}}\mathrm{T}^{(+)}$, with the corresponding decomposition set $\boldsymbol{Q}(T, \boldsymbol{R}) = \{R^3, R^6, R^{2t-1}\}$.
(b) The vertex $T := {}_{-\{2,4,t\}}\mathrm{T}^{(+)}$, with the corresponding decomposition set $\boldsymbol{Q}(T, \boldsymbol{R}) = \{R^2, R^6, R^8\}$.

see Figure 5.8.

(ii) By Lemma 5.13(ii), in the set (5.25) there are

$$\binom{j-1}{(\ell-1)/2}\binom{t-j-1}{(\ell-3)/2} := \binom{3-1}{(3-1)/2}\binom{5-3-1}{(3-3)/2} = 2$$

vertices T whose negative parts $\mathfrak{n}(T)$, of cardinality $j := 3$, are disjoint unions (5.23) of $\frac{\ell+1}{2} := \frac{3+1}{2} = 2$ inclusion-maximal

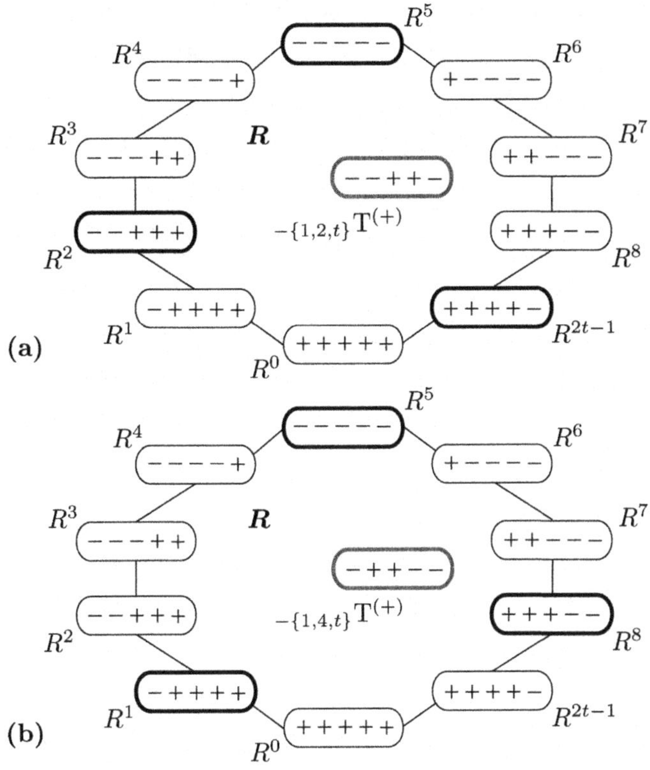

Figure 5.9 Two instances of the same distinguished symmetric cycle $\boldsymbol{R} := (R^0, R^1, \ldots, R^{2t-1}, R^0)$ in the hypercube graph $\boldsymbol{H}((t := 5), 2)$, defined by (5.2)(5.3), and the two vertices T (that admit decompositions of size $\ell := 3$) of the graph, with the negative parts of cardinality $j := 3$, such that $\{1, t\} \cap \mathfrak{n}(T) = \{1, t\}$.
(a) The vertex $T := {}_{-\{1,2,t\}}\mathrm{T}^{(+)}$, with the corresponding decomposition set $\boldsymbol{Q}(T, \boldsymbol{R}) = \{R^2, R^5, R^{2t-1}\}$.
(b) The vertex $T := {}_{-\{1,4,t\}}\mathrm{T}^{(+)}$, with the corresponding decomposition set $\boldsymbol{Q}(T, \boldsymbol{R}) = \{R^1, R^5, R^8\}$.

intervals of E_t, such that $\{1, t\} \cap \mathfrak{n}(T) = \{i_1, j_{(\ell+1)/2}\} := \{i_1, j_{(3+1)/2}\} = \{i_1, j_2\} = \{1, t\}$.

These negative parts $\mathfrak{n}(T)$ are the sets (5.20) and (5.21):

$$\{1,\ 2,\ \ ,\ \ ,\ t\} \subset E_t\,,$$

$$\{1,\ \ ,\ \ ,\ 4,\ t\}\,,$$

and the two corresponding vertices T are

$$_{-\{1,2,t\}}\mathrm{T}^{(+)} := (-1, -1, \quad 1, \quad 1, -1) \in \{1, -1\}^t ,$$

and

$$_{-\{1,4,t\}}\mathrm{T}^{(+)} := (-1, \quad 1, \quad 1, -1, -1) .$$

Using Proposition 5.9(ii), we have

$$\boldsymbol{x}(\{1, 2, t\}) = (-1, \quad 0, \quad 1, \quad 0, -1) ,$$

and

$$\boldsymbol{x}(\{1, 4, t\}) = (-1, \quad 1, \quad 0, -1, \quad 0) ;$$

see Figure 5.9.

(iii) Since

$$\binom{j-1}{(\ell-3)/2}\binom{t-j-1}{(\ell-1)/2} := \binom{3-1}{(3-3)/2}\binom{5-3-1}{(3-1)/2} = 1 ,$$

Lemma 5.13(iii) implies that in the set (5.25) there is only one vertex T whose negative part $\mathfrak{n}(T)$, of cardinality $j := 3$, is an interval (because $\frac{\ell-1}{2} := \frac{3-1}{2} = 1$) of E_t, such that $|\{1, t\} \cap \mathfrak{n}(T)| = 0$.

This is the vertex

$$_{-\{2,3,4\}}\mathrm{T}^{(+)} := (1, -1, -1, -1, \quad 1) ,$$

with its negative part

$$\{ \ , 2, 3, 4, \ \} \subset E_t .$$

Formula (5.15) derived from Proposition 5.9(iii) shows that

$$\boldsymbol{x}(\{2, 3, 4\}) = (1, -1, \quad 0, \quad 0, \quad 1) ;$$

see Figure 5.10.

We can now give a refined statistic on the decompositions of vertices with respect to the distinguished symmetric cycle.

Theorem 5.15. *Pick some $j \in E_t$, and let ℓ be an odd integer such that $3 \le \ell \le t$. Consider the symmetric cycle $\boldsymbol{R}$ in the hypercube graph $\boldsymbol{H}(t, 2)$, defined by* (5.2)(5.3)*.*

(i) *If*

$$j < \frac{\ell-1}{2} \quad or \quad j > t - \frac{\ell-1}{2} ,$$

then

$$|\{T \in \{1, -1\}^t : |\mathfrak{n}(T)| = j, \ \mathfrak{q}(T) = \ell\}| = 0 .$$

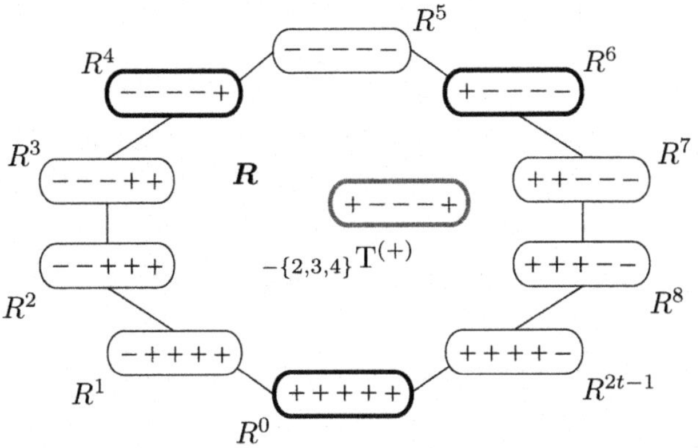

Figure 5.10 The distinguished symmetric cycle $\boldsymbol{R} := (R^0, R^1, \ldots, R^{2t-1}, R^0)$ in the hypercube graph $\boldsymbol{H}((t := 5), 2)$, defined by (5.2)(5.3), and the vertex $T := {}_{-\{2,3,4\}}\mathrm{T}^{(+)}$ (that admits the decomposition of size $\ell := 3$) of the graph, with the negative part of cardinality $j := 3$, such that $|\{1, t\} \cap \mathfrak{n}(T)| = 0$.
The corresponding decomposition set is $\boldsymbol{Q}(T, \boldsymbol{R}) = \{R^0, R^4, R^6\}$.

(ii) *If*

$$\frac{\ell-1}{2} \leq j \leq t - \frac{\ell-1}{2},$$

then

$$\begin{aligned}
&|\{T \in \{1,-1\}^t : |\mathfrak{n}(T)| = j,\ \mathfrak{q}(T) = \ell\}| \\
&= |\{T \in \{1,-1\}^t : |\mathfrak{n}(T)| = t - j,\ \mathfrak{q}(T) = \ell\}| \\
&= 2\mathrm{c}\left(\frac{\ell+1}{2}; j\right) \cdot \mathrm{c}\left(\frac{\ell+1}{2}; t-j\right) + \mathrm{c}\left(\frac{\ell+1}{2}; j\right) \\
&\times \mathrm{c}\left(\frac{\ell-1}{2}; t-j\right) + \mathrm{c}\left(\frac{\ell-1}{2}; j\right) \cdot \mathrm{c}\left(\frac{\ell+1}{2}; t-j\right) \\
&= \binom{j-1}{\frac{\ell-1}{2}}\binom{t-j}{\frac{\ell-1}{2}} + \binom{t-j-1}{\frac{\ell-1}{2}}\binom{j}{\frac{\ell-1}{2}} \qquad (5.26) \\
&= \mathrm{c}\left(\frac{\ell+1}{2}; j\right) \cdot \mathrm{c}\left(\frac{\ell+1}{2}; t-j+1\right) + \mathrm{c}\left(\frac{\ell+1}{2}; t-j\right) \\
&\times \mathrm{c}\left(\frac{\ell+1}{2}; j+1\right).
\end{aligned}$$

In particular, we have

$$|\{T \in \{1,-1\}^t \colon |\mathfrak{n}(T)| = \frac{\ell-1}{2},\ \mathfrak{q}(T) = \ell\}| = \binom{t-\frac{\ell+1}{2}}{\frac{\ell-1}{2}} = \binom{t-\frac{\ell+1}{2}}{t-\ell}, \quad (5.27)$$

and for j such that $1 \le j \le t-1$, we have

$$|\{T \in \{1,-1\}^t \colon |\mathfrak{n}(T)| = j,\ \mathfrak{q}(T) = 3\}| = 2j(t-j) - t. \quad (5.28)$$

Example 5.16. Suppose $t := 5$, $j := 3$, and $\ell := 3$. By formula (5.26) we have

$$\begin{aligned} &|\{T \in \{1,-1\}^t \colon |\mathfrak{n}(T)| = j,\ \mathfrak{q}(T) = \ell\}| \\ &= \binom{j-1}{\frac{\ell-1}{2}}\binom{t-j}{\frac{\ell-1}{2}} + \binom{t-j-1}{\frac{\ell-1}{2}}\binom{j}{\frac{\ell-1}{2}} \\ &:= \binom{3-1}{\frac{3-1}{2}}\binom{5-3}{\frac{3-1}{2}} + \binom{5-3-1}{\frac{3-1}{2}}\binom{3}{\frac{3-1}{2}} = 7. \end{aligned} \quad (5.29)$$

Indeed, in Example 5.14 we enumerated all seven vertices composing the set $\{T \in \{1,-1\}^t \colon |\mathfrak{n}(T)| = j,\ \mathfrak{q}(T) = \ell\}$.

Formula (5.27) yields

$$\begin{aligned} &|\{T \in \{1,-1\}^t \colon |\mathfrak{n}(T)| = \frac{\ell-1}{2},\ \mathfrak{q}(T) = \ell\}| \quad (5.30) \\ &= \binom{t-\frac{\ell+1}{2}}{t-\ell} := \binom{5-\frac{3+1}{2}}{5-3} = 3. \end{aligned}$$

Indeed, the set (5.30) is the set $\{{}_{-2}\mathrm{T}^{(+)}, {}_{-3}\mathrm{T}^{(+)}, {}_{-4}\mathrm{T}^{(+)}\}$.

Using formula (5.28), we have

$$\begin{aligned} &|\{T \in \{1,-1\}^t \colon |\mathfrak{n}(T)| = j,\ \mathfrak{q}(T) = 3\}| \\ &= 2j(t-j) - t := 2 \cdot 3(5-3) - 5 = 7; \end{aligned}$$

cf. the result (5.29).

5.3 The Sizes of Decompositions

For the symmetric cycle $\boldsymbol{R}$ in the hypercube graph $\boldsymbol{H}(t,2)$, defined by (5.2)(5.3), the (i,j)th entry of the symmetric Toeplitz matrix $\mathbf{M} \cdot \mathbf{M}^\top$, where $\mathbf{M} := \mathbf{M}(\boldsymbol{R})$ is given in (5.5), is

$$t - 2|j-i|.$$

The ith row $(\mathbf{M}^{-1} \cdot (\mathbf{M}^{-1})^\top)_i$ of the symmetric Toeplitz matrix $\mathbf{M}^{-1} \cdot (\mathbf{M}^{-1})^\top$ is

$$
\left(\mathbf{M}^{-1} \cdot (\mathbf{M}^{-1})^\top\right)_i =
\begin{cases}
\frac{1}{4} \cdot \left(2\sigma(1) - \sigma(2) + \sigma(t)\right), & \text{if } i = 1, \\
\frac{1}{4} \cdot \left(-\sigma(i-1) + 2\sigma(i) + \sigma(i+1)\right), & \text{if } 2 \le i \le t-1, \\
\frac{1}{4} \cdot \left(\sigma(1) - \sigma(t-1) + 2\sigma(t)\right), & \text{if } i = t.
\end{cases}
\tag{5.31}
$$

See, e.g., Ref. [23] on Toeplitz matrices.

Example 5.17. Suppose $t := 6$. Let $\boldsymbol{R}$ be the distinguished symmetric cycle in the hypercube graph $\boldsymbol{H}(t, 2)$, defined by (5.2)(5.3). We have

$$
\mathbf{M} \cdot \mathbf{M}^\top = \begin{pmatrix}
6 & 4 & 2 & 0 & -2 & -4 \\
4 & 6 & 4 & 2 & 0 & -2 \\
2 & 4 & 6 & 4 & 2 & 0 \\
0 & 2 & 4 & 6 & 4 & 2 \\
-2 & 0 & 2 & 4 & 6 & 4 \\
-4 & -2 & 0 & 2 & 4 & 6
\end{pmatrix},
$$

and

$$
\mathbf{M}^{-1} \cdot (\mathbf{M}^{-1})^\top = \frac{1}{4} \begin{pmatrix}
2 & -1 & 0 & 0 & 0 & 1 \\
-1 & 2 & -1 & 0 & 0 & 0 \\
0 & -1 & 2 & -1 & 0 & 0 \\
0 & 0 & -1 & 2 & -1 & 0 \\
0 & 0 & 0 & -1 & 2 & -1 \\
1 & 0 & 0 & 0 & -1 & 2
\end{pmatrix}.
$$

Recall that for a vertex $T \in \{1, -1\}^t$ we have

$$
\begin{aligned}
\mathfrak{q}(T) :&= \mathfrak{q}(T, \boldsymbol{R}) := |\boldsymbol{Q}(T, \boldsymbol{R})| = \|\boldsymbol{x}(T)\|^2 \\
&= T \cdot \mathbf{M}^{-1} \cdot (\mathbf{M}^{-1})^\top \cdot T^\top .
\end{aligned}
$$

If $T', T'' \in \{1, -1\}^t$, then (5.10) implies that

$$
\begin{aligned}
\mathfrak{q}(T'') = \mathfrak{q}(T') - 4\Big\langle T'\mathbf{M}^{-1}, \Big(\sum_{s \in \mathbf{S}(T', T'')} T'(s)\sigma(s)\Big) \cdot \mathbf{M}^{-1}\Big\rangle \\
+ 4\Big\|\Big(\sum_{s \in \mathbf{S}(T', T'')} T'(s)\sigma(s)\Big) \cdot \mathbf{M}^{-1}\Big\|^2 ,
\end{aligned}
$$

and, as a consequence, we have

$$\mathfrak{q}(T') - \mathfrak{q}(T'') = 4\Big\langle \Big(T' - \Big(\sum_{s\in \mathbf{S}(T',T'')} T'(s)\sigma(s)\Big)\Big)\cdot \mathbf{M}^{-1},\ \Big(\sum_{s\in \mathbf{S}(T',T'')} T'(s)\sigma(s)\Big)\cdot \mathbf{M}^{-1}\Big\rangle . \qquad (5.32)$$

5.4 Equinumerous Decompositions of Vertices

If $T', T'' \in \{1, -1\}^t$ are two vertices of the hypercube graph $\boldsymbol{H}(t, 2)$ with its distinguished symmetric cycle $\boldsymbol{R}$, defined by (5.2)(5.3), then it follows from (5.32) that

$$\mathfrak{q}(T') := |\boldsymbol{Q}(T', \boldsymbol{R})| = |\boldsymbol{Q}(T'', \boldsymbol{R})| =: \mathfrak{q}(T'') \qquad (5.33)$$

if and only if

$$\Big\langle \Big(T' - \Big(\sum_{s\in \mathbf{S}(T',T'')} T'(s)\sigma(s)\Big)\Big)\cdot \mathbf{M}^{-1},\ \Big(\sum_{s\in \mathbf{S}(T',T'')} T'(s)\sigma(s)\Big)\cdot \mathbf{M}^{-1}\Big\rangle = 0 .$$

Let us denote by $\omega(i, j)$ the (i, j)th entry of the matrix $\mathbf{M}^{-1}\cdot(\mathbf{M}^{-1})^\top$ whose rows are given in (5.31). We see that (5.33) holds if and only if

$$\sum_{i\in E_t-\mathbf{S}(T',T'')}\ \sum_{j\in \mathbf{S}(T',T'')} T'(i)\cdot T'(j)\cdot \omega(i, j) = \sum_{i\in \mathbf{S}(T',T'')}\ \sum_{j\in E_t-\mathbf{S}(T',T'')} T'(i)\cdot T'(j)\cdot \omega(i, j) = 0$$

or, equivalently,

$$\sum_{i\in[t-1]}\ \sum_{\substack{j\in[i+1,t]:\\ |\{i,j\}\cap \mathbf{S}(T',T'')|=1}} T'(i)\cdot T'(j)\cdot \omega(i, j) = 0 . \qquad (5.34)$$

Proposition 5.18. *Let $\boldsymbol{R}$ be the symmetric cycle in the hypercube graph $\boldsymbol{H}(t, 2)$, defined by* (5.2)(5.3).

Let $T \in \{1, -1\}^t$ be a vertex of $\boldsymbol{H}(t, 2)$, and let A be a proper subset of the set E_t.

(i) *If* $|\{1, t\} \cap A| = 1$, *then*

$$\mathfrak{q}(T) = \mathfrak{q}({}_{-A}T) \iff \sum_{\substack{i\in[t-1]:\\ |\{i,i+1\}\cap A|=1}} T(i)\cdot T(i+1) = T(1)\cdot T(t)\,.$$

(ii) *If* $|\{1, t\} \cap A| \neq 1$, *then*

$$\mathfrak{q}(T) = \mathfrak{q}({}_{-A}T) \iff \sum_{\substack{i\in[t-1]:\\ |\{i,i+1\}\cap A|=1}} T(i)\cdot T(i+1) = 0\,.$$

Proof. Consider relation (5.34) for the vertices $T' := T$ and $T'' := {}_{-A}T'$ with their separation set $\mathbf{S}(T', T'') = A$.

(i) Since $|\{1, t\} \cap \mathbf{S}(T', T'')| = 1$, we have

$$\mathfrak{q}(T') = \mathfrak{q}(T'') \iff T'(1)\cdot T'(t)\cdot 1 + \sum_{i\in[t-1]} \sum_{\substack{j\in[i+1,t]:\\ |\{i,j\}\cap\mathbf{S}(T',T'')|=1}} T'(i)\cdot T'(j)\cdot(-1)$$

$$= T'(1)\cdot T'(t)\cdot 1 + \sum_{\substack{i\in[t-1]:\\ |\{i,i+1\}\cap\mathbf{S}(T',T'')|=1}} T'(i)\cdot T'(i+1)\cdot(-1) = 0\,.$$

(ii) Since $|\{1, t\} \cap \mathbf{S}(T', T'')| \neq 1$, we have

$$\mathfrak{q}(T') = \mathfrak{q}(T'') \iff \sum_{i\in[t-1]} \sum_{\substack{j\in[i+1,t]:\\ |\{i,j\}\cap\mathbf{S}(T',T'')|=1}} T'(i)\cdot T'(j)\cdot(-1)$$

$$= \sum_{\substack{i\in[t-1]:\\ |\{i,i+1\}\cap\mathbf{S}(T',T'')|=1}} T'(i)\cdot T'(i+1)\cdot(-1) = 0\,. \quad \square$$

We conclude this section with simple structural criteria (derived from Proposition 5.9) of the equicardinality of decompositions.

Corollary 5.19. *Let $\boldsymbol{R}$ be the symmetric cycle in the hypercube graph $\boldsymbol{H}(t, 2)$, defined by* (5.2)(5.3).

Let A and B be two nonempty subsets of the set E_t, and let

$$A = [i'_1, j'_1] \,\dot{\cup}\, [i'_2, j'_2] \,\dot{\cup}\, \cdots \,\dot{\cup}\, [i'_{\varrho(A)}, j'_{\varrho(A)}]$$

and

$$B = [i''_1, j''_1] \,\dot{\cup}\, [i''_2, j''_2] \,\dot{\cup}\, \cdots \,\dot{\cup}\, [i''_{\varrho(B)}, j''_{\varrho(B)}]$$

be their partitions into inclusion-maximal intervals such that

$$j'_1 + 2 \leq i'_2, \quad j'_2 + 2 \leq i'_3, \quad \ldots, \quad j'_{\varrho(A)-1} + 2 \leq i'_{\varrho(A)}$$

and

$$j''_1 + 2 \leq i''_2, \quad j''_2 + 2 \leq i''_3, \quad \ldots, \quad j''_{\varrho(B)-1} + 2 \leq i''_{\varrho(B)} \,.$$

(i) *If*

$$|\{1, t\} \cap A| > 0\,, \quad \textit{and} \quad |\{1, t\} \cap B| > 0\,,$$

or

$$|\{1, t\} \cap A| = |\{1, t\} \cap B| = 0\,,$$

then

$$\mathfrak{q}(A) := |\boldsymbol{Q}({}_{-A}\mathrm{T}^{(+)}, \boldsymbol{R})| = |\boldsymbol{Q}({}_{-B}\mathrm{T}^{(+)}, \boldsymbol{R})| =: \mathfrak{q}(B)$$
$$\Longleftrightarrow \quad \varrho(B) = \varrho(A)\,.$$

(ii) *If*

$$|\{1, t\} \cap A| > 0\,, \quad \textit{and} \quad |\{1, t\} \cap B| = 0\,,$$

then

$$\mathfrak{q}(A) = \mathfrak{q}(B) \quad \Longleftrightarrow \quad \varrho(B) = \varrho(A) - 1\,.$$

5.5 Decompositions, Inclusion–Exclusion and Valuations

Let $\boldsymbol{R}$ be the distinguished symmetric cycle in a hypercube graph $\boldsymbol{H}(t, 2)$, defined by (5.2)(5.3).

As will be noted in Remark 6.1 of Chapter 6, for *disjoint* subsets A and B of the ground set E_t, we have

$$\boldsymbol{x}({}_{-A}\mathrm{T}^{(+)}, \boldsymbol{R}) + \boldsymbol{x}({}_{-B}\mathrm{T}^{(+)}, \boldsymbol{R}) = \boldsymbol{\sigma}(1) + \boldsymbol{x}({}_{-(A\dot{\cup}B)}\mathrm{T}^{(+)}, \boldsymbol{R}) .$$

In fact, this is an *Inclusion–Exclusion* type relation, since $\boldsymbol{\sigma}(1) = \boldsymbol{x}(\mathrm{T}^{(+)}, \boldsymbol{R}) = \boldsymbol{x}({}_{-(A\cap B)}\mathrm{T}^{(+)}, \boldsymbol{R})$.

Turning to arbitrary vertices and symmetric cycles, we arrive at a general conclusion (see page 265 of Ref. [7] on the poset-theoretic context):

Proposition 5.20. *Let* $\boldsymbol{D}$ *be an arbitrary symmetric cycle in a hypercube graph* $\boldsymbol{H}(t, 2)$*, defined by* (5.1)*.*

(i) *The map*

$$\mathbb{B}(t) \to \mathbb{Z}^t , \quad A \mapsto \boldsymbol{x}({}_{-A}\mathrm{T}^{(+)}, \boldsymbol{D}) =: \boldsymbol{x}(A, \boldsymbol{D}) ,$$

is a valuation *on the* Boolean lattice $\mathbb{B}(t)$ *of subsets of the set* E_t*, since for any two sets* $A, B \in \mathbb{B}(t)$*, we have*

$$\boldsymbol{x}(A, \boldsymbol{D}) + \boldsymbol{x}(B, \boldsymbol{D}) = \boldsymbol{x}(A \cap B, \boldsymbol{D}) + \boldsymbol{x}(A \cup B, \boldsymbol{D}) .$$

As a consequence, we have

$$\boldsymbol{x}(A \cap B, \boldsymbol{D}) + \boldsymbol{x}(A\triangle B, \boldsymbol{D}) = \underbrace{\boldsymbol{x}(\hat{0}, \boldsymbol{D})}_{\boldsymbol{x}(\mathrm{T}^{(+)}, \boldsymbol{D})} + \boldsymbol{x}(A \cup B, \boldsymbol{D}) ,$$

where $A\triangle B$ *denotes the symmetric difference of the sets* A *and* B*.*

(ii) *Let* $\mathcal{A} := \{A_1, \ldots, A_\alpha\}$ *be a nonempty family of subsets of the set* E_t*. Let* $\mu(\cdot, \cdot) := \mu_{P(\mathcal{A})}(\cdot, \cdot)$ *denote the* Möbius function *of the poset* $P(\mathcal{A})$ *of all intersections* $\bigcap_{s\in S} A_s$*,* $S \subseteq [\alpha]$*, ordered by inclusion; the greatest element* $\hat{1} := \bigcup_{A\in\mathcal{A}} A$ *of* $P(\mathcal{A})$ *represents the empty intersection. We have*

$$\begin{aligned} \boldsymbol{x}(\bigcup_{A\in\mathcal{A}} A, \boldsymbol{D}) &= - \sum_{S\subseteq[\alpha]:\ |S|>0} (-1)^{|S|} \cdot \boldsymbol{x}(\bigcap_{s\in S} A_s, \boldsymbol{D}) \\ &= - \sum_{B\in P(\mathcal{A})-\{\hat{1}\}} \mu(B, \hat{1}) \cdot \boldsymbol{x}(B, \boldsymbol{D}) . \end{aligned}$$

Chapter 6

Distinguished Symmetric Cycles in Hypercube Graphs and Pairwise Decompositions of Vertices: Two-member Families of Disjoint Sets

In this chapter, we continue to investigate the decompositions of vertices $T \in \{1, -1\}^t$ of the *hypercube graph* $\boldsymbol{H}(t, 2)$ with respect to its distinguished symmetric cycle $\boldsymbol{R} := (R^0, R^1, \ldots, R^{2t-1}, R^0)$ whose vertex sequence $\vec{\mathrm{V}}(\boldsymbol{R}) := (R^0, R^1, \ldots, R^{2t-1})$ by definition is as follows:

$$R^0 := \mathrm{T}^{(+)} \,, \qquad R^s := {}_{-[s]}R^0 \,, \quad 1 \le s \le t-1 \,, \tag{6.1}$$

and

$$R^{k+t} := -R^k \,, \quad 0 \le k \le t-1 \,. \tag{6.2}$$

We consider partitions $\mathfrak{n}(T) = A \dot{\cup} B$ of the negative parts of vertices T of the hypercube graph $\boldsymbol{H}(t, 2)$ into two nonempty subsets A and B, and we describe the decomposition sets $\boldsymbol{Q}({}_{-A}\mathrm{T}^{(+)}, \boldsymbol{R}) \subset \mathrm{V}(\boldsymbol{R})$, $\boldsymbol{Q}({}_{-B}\mathrm{T}^{(+)}, \boldsymbol{R})$ and $\boldsymbol{Q}(T, \boldsymbol{R})$ of the vertices ${}_{-A}\mathrm{T}^{(+)}$, ${}_{-B}\mathrm{T}^{(+)}$ and T with respect to the symmetric cycle $\boldsymbol{R}$.

Symmetric Cycles
Andrey O. Matveev

ISBN 978-981-4968-81-2 (Hardcover), 978-1-003-43832-8 (eBook)
www.jennystanford.com

In Section 6.1 we show how a vector description of the decomposition set $\boldsymbol{Q}(T, \boldsymbol{R})$ can be restored from vector descriptions of the sets $\boldsymbol{Q}({}_{-A}\mathrm{T}^{(+)}, \boldsymbol{R})$ and $\boldsymbol{Q}({}_{-B}\mathrm{T}^{(+)}, \boldsymbol{R})$. In Section 6.2 we touch on the related question on a structural connection between the decomposition sets for vertices whose negative parts are comparable by inclusion. Enumerative results of Section 6.3 concern statistics on partitions of the negative parts of vertices of hypercube graphs and on decompositions of vertices. The key computational tool that allows us to present quite fine statistics is an approach to enumeration of ternary Smirnov words (i.e., words over a three-letter alphabet, such that adjacent letters in the words never coincide) discussed in Appendix A.

6.1 Partitions of the Negative Parts of Vertices into Two Subsets, and Decompositions of Vertices

Recall that for any vertex ${}_{-A}\mathrm{T}^{(+)} \in \{1, -1\}^t$ of the hypercube graph $\boldsymbol{H}(t, 2)$ with its distinguished symmetric cycle $\boldsymbol{R}$, defined by (6.1)(6.2), there exists a vector

$$\boldsymbol{x}(A) := \boldsymbol{x}({}_{-A}\mathrm{T}^{(+)}) := \boldsymbol{x}({}_{-A}\mathrm{T}^{(+)}, \boldsymbol{R}) = (x_1, \ldots, x_t) \in \{-1, 0, 1\}^t$$

such that the set

$$\boldsymbol{Q}({}_{-A}\mathrm{T}^{(+)}, \boldsymbol{R}) := \{x_i \cdot R^{i-1} : x_i \neq 0\}$$

is the unique inclusion-minimal subset of the vertex set $\mathrm{V}(\boldsymbol{R})$ of the cycle $\boldsymbol{R}$ for which we have

$${}_{-A}\mathrm{T}^{(+)} = \sum_{Q \in \boldsymbol{Q}({}_{-A}\mathrm{T}^{(+)}, \boldsymbol{R})} Q\,;$$

see Subsection 2.1.4. As earlier, we denote the cardinality of the decomposition set $\boldsymbol{Q}({}_{-A}\mathrm{T}^{(+)}, \boldsymbol{R})$ by

$$\mathfrak{q}(A) := \mathfrak{q}({}_{-A}\mathrm{T}^{(+)}) := |\boldsymbol{Q}({}_{-A}\mathrm{T}^{(+)}, \boldsymbol{R})|\,.$$

Remark 6.1. (See also Section 5.5.) It follows from Remark 5.6 that if A and B are *disjoint* subsets of the ground set E_t, and if we define, as earlier in Section 5.2, row vectors $\boldsymbol{y}(s) \in \{-1, 0, 1\}^t$, where $s \in E_t$, by

$$\boldsymbol{y}(s) := \boldsymbol{x}(\{s\}) := \boldsymbol{x}({}_{-s}\mathrm{T}^{(+)}) ,$$

then we have

$$\begin{aligned}\boldsymbol{x}(A\dot{\cup}B) &= (1 - |A| - |B|) \cdot \boldsymbol{\sigma}(1) + \sum_{s\in A} \boldsymbol{y}(s) + \sum_{s\in B} \boldsymbol{y}(s)\\ &= \boldsymbol{x}(A) - |B| \cdot \boldsymbol{\sigma}(1) + \sum_{s\in B} \boldsymbol{y}(s)\\ &= -|A| \cdot \boldsymbol{\sigma}(1) + \sum_{s\in A} \boldsymbol{y}(s) + \boldsymbol{x}(B) ,\end{aligned}$$

that is,

$$\boldsymbol{x}(A\dot{\cup}B) = -\boldsymbol{\sigma}(1) + \boldsymbol{x}(A) + \boldsymbol{x}(B) . \tag{6.3}$$

Example 6.2. Suppose $t := 6$, and consider the disjoint subsets $A := \{1, 2, 5\}$ and $B := \{4\}$ of the set E_t. For the corresponding vertices ${}_{-A}\mathrm{T}^{(+)}$ and ${}_{-B}\mathrm{T}^{(+)}$ of the hypercube graph $\boldsymbol{H}(t, 2)$, with its distinguished symmetric cycle $\boldsymbol{R}$ defined by (6.1)(6.2), we have

$$\begin{aligned}\boldsymbol{x}(A) &= (0,\ \ 0,\ \ 1,\ \ 0, -1,\ \ 1) ,\\ \boldsymbol{x}(B) &= (1,\ \ 0,\ \ 0, -1,\ \ 1,\ \ 0) ,\end{aligned}$$

see Proposition 5.9(i) and formula (5.15). Formula (6.3) yields

$$\begin{aligned}\boldsymbol{x}(A\dot{\cup}B) &= -\boldsymbol{\sigma}(1) + \boldsymbol{x}(A) + \boldsymbol{x}(B)\\ &:= -(1,\ \ 0,\ \ 0,\ \ 0,\ \ 0,\ \ 0) + (0,\ \ 0,\ \ 1,\ \ 0, -1,\ \ 1)\\ &\quad + (1,\ \ 0,\ \ 0, -1,\ \ 1,\ \ 0) = (0,\ \ 0,\ \ 1, -1,\ \ 0,\ \ 1) ,\end{aligned}$$

that is, $\boldsymbol{Q}({}_{-(A\dot{\cup}B)}\mathrm{T}^{(+)}, \boldsymbol{R}) = \{R^2, R^5, R^9\}$; see Figure 6.1.

As a consequence, we see that

$$\begin{aligned}&\|\boldsymbol{x}(A\dot{\cup}B)\|^2 = \|\boldsymbol{x}(A) + \boldsymbol{x}(B)\|^2 - 2\big(x_1(A) + x_1(B)\big) + 1\\ &= \|\boldsymbol{x}(A)\|^2 + \|\boldsymbol{x}(B)\|^2 + 2\langle \boldsymbol{x}(A), \boldsymbol{x}(B)\rangle - 2x_1(A) - 2x_1(B) + 1 .\end{aligned}$$

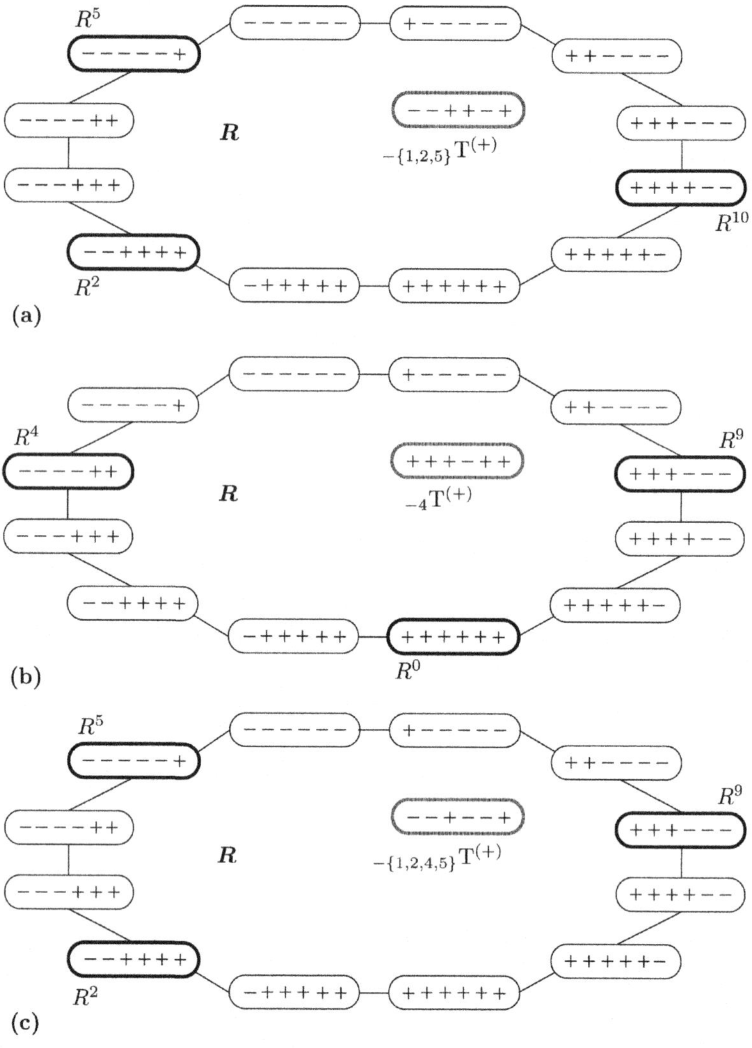

Figure 6.1 Three instances of the same distinguished symmetric cycle $\boldsymbol{R}$, defined by (6.1)(6.2), in the hypercube graph $\boldsymbol{H}((t := 6), 2)$, and the decomposition sets $\boldsymbol{Q}(T, \boldsymbol{R}) \subset \mathrm{V}(\boldsymbol{R})$ for three vertices $T \in \{1, -1\}^t$ of the graph $\boldsymbol{H}(t, 2)$, whose negative parts are some sets A, B, and $A \dot{\cup} B$.
(a) $\boldsymbol{Q}({}_{-\{1,2,5\}}\mathrm{T}^{(+)}, \boldsymbol{R}) = \{R^2, R^5, R^{10}\}$.
(b) $\boldsymbol{Q}({}_{-4}\mathrm{T}^{(+)}, \boldsymbol{R}) = \{R^0, R^4, R^9\}$.
(c) $\boldsymbol{Q}({}_{-\{1,2,4,5\}}\mathrm{T}^{(+)}, \boldsymbol{R}) = \{R^2, R^5, R^9\}$.

6.2 Inclusion of the Negative Parts of Vertices, and Decompositions of Vertices

Let us consider the vectors $\mathbf{x}(T')$ and $\mathbf{x}(T'')$ associated with vertices $T', T'' \in \{1, -1\}^t$ of the graph $\boldsymbol{H}(t, 2)$, whose negative parts $\mathfrak{n}(T')$ and $\mathfrak{n}(T'')$ are *comparable by inclusion*.

Proposition 6.3. *Let A and C be two sets such that*

$$A \subseteq C \subseteq E_t \, .$$

(i) *If*

$$\{1, t\} \cap A = \{1, t\} \cap C \, ,$$

then

$$\mathbf{x}(C) = \mathbf{x}(A) - \sum_{i \in C - A} (\, \sigma(i) - \sigma(i+1) \,) \, .$$

(ii) *If*

$$\{1, t\} \cap A = \{1\} \, , \quad \textit{and} \quad \{1, t\} \cap C = \{1, t\} \, ,$$

or

$$|\{1, t\} \cap A| = 0 \, , \quad \textit{and} \quad \{1, t\} \cap C = \{t\} \, ,$$

then

$$\mathbf{x}(C) = \mathbf{x}(A) - \sigma(1) - \sigma(t) - \sum_{i \in C - (A \dot{\cup} \{t\})} (\, \sigma(i) - \sigma(i+1) \,) \, .$$

(iii) *If*

$$|\{1, t\} \cap A| = 0 \, , \quad \textit{and} \quad \{1, t\} \cap C = \{1\} \, ,$$

or

$$\{1, t\} \cap A = \{t\} \, , \quad \textit{and} \quad \{1, t\} \cap C = \{1, t\} \, ,$$

then

$$\mathbf{x}(C) = \mathbf{x}(A) - \sigma(1) + \sigma(2) - \sum_{i \in C - (A \dot{\cup} \{1\})} (\, \sigma(i) - \sigma(i+1) \,) \, .$$

(iv) *If*

$$|\{1,\, t\} \cap A| = 0\,, \quad \textit{and} \quad \{1,\, t\} \cap C = \{1,\, t\}\,,$$

then

$$\begin{aligned}\mathbf{x}(C) = \mathbf{x}(A) - 2\sigma(1) + \sigma(2) - \sigma(t) \\ - \sum_{i \in C - (A \dot{\cup} \{1,t\})} (\, \sigma(i) - \sigma(i+1)\,)\,.\end{aligned}$$

Proof. We will use Remark 5.8:
(a) If

$$\{1,\, t\} \cap C = \{1\}\,,$$

then

$$\begin{aligned} A \subseteq C \quad \Longrightarrow \quad & \mathbf{x}(A) \\ & = \begin{cases} \sigma(2) - \sum_{i \in A - \{1\}}(\, \sigma(i) - \sigma(i+1)\,)\,, & \text{if } 1 \in A\,, \\ \sigma(1) - \sum_{i \in A}(\, \sigma(i) - \sigma(i+1)\,)\,, & \text{if } 1 \notin A \end{cases} \\ = & \begin{cases} \mathbf{x}(C) + \sum_{i \in C - A}(\, \sigma(i) - \sigma(i+1)\,)\,, & \text{if } 1 \in A\,, \\ \mathbf{x}(C) + \sigma(1) - \sigma(2) + \sum_{i \in C - (A \dot{\cup} \{1\})}(\, \sigma(i) - \sigma(i+1)\,)\,, & \\ & \text{if } 1 \notin A\,. \end{cases} \end{aligned}$$

(b) If

$$\{1,\, t\} \cap C = \{1,\, t\}\,,$$

then

$$\begin{aligned} A \subseteq C \quad \Longrightarrow \quad & \mathbf{x}(A) \\ = & \begin{cases} \sigma(2) - \sum_{i \in A - \{1\}}(\, \sigma(i) - \sigma(i+1)\,)\,, & \\ & \text{if } \{1,\, t\} \cap A = \{1\}\,, \\ -\sigma(1) + \sigma(2) - \sigma(t) - \sum_{i \in A - \{1,t\}}(\, \sigma(i) - \sigma(i+1)\,)\,, & \\ & \text{if } \{1,\, t\} \cap A = \{1,\, t\}\,, \\ \sigma(1) - \sum_{i \in A}(\, \sigma(i) - \sigma(i+1)\,)\,, \ \text{if } |\{1,\, t\} \cap A| = 0\,, & \\ -\sigma(t) - \sum_{i \in A - \{t\}}(\, \sigma(i) - \sigma(i+1)\,)\,, & \\ & \text{if } \{1,\, t\} \cap A = \{t\} \end{cases} \end{aligned}$$

$$= \begin{cases} \mathbf{x}(C) + \sigma(1) + \sigma(t) + \sum_{i \in C-(A \dot\cup \{t\})} (\,\sigma(i) - \sigma(i+1)\,)\,, \\ \qquad\qquad\qquad\qquad \text{if } \{1, t\} \cap A = \{1\}\,, \\ \mathbf{x}(C) + \sum_{i \in C-A} (\,\sigma(i) - \sigma(i+1)\,)\,, \\ \qquad\qquad\qquad\qquad \text{if } \{1, t\} \cap A = \{1, t\}\,, \\ \mathbf{x}(C) + 2\sigma(1) - \sigma(2) + \sigma(t) \\ \quad + \sum_{i \in C-(A \dot\cup \{1,t\})} (\,\sigma(i) - \sigma(i+1)\,)\,, \\ \qquad\qquad\qquad\qquad \text{if } |\{1, t\} \cap A| = 0\,, \\ \mathbf{x}(C) + \sigma(1) - \sigma(2) + \sum_{i \in C-(A \dot\cup \{1\})} (\,\sigma(i) - \sigma(i+1)\,)\,, \\ \qquad\qquad\qquad\qquad \text{if } \{1, t\} \cap A = \{t\}\,. \end{cases}$$

(c) If

$$|\{1, t\} \cap C| = 0\,,$$

then

$$A \subseteq C \implies \mathbf{x}(A) = \sigma(1) - \sum_{i \in A} (\,\sigma(i) - \sigma(i+1)\,)$$
$$= \mathbf{x}(C) + \sum_{i \in C-A} (\,\sigma(i) - \sigma(i+1)\,)\,.$$

(d) If

$$\{1, t\} \cap C = \{t\}\,,$$

then

$$A \subseteq C \implies \mathbf{x}(A)$$
$$= \begin{cases} \sigma(1) - \sum_{i \in A} (\,\sigma(i) - \sigma(i+1)\,)\,, & \text{if } t \notin A\,, \\ -\sigma(t) - \sum_{i \in A-\{t\}} (\,\sigma(i) - \sigma(i+1)\,)\,, & \text{if } t \in A \end{cases}$$
$$= \begin{cases} \mathbf{x}(C) + \sigma(1) + \sigma(t) + \sum_{i \in C-(A \dot\cup \{t\})} (\,\sigma(i) - \sigma(i+1)\,)\,, \\ \qquad\qquad\qquad\qquad \text{if } t \notin A\,, \\ \mathbf{x}(C) + \sum_{i \in C-A} (\,\sigma(i) - \sigma(i+1)\,)\,, \qquad \text{if } t \in A\,. \end{cases} \quad \square$$

6.3 Statistics on Partitions of the Negative Parts of Vertices of Hypercube Graphs and on Decompositions of Vertices

Given *odd* integers $\ell', \ell'', \ell \in E_t$, and positive integers j' and j'', in this section we obtain statistics related to the following family of *ordered pairs* (A, B) of disjoint *unordered subsets* A and B of the ground set E_t:

$$\begin{aligned}\{(A, B) \in \mathbf{2}^{[t]} \times \mathbf{2}^{[t]}:\ & |A \cap B| = 0,\\ & 0 < |A| = j',\quad 0 < |B| = j'',\quad j' + j'' < t,\\ & \mathfrak{q}(A) = \ell',\quad \mathfrak{q}(B) = \ell'',\quad \mathfrak{q}(A \dot{\cup} B) = \ell\}\,. \end{aligned} \tag{6.4}$$

Theorem 6.4.

(i) *In the family* (6.4) *there are*

$$\binom{t-(j'+j'')-1}{\frac{\ell-1}{2}}\binom{j'-1}{\frac{\ell'-3}{2}}\binom{j''-1}{\frac{\ell''-3}{2}} \times \begin{cases} \binom{\frac{\ell-1}{2}}{\frac{\ell+\ell'-\ell''-1}{4}}\binom{\frac{\ell+\ell'+\ell''-7}{4}}{\frac{\ell-3}{2}}\,, & \text{if } \frac{\ell+\ell'+\ell''-1}{2} \text{ odd}\,,\\ \frac{\ell+\ell'-\ell''+1}{2}\binom{\frac{\ell-1}{2}}{\frac{\ell+\ell'-\ell''+1}{4}}\binom{\frac{\ell+\ell'+\ell''-9}{4}}{\frac{\ell-3}{2}}\,, & \text{if } \frac{\ell+\ell'+\ell''-1}{2} \text{ even}\,, \end{cases} \tag{6.5}$$

pairs (A, B) *of sets* A *and* B *such that*

$$|\{1, t\} \cap A| = |\{1, t\} \cap B| = 0\,. \tag{6.6}$$

(ii) *In the family* (6.4) *there are*

$$\binom{t-(j'+j'')-1}{\frac{\ell-3}{2}}\binom{j'-1}{\frac{\ell'-1}{2}}\binom{j''-1}{\frac{\ell''-3}{2}} \times \begin{cases} \binom{\frac{\ell'-1}{2}}{\frac{\ell+\ell'-\ell''-1}{4}}\binom{\frac{\ell+\ell'+\ell''-7}{4}}{\frac{\ell'-3}{2}}\,, & \text{if } \frac{\ell+\ell'+\ell''-1}{2} \text{ odd}\,,\\ \frac{\ell+\ell'-\ell''+1}{2}\binom{\frac{\ell'-1}{2}}{\frac{\ell+\ell'-\ell''+1}{4}}\binom{\frac{\ell+\ell'+\ell''-9}{4}}{\frac{\ell'-3}{2}}\,, & \text{if } \frac{\ell+\ell'+\ell''-1}{2} \text{ even}\,, \end{cases} \tag{6.7}$$

pairs (A, B) *of sets* A *and* B *such that*

$$\{1, t\} \cap A = \{1, t\} \quad \textit{and} \quad |\{1, t\} \cap B| = 0\,. \tag{6.8}$$

(iii) *In the family* (6.4) *there are*

$$\binom{t-(j'+j'')-1}{\frac{\ell-1}{2}}\binom{j'-1}{\frac{\ell'-3}{2}}\binom{j''-1}{\frac{\ell''-1}{2}}$$
$$\times\begin{cases}\frac{\ell+\ell''-\ell'+3}{2}\binom{\frac{\ell-1}{2}}{\frac{\ell+\ell''-\ell'+1}{4}}\binom{\frac{\ell+\ell'+\ell''-5}{4}}{\frac{\ell-1}{2}}\ , \textit{if}\ \frac{\ell+\ell'+\ell''+1}{2}\ \textit{odd}\ ,\\ \binom{\frac{\ell-1}{2}}{\frac{\ell+\ell''-\ell'-1}{4}}\binom{\frac{\ell+\ell'+\ell''-3}{4}}{\frac{\ell-1}{2}}\\ \quad+\frac{\ell+\ell''-\ell'+3}{2}\binom{\frac{\ell-1}{2}}{\frac{\ell+\ell''-\ell'+3}{4}}\binom{\frac{\ell+\ell'+\ell''-7}{4}}{\frac{\ell-1}{2}}\ , \textit{if}\ \frac{\ell+\ell'+\ell''+1}{2}\ \textit{even}\ ,\end{cases} \tag{6.9}$$

pairs (A, B) *of sets* A *and* B *such that*

$$|\{1, t\}\cap A| = 0 \quad \textit{and} \quad \{1, t\}\cap B = \{t\}\,. \tag{6.10}$$

(iv) *In the family* (6.4) *there are*

$$\binom{t-(j'+j'')-1}{\frac{\ell-1}{2}}\binom{j'-1}{\frac{\ell'-1}{2}}\binom{j''-1}{\frac{\ell''-3}{2}}$$
$$\times\begin{cases}\frac{\ell+\ell'-\ell''+3}{2}\binom{\frac{\ell'-1}{2}}{\frac{\ell+\ell'-\ell''+1}{4}}\binom{\frac{\ell+\ell'+\ell''-5}{4}}{\frac{\ell'-1}{2}}\ , \textit{if}\ \frac{\ell+\ell'+\ell''+1}{2}\ \textit{odd}\ ,\\ \binom{\frac{\ell'-1}{2}}{\frac{\ell+\ell'-\ell''-1}{4}}\binom{\frac{\ell+\ell'+\ell''-3}{4}}{\frac{\ell'-1}{2}}\\ \quad+\frac{\ell+\ell'-\ell''+3}{2}\binom{\frac{\ell'-1}{2}}{\frac{\ell+\ell'-\ell''+3}{4}}\binom{\frac{\ell+\ell'+\ell''-7}{4}}{\frac{\ell'-1}{2}}\ , \textit{if}\ \frac{\ell+\ell'+\ell''+1}{2}\ \textit{even}\ ,\end{cases} \tag{6.11}$$

pairs (A, B) *of sets* A *and* B *such that*

$$\{1, t\}\cap A = \{1\} \quad \textit{and} \quad |\{1, t\}\cap B| = 0\,. \tag{6.12}$$

(v) *In the family* (6.4) *there are*

$$\binom{t-(j'+j'')-1}{\frac{\ell-3}{2}}\binom{j'-1}{\frac{\ell'-1}{2}}\binom{j''-1}{\frac{\ell''-1}{2}}$$
$$\times\begin{cases}\frac{\ell'+\ell''-\ell+3}{2}\binom{\frac{\ell'-1}{2}}{\frac{\ell'+\ell''-\ell+1}{4}}\binom{\frac{\ell+\ell'+\ell''-5}{4}}{\frac{\ell'-1}{2}}\ , \textit{if}\ \frac{\ell+\ell'+\ell''+1}{2}\ \textit{odd}\ ,\\ \binom{\frac{\ell'-1}{2}}{\frac{\ell'+\ell''-\ell-1}{4}}\binom{\frac{\ell+\ell'+\ell''-3}{4}}{\frac{\ell'-1}{2}}\\ \quad+\frac{\ell'+\ell''-\ell+3}{2}\binom{\frac{\ell'-1}{2}}{\frac{\ell'+\ell''-\ell+3}{4}}\binom{\frac{\ell+\ell'+\ell''-7}{4}}{\frac{\ell'-1}{2}}\ , \textit{if}\ \frac{\ell+\ell'+\ell''+1}{2}\ \textit{even}\ ,\end{cases} \tag{6.13}$$

pairs (A, B) of sets A and B such that

$$\{1, t\} \cap A = \{1\} \quad \textit{and} \quad \{1, t\} \cap B = \{t\} . \tag{6.14}$$

Before proceeding to the proof of the theorem, recall that the *Smirnov words* (see Appendix A on page 281) are defined to be the words, any two consecutive letters of which are distinct.

Let (θ, α, β) be an ordered three-letter alphabet.

Given two letters $\mathfrak{s}' \in (\theta, \alpha, \beta)$ and $\mathfrak{s}'' \in (\theta, \alpha, \beta)$, we denote by

$$\mathfrak{T}(\mathfrak{s}', \mathfrak{s}''; n(\theta), n(\alpha), n(\beta))$$

the number of ternary Smirnov words (that start with $\mathfrak{s}'$ and end with $\mathfrak{s}''$) with exactly $n(\theta)$ letters θ, with $n(\alpha)$ letters α, and with $n(\beta)$ letters β.

As earlier, we denote by $\mathrm{c}(m; n)$, where $\mathrm{c}(m; n) = \binom{n-1}{m-1}$, the number of *compositions* of a positive integer n with m positive parts.

We will regard sets A and B composing pairs of the family (6.4) as disjoint unions

$$A = [i'_1, k'_1] \dot\cup [i'_2, k'_2] \dot\cup \cdots \dot\cup [i'_{\varrho(A)}, k'_{\varrho(A)}]$$

and

$$B = [i''_1, k''_1] \dot\cup [i''_2, k''_2] \dot\cup \cdots \dot\cup [i''_{\varrho(B)}, k''_{\varrho(B)}]$$

of inclusion-maximal intervals $[i, k]$ such that

$$k'_1 + 2 \le i'_2, \quad k'_2 + 2 \le i'_3, \quad \ldots, \quad k'_{\varrho(A)-1} + 2 \le i'_{\varrho(A)}$$

and

$$k''_1 + 2 \le i''_2, \quad k''_2 + 2 \le i''_3, \quad \ldots, \quad k''_{\varrho(B)-1} + 2 \le i''_{\varrho(B)} .$$

Proof. (i) Let us count the number of pairs (A, B) in the family (6.4) such that

$$|\{1, t\} \cap A| = |\{1, t\} \cap B| = 0 .$$

We know from Lemma 5.13(iii) that for any such a pair we have

$$\varrho(A) = \tfrac{\ell'-1}{2} , \quad \text{and} \quad \varrho(B) = \tfrac{\ell''-1}{2} ,$$

and the set $E_t - (A \dot\cup B)$ is a disjoint union of $\frac{\ell+1}{2}$ intervals.

For each pair (A, B), pick an arbitrary *system* (arranged in *ascending order*) of *distinct representatives* $(e_1 < e_2 < \cdots <$

$e_{(\ell+\ell'+\ell''-1)/2})$ of the intervals composing the sets A, B and $E_t - (A\dot{\cup}B)$. By making the substitutions

$$e_i \mapsto \begin{cases} \theta\,, & \text{if } e_i \in E_t - (A\dot{\cup}B)\,, \\ \alpha\,, & \text{if } e_i \in A\,, \\ \beta\,, & \text{if } e_i \in B\,, \end{cases} \qquad 1 \le i \le (\ell+\ell'+\ell''-1)/2\,,$$

for all of the pairs, we get

$$\mathfrak{T}\left(\theta, \theta; \frac{\ell+1}{2}, \frac{\ell'-1}{2}, \frac{\ell''-1}{2}\right)$$

different *ternary Smirnov words*, of length $(\ell+\ell'+\ell''-1)/2$, that start with θ, end with θ, and contain exactly $\frac{\ell+1}{2}$ letters θ, $\frac{\ell'-1}{2}$ letters α, and $\frac{\ell''-1}{2}$ letters β.

By Remark A.1(i), we see that

$$\mathfrak{T}\left(\theta, \theta; \tfrac{\ell+1}{2}, \tfrac{\ell'-1}{2}, \tfrac{\ell''-1}{2}\right) = \begin{cases} \dbinom{\frac{\ell+1}{2}-1}{\frac{\frac{\ell+1}{2}+\frac{\ell'-1}{2}-\frac{\ell''-1}{2}-1}{2}}\dbinom{\frac{\frac{\ell+1}{2}+\frac{\ell'-1}{2}+\frac{\ell''-1}{2}-3}{2}}{\frac{\ell+1}{2}-2}\,, \text{ if } \frac{\ell+\ell'+\ell''-1}{2} \text{ odd}\,, \\ \left(\frac{\ell+1}{2}+\frac{\ell'-1}{2}-\frac{\ell''-1}{2}\right)\cdot\dbinom{\frac{\ell+1}{2}-1}{\frac{\frac{\ell+1}{2}+\frac{\ell'-1}{2}-\frac{\ell''-1}{2}}{2}} \\ \quad\times\dbinom{\frac{\frac{\ell+1}{2}+\frac{\ell'-1}{2}+\frac{\ell''-1}{2}}{2}-2}{\frac{\ell+1}{2}-2}\,, \quad \text{if } \frac{\ell+\ell'+\ell''-1}{2} \text{ even}\,, \end{cases}$$

that is,

$$\mathfrak{T}\left(\theta, \theta; \tfrac{\ell+1}{2}, \tfrac{\ell'-1}{2}, \tfrac{\ell''-1}{2}\right) = \begin{cases} \dbinom{\frac{\ell-1}{2}}{\frac{\ell+\ell'-\ell''-1}{4}}\dbinom{\frac{\ell+\ell'+\ell''-7}{4}}{\frac{\ell-3}{2}}\,, & \text{if } \frac{\ell+\ell'+\ell''-1}{2} \text{ odd}\,, \\ \frac{\ell+\ell'-\ell''+1}{2}\dbinom{\frac{\ell-1}{2}}{\frac{\ell+\ell'-\ell''+1}{4}}\dbinom{\frac{\ell+\ell'+\ell''-9}{4}}{\frac{\ell-3}{2}}\,, & \text{if } \frac{\ell+\ell'+\ell''-1}{2} \text{ even}\,. \end{cases}$$

Since there are

$$\mathfrak{T}\left(\theta, \theta; \frac{\ell+1}{2}, \frac{\ell'-1}{2}, \frac{\ell''-1}{2}\right)\cdot \mathrm{c}\left(\frac{\ell+1}{2}; t-(j'+j'')\right) \\ \times \mathrm{c}\left(\frac{\ell'-1}{2}; j'\right)\cdot \mathrm{c}\left(\frac{\ell''-1}{2}; j''\right)$$

pairs (A, B) of sets A and B with the properties given in (6.6), we see that the number of these pairs in the family (6.4) can be calculated by means of (6.5).

(ii) Let us count the number of pairs $(A,\ B)$ in the family (6.4) such that

$$\{1,\, t\} \cap A = \{1,\, t\}\,, \quad \text{and} \quad |\{1,\, t\} \cap B| = 0\,.$$

We know from Lemma 5.13(ii) and Lemma 5.13(iii) that for any such a pair we have

$$\varrho(A) = \frac{\ell' + 1}{2}\,, \quad \text{and} \quad \varrho(B) = \frac{\ell'' - 1}{2}\,,$$

and the set $E_t - (A \dot{\cup} B)$ is a disjoint union of $\frac{\ell-1}{2}$ intervals. We denote by

$$\mathfrak{T}\left(\alpha,\, \alpha;\, \frac{\ell - 1}{2},\, \frac{\ell' + 1}{2},\, \frac{\ell'' - 1}{2}\right)$$

the number of ternary Smirnov words, of length $(\ell + \ell' + \ell'' - 1)/2$, that start with α, end with α, and contain $\frac{\ell-1}{2}$ letters θ, $\frac{\ell'+1}{2}$ letters α, and $\frac{\ell''-1}{2}$ letters β. In the family (6.4) there are

$$\mathfrak{T}\left(\alpha,\, \alpha;\, \frac{\ell - 1}{2},\, \frac{\ell' + 1}{2},\, \frac{\ell'' - 1}{2}\right) \cdot \mathrm{c}\left(\frac{\ell - 1}{2};\, t - (j' + j'')\right) \times \mathrm{c}\left(\frac{\ell' + 1}{2};\, j'\right) \cdot \mathrm{c}\left(\frac{\ell'' - 1}{2};\, j''\right)$$

pairs of sets A and B with the properties given in (6.8).

By analogy with expression (A.7), the number $\mathfrak{T}(\alpha,\, \alpha;\, n(\theta),\, n(\alpha),\, n(\beta))$ of ternary Smirnov words that start with α, end with α, and contain $n(\theta)$ letters θ, $n(\alpha)$ letters α, and $n(\beta)$ letters β, is

$$\mathfrak{T}(\alpha,\, \alpha;\, n(\theta),\, n(\alpha),\, n(\beta)) = \begin{cases} \binom{n(\alpha)-1}{\frac{n(\alpha)+n(\theta)-n(\beta)-1}{2}} \binom{\frac{n(\alpha)+n(\theta)+n(\beta)-3}{2}}{n(\alpha)-2}\,, & \text{if } n(\alpha) + n(\theta) + n(\beta) \text{ odd}\,, \\ (n(\alpha) + n(\theta) - n(\beta)) \cdot \binom{n(\alpha)-1}{\frac{n(\alpha)+n(\theta)-n(\beta)}{2}} \binom{\frac{n(\alpha)+n(\theta)+n(\beta)}{2}-2}{n(\alpha)-2}\,, & \text{if } n(\alpha) + n(\theta) + n(\beta) \text{ even}\,. \end{cases} \tag{6.15}$$

As a consequence,

$$\mathfrak{T}\left(\alpha, \alpha; \tfrac{\ell-1}{2}, \tfrac{\ell'+1}{2}, \tfrac{\ell''-1}{2}\right)$$
$$= \begin{cases} \dbinom{\frac{\ell'+1}{2}-1}{\frac{\frac{\ell'+1}{2}+\frac{\ell-1}{2}-\frac{\ell''-1}{2}-1}{2}} \dbinom{\frac{\frac{\ell'+1}{2}+\frac{\ell-1}{2}+\frac{\ell''-1}{2}-3}{2}}{\frac{\ell'+1}{2}-2}, \\ \qquad\qquad \text{if } \frac{\ell'+1}{2}+\frac{\ell-1}{2}+\frac{\ell''-1}{2} \text{ odd }, \\ \left(\frac{\ell'+1}{2}+\frac{\ell-1}{2}-\frac{\ell''-1}{2}\right) \cdot \dbinom{\frac{\ell'+1}{2}-1}{\frac{\frac{\ell'+1}{2}+\frac{\ell-1}{2}-\frac{\ell''-1}{2}}{2}} \\ \quad \times \dbinom{\frac{\frac{\ell'+1}{2}+\frac{\ell-1}{2}+\frac{\ell''-1}{2}}{2}-2}{\frac{\ell'+1}{2}-2}, \\ \qquad\qquad \text{if } \frac{\ell'+1}{2}+\frac{\ell-1}{2}+\frac{\ell''-1}{2} \text{ even }, \end{cases}$$

that is,

$$\mathfrak{T}\left(\alpha, \alpha; \tfrac{\ell-1}{2}, \tfrac{\ell'+1}{2}, \tfrac{\ell''-1}{2}\right)$$
$$= \begin{cases} \dbinom{\frac{\ell'-1}{2}}{\frac{\ell+\ell'-\ell''-1}{4}} \dbinom{\frac{\ell+\ell'+\ell''-7}{4}}{\frac{\ell'-3}{2}}, & \text{if } \frac{\ell+\ell'+\ell''-1}{2} \text{ odd }, \\ \frac{\ell+\ell'-\ell''+1}{2} \dbinom{\frac{\ell'-1}{2}}{\frac{\ell+\ell'-\ell''+1}{4}} \dbinom{\frac{\ell+\ell'+\ell''-9}{4}}{\frac{\ell'-3}{2}}, & \text{if } \frac{\ell+\ell'+\ell''-1}{2} \text{ even }. \end{cases}$$

Thus, the number of pairs (A, B) of sets A and B in the family (6.4), with the properties given in (6.8), can be calculated by means of (6.7).

(iii) Let us consider the pairs (A, B) in the family (6.4) such that

$$|\{1, t\} \cap A| = 0\,, \quad \text{and} \quad \{1, t\} \cap B = \{t\}\,.$$

We know from Lemma 5.13(iii) and Lemma 5.13(i)(b) that for any such a pair we have

$$\varrho(A) = \tfrac{\ell'-1}{2}\,, \quad \text{and} \quad \varrho(B) = \tfrac{\ell''+1}{2}\,,$$

and the set $E_t - (A \dot{\cup} B)$ is a disjoint union of $\frac{\ell+1}{2}$ intervals. We denote by

$$\mathfrak{T}\left(\theta, \beta; \tfrac{\ell+1}{2}, \tfrac{\ell'-1}{2}, \tfrac{\ell''+1}{2}\right)$$

the number of ternary Smirnov words, of length $(\ell + \ell' + \ell'' + 1)/2$, that start with θ, end with β, and contain $\frac{\ell+1}{2}$ letters θ, $\frac{\ell'-1}{2}$ letters α, and $\frac{\ell''+1}{2}$ letters β; in the family (6.4) there are

$$\mathfrak{T}\left(\theta, \beta; \frac{\ell+1}{2}, \frac{\ell'-1}{2}, \frac{\ell''+1}{2}\right) \cdot \mathrm{c}\left(\frac{\ell+1}{2}; t-(j'+j'')\right)$$
$$\times \mathrm{c}\left(\frac{\ell'-1}{2}; j'\right) \cdot \mathrm{c}\left(\frac{\ell''+1}{2}; j''\right)$$

pairs of sets A and B with the properties given in (6.10).

By Remark A.1(ii), we have

$$\mathfrak{T}\left(\theta, \beta; \frac{\ell+1}{2}, \frac{\ell'-1}{2}, \frac{\ell''+1}{2}\right)$$
$$= \begin{cases} \left(\frac{\ell+1}{2} + \frac{\ell''+1}{2} - \frac{\ell'-1}{2}\right) \cdot \binom{\frac{\ell+1}{2}-1}{\frac{\frac{\ell+1}{2}+\frac{\ell''+1}{2}-\frac{\ell'-1}{2}-1}{2}} \binom{\frac{\frac{\ell+1}{2}+\frac{\ell'-1}{2}+\frac{\ell''+1}{2}-3}{2}}{\frac{\ell+1}{2}-1}, \\ \qquad \text{if } \frac{\ell+1}{2} + \frac{\ell'-1}{2} + \frac{\ell''+1}{2} \text{ odd}, \\ \binom{\frac{\ell+1}{2}-1}{\frac{\frac{\ell+1}{2}+\frac{\ell''+1}{2}-\frac{\ell'-1}{2}}{2}-1} \binom{\frac{\frac{\ell+1}{2}+\frac{\ell'-1}{2}+\frac{\ell''+1}{2}}{2}-1}{\frac{\ell+1}{2}-1} \\ + \left(\frac{\ell+1}{2} + \frac{\ell''+1}{2} - \frac{\ell'-1}{2}\right) \cdot \binom{\frac{\ell+1}{2}-1}{\frac{\frac{\ell+1}{2}+\frac{\ell''+1}{2}-\frac{\ell'-1}{2}}{2}} \binom{\frac{\frac{\ell+1}{2}+\frac{\ell'-1}{2}+\frac{\ell''+1}{2}}{2}-2}{\frac{\ell+1}{2}-1}, \\ \qquad \text{if } \frac{\ell+1}{2} + \frac{\ell'-1}{2} + \frac{\ell''+1}{2} \text{ even}, \end{cases}$$

that is,

$$\mathfrak{T}\left(\theta, \beta; \tfrac{\ell+1}{2}, \tfrac{\ell'-1}{2}, \tfrac{\ell''+1}{2}\right)$$
$$= \begin{cases} \frac{\ell+\ell''-\ell'+3}{2} \binom{\frac{\ell-1}{2}}{\frac{\ell+\ell''-\ell'+1}{4}} \binom{\frac{\ell+\ell'+\ell''-5}{4}}{\frac{\ell-1}{2}}, & \text{if } \frac{\ell+\ell'+\ell''+1}{2} \text{ odd}, \\ \binom{\frac{\ell-1}{2}}{\frac{\ell+\ell''-\ell'-1}{4}} \binom{\frac{\ell+\ell'+\ell''-3}{4}}{\frac{\ell-1}{2}} \\ + \frac{\ell+\ell''-\ell'+3}{2} \binom{\frac{\ell-1}{2}}{\frac{\ell+\ell''-\ell'+3}{4}} \binom{\frac{\ell+\ell'+\ell''-7}{4}}{\frac{\ell-1}{2}}, & \text{if } \frac{\ell+\ell'+\ell''+1}{2} \text{ even}. \end{cases}$$

We see that the number of pairs (A, B) of sets A and B in the family (6.4), with the properties given in (6.10), can be calculated by means of (6.9).

(iv) Let us consider the pairs (A, B) in the family (6.4) such that

$$\{1, t\} \cap A = \{1\}, \quad \text{and} \quad |\{1, t\} \cap B| = 0 .$$

We know from Lemma 5.13(i)(a) and Lemma 5.13(iii) that we have

$$\varrho(A) = \frac{\ell'+1}{2}, \quad \text{and} \quad \varrho(B) = \frac{\ell''-1}{2};$$

note also that the set $E_t - (A \dot{\cup} B)$ is a disjoint union of $\frac{\ell+1}{2}$ intervals.

We denote by

$$\mathfrak{T}\left(\alpha, \theta; \frac{\ell+1}{2}, \frac{\ell'+1}{2}, \frac{\ell''-1}{2}\right)$$

the number of ternary Smirnov words, of length $(\ell + \ell' + \ell'' + 1)/2$, that start with α, end with θ, and contain exactly $\frac{\ell+1}{2}$ letters θ, $\frac{\ell'+1}{2}$ letters α, and $\frac{\ell''-1}{2}$ letters β; in the family (6.4) there are

$$\mathfrak{T}\left(\alpha,\theta;\frac{\ell+1}{2},\frac{\ell'+1}{2},\frac{\ell''-1}{2}\right)\cdot \mathrm{c}\left(\frac{\ell+1}{2};t-(j'+j'')\right)$$
$$\times \mathrm{c}\left(\frac{\ell'+1}{2};j'\right)\cdot \mathrm{c}\left(\frac{\ell''-1}{2};j''\right)$$

pairs of sets A and B with the properties given in (6.12).

By analogy with expression (A.8), the number $\mathfrak{T}(\alpha,\theta;n(\theta),n(\alpha),n(\beta))$ of ternary Smirnov words that start with α, end with θ, and contain $n(\theta)$ letters θ, $n(\alpha)$ letters α, and $n(\beta)$ letters β, is

$$\mathfrak{T}(\alpha,\theta;n(\theta),n(\alpha),n(\beta))$$
$$=\begin{cases}(n(\alpha)+n(\theta)-n(\beta))\cdot\binom{n(\alpha)-1}{\frac{n(\alpha)+n(\theta)-n(\beta)-1}{2}}\\ \quad\times\binom{\frac{n(\alpha)+n(\beta)+n(\theta)-3}{2}}{n(\alpha)-1}, \quad \text{if } n(\alpha)+n(\beta)+n(\theta) \text{ odd },\\ \binom{n(\alpha)-1}{\frac{n(\alpha)+n(\theta)-n(\beta)}{2}-1}\binom{\frac{n(\alpha)+n(\beta)+n(\theta)}{2}-1}{n(\alpha)-1}\\ +(n(\alpha)+n(\theta)-n(\beta))\cdot\binom{n(\alpha)-1}{\frac{n(\alpha)+n(\theta)-n(\beta)}{2}}\binom{\frac{n(\alpha)+n(\beta)+n(\theta)}{2}-2}{n(\alpha)-1},\\ \quad \text{if } n(\alpha)+n(\beta)+n(\theta) \text{ even }.\end{cases} \tag{6.16}$$

As a consequence,

$$\mathfrak{T}\left(\alpha,\theta;\tfrac{\ell+1}{2},\tfrac{\ell'+1}{2},\tfrac{\ell''-1}{2}\right)$$
$$=\begin{cases}\left(\frac{\ell'+1}{2}+\frac{\ell+1}{2}-\frac{\ell''-1}{2}\right)\cdot\binom{\frac{\ell'+1}{2}-1}{\frac{\frac{\ell'+1}{2}+\frac{\ell+1}{2}-\frac{\ell''-1}{2}-1}{2}}\\ \quad\times\binom{\frac{\frac{\ell'+1}{2}+\frac{\ell''-1}{2}+\frac{\ell+1}{2}-3}{2}}{\frac{\ell'+1}{2}-1}, \quad \text{if } \frac{\ell'+1}{2}+\frac{\ell''-1}{2}+\frac{\ell+1}{2} \text{ odd },\\ \binom{\frac{\ell'+1}{2}-1}{\frac{\frac{\ell'+1}{2}+\frac{\ell+1}{2}-\frac{\ell''-1}{2}}{2}-1}\binom{\frac{\frac{\ell'+1}{2}+\frac{\ell''-1}{2}+\frac{\ell+1}{2}}{2}-1}{\frac{\ell'+1}{2}-1}\\ +\left(\frac{\ell'+1}{2}+\frac{\ell+1}{2}-\frac{\ell''-1}{2}\right)\cdot\binom{\frac{\ell'+1}{2}-1}{\frac{\frac{\ell'+1}{2}+\frac{\ell+1}{2}-\frac{\ell''-1}{2}}{2}}\\ \quad\times\binom{\frac{\frac{\ell'+1}{2}+\frac{\ell''-1}{2}+\frac{\ell+1}{2}}{2}-2}{\frac{\ell'+1}{2}-1}, \quad \text{if } \frac{\ell'+1}{2}+\frac{\ell''-1}{2}+\frac{\ell+1}{2} \text{ even },\end{cases}$$

that is,

$$\mathfrak{T}\left(\alpha,\theta;\tfrac{\ell+1}{2},\tfrac{\ell'+1}{2},\tfrac{\ell''-1}{2}\right)$$

$$=\begin{cases}\frac{\ell+\ell'-\ell''+3}{2}\binom{\frac{\ell'-1}{2}}{\frac{\ell+\ell'-\ell''+1}{4}}\binom{\frac{\ell+\ell'+\ell''-5}{4}}{\frac{\ell'-1}{2}}\,, & \text{if } \frac{\ell+\ell'+\ell''+1}{2} \text{ odd}\,,\\ \binom{\frac{\ell'-1}{2}}{\frac{\ell+\ell'-\ell''-1}{4}}\binom{\frac{\ell+\ell'+\ell''-3}{4}}{\frac{\ell'-1}{2}} & \\ \quad+\frac{\ell+\ell'-\ell''+3}{2}\binom{\frac{\ell'-1}{2}}{\frac{\ell+\ell'-\ell''+3}{4}}\binom{\frac{\ell+\ell'+\ell''-7}{4}}{\frac{\ell'-1}{2}}\,, & \text{if } \frac{\ell+\ell'+\ell''+1}{2} \text{ even}\,.\end{cases}$$

We see that the number of pairs (A, B) of sets A and B in the family (6.4), with the properties given in (6.12), can be calculated by means of (6.11).

(v) Let us consider the pairs (A, B) in the family (6.4) such that

$$\{1, t\}\cap A=\{1\}\,,\quad\text{and}\quad\{1, t\}\cap B=\{t\}\,.$$

We know from Lemma 5.13(i)(a), Lemma 5.13(i)(b) and Lemma 5.13(iii) that we have

$$\varrho(A)=\tfrac{\ell'+1}{2}\,,\quad\text{and}\quad\varrho(B)=\tfrac{\ell''+1}{2}\,,$$

and the set $E_t-(A\dot{\cup}B)$ is a disjoint union of $\frac{\ell-1}{2}$ intervals.

We denote by

$$\mathfrak{T}\left(\alpha,\beta;\frac{\ell-1}{2},\frac{\ell'+1}{2},\frac{\ell''+1}{2}\right)$$

the number of ternary Smirnov words, of length $(\ell+\ell'+\ell''+1)/2$, that start with α, end with β, and contain $\frac{\ell-1}{2}$ letters θ, $\frac{\ell'+1}{2}$ letters α, and $\frac{\ell''+1}{2}$ letters β; in the family (6.4) there are

$$\mathfrak{T}\left(\alpha,\beta;\frac{\ell-1}{2},\frac{\ell'+1}{2},\frac{\ell''+1}{2}\right)\cdot\mathrm{c}\left(\frac{\ell-1}{2};t-(j'+j'')\right)$$
$$\times\mathrm{c}\left(\frac{\ell'+1}{2};j'\right)\cdot\mathrm{c}\left(\frac{\ell''+1}{2};j''\right)$$

pairs of sets A and B with the properties given in (6.14).

By analogy with expression (A.8), the number $\mathfrak{T}(\alpha,\beta;n(\theta), n(\alpha), n(\beta))$ of ternary Smirnov words that start with α, end with β, and contain $n(\theta)$ letters θ, $n(\alpha)$ letters α and $n(\beta)$ letters β, is

$$\mathfrak{T}(\alpha,\beta;n(\theta),n(\alpha),n(\beta))$$

$$= \begin{cases} (n(\alpha)+n(\beta)-n(\theta))\cdot\binom{n(\alpha)-1}{\frac{n(\alpha)+n(\beta)-n(\theta)-1}{2}}\binom{\frac{n(\alpha)+n(\theta)+n(\beta)-3}{2}}{n(\alpha)-1}\,, \\ \qquad\qquad \text{if } n(\alpha)+n(\theta)+n(\beta) \text{ odd}\,, \\ \binom{n(\alpha)-1}{\frac{n(\alpha)+n(\beta)-n(\theta)}{2}-1}\binom{\frac{n(\alpha)+n(\theta)+n(\beta)}{2}-1}{n(\alpha)-1} \\ +(n(\alpha)+n(\beta)-n(\theta))\cdot\binom{n(\alpha)-1}{\frac{n(\alpha)+n(\beta)-n(\theta)}{2}}\binom{\frac{n(\alpha)+n(\theta)+n(\beta)}{2}-2}{n(\alpha)-1}\,, \\ \qquad\qquad \text{if } n(\alpha)+n(\theta)+n(\beta) \text{ even}\,. \end{cases} \tag{6.17}$$

As a consequence,

$$\mathfrak{T}\left(\alpha,\beta;\tfrac{\ell-1}{2},\tfrac{\ell'+1}{2},\tfrac{\ell''+1}{2}\right)$$

$$= \begin{cases} \left(\frac{\ell'+1}{2}+\frac{\ell''+1}{2}-\frac{\ell-1}{2}\right)\cdot\binom{\frac{\ell'+1}{2}-1}{\frac{\frac{\ell'+1}{2}+\frac{\ell''+1}{2}-\frac{\ell-1}{2}-1}{2}} \\ \quad\times\binom{\frac{\frac{\ell'+1}{2}+\frac{\ell-1}{2}+\frac{\ell''+1}{2}-3}{2}}{\frac{\ell'+1}{2}-1}\,, \quad \text{if } \frac{\ell'+1}{2}+\frac{\ell-1}{2}+\frac{\ell''+1}{2} \text{ odd}\,, \\ \binom{\frac{\ell'+1}{2}-1}{\frac{\frac{\ell'+1}{2}+\frac{\ell''+1}{2}-\frac{\ell-1}{2}}{2}-1}\binom{\frac{\frac{\ell'+1}{2}+\frac{\ell-1}{2}+\frac{\ell''+1}{2}}{2}-1}{\frac{\ell'+1}{2}-1} \\ +\left(\frac{\ell'+1}{2}+\frac{\ell''+1}{2}-\frac{\ell-1}{2}\right)\cdot\binom{\frac{\ell'+1}{2}-1}{\frac{\frac{\ell'+1}{2}+\frac{\ell''+1}{2}-\frac{\ell-1}{2}}{2}} \\ \quad\times\binom{\frac{\frac{\ell'+1}{2}+\frac{\ell-1}{2}+\frac{\ell''+1}{2}}{2}-2}{\frac{\ell'+1}{2}-1}\,, \quad \text{if } \frac{\ell'+1}{2}+\frac{\ell-1}{2}+\frac{\ell''+1}{2} \text{ even}\,, \end{cases}$$

that is,

$$\mathfrak{T}\left(\alpha,\beta;\tfrac{\ell-1}{2},\tfrac{\ell'+1}{2},\tfrac{\ell''+1}{2}\right)$$

$$= \begin{cases} \frac{\ell'+\ell''-\ell+3}{2}\binom{\frac{\ell'-1}{2}}{\frac{\ell'+\ell''-\ell+1}{4}}\binom{\frac{\ell+\ell'+\ell''-5}{4}}{\frac{\ell'-1}{2}}\,, \qquad \text{if } \frac{\ell+\ell'+\ell''+1}{2} \text{ odd}\,, \\ \binom{\frac{\ell'-1}{2}}{\frac{\ell'+\ell''-\ell-1}{4}}\binom{\frac{\ell+\ell'+\ell''-3}{4}}{\frac{\ell'-1}{2}} \\ +\frac{\ell'+\ell''-\ell+3}{2}\binom{\frac{\ell'-1}{2}}{\frac{\ell'+\ell''-\ell+3}{4}}\binom{\frac{\ell+\ell'+\ell''-7}{4}}{\frac{\ell'-1}{2}}\,, \quad \text{if } \frac{\ell+\ell'+\ell''+1}{2} \text{ even}\,. \end{cases}$$

We see that the number of pairs (A, B) of sets A and B in the family (6.4), with the properties given in (6.14), can be calculated by means of (6.13). $\square$

Example 6.5. Suppose $t := 5$.
(i) Consider the family of pairs

$$\big\{(A, B) \in \mathbf{2}^{[t]} \times \mathbf{2}^{[t]} : \ |A \cap B| = 0, \ \ j' := |A| = 2, \ \ j'' := |B| = 1, \\ \ell' := \mathfrak{q}(A) = 3, \ \ \ell'' := \mathfrak{q}(B) = 3, \ \ \ell := \mathfrak{q}(A \dot{\cup} B) = 3\big\} . \quad (6.18)$$

Since the quantity $\frac{\ell+\ell'+\ell''-1}{2} := \frac{3+3+3-1}{2} = 4$ is even, Theorem 6.4(i) implies that the family (6.18) contains

$$\binom{t-(j'+j'')-1}{\frac{\ell-1}{2}}\binom{j'-1}{\frac{\ell'-3}{2}}\binom{j''-1}{\frac{\ell''-3}{2}} \times \frac{\ell+\ell'-\ell''+1}{2}\binom{\frac{\ell-1}{2}}{\frac{\ell+\ell'-\ell''+1}{4}}\binom{\frac{\ell+\ell'+\ell''-9}{4}}{\frac{\ell-3}{2}}$$
$$:= \binom{5-(2+1)-1}{\frac{3-1}{2}}\binom{2-1}{\frac{3-3}{2}}\binom{1-1}{\frac{3-3}{2}} \times \frac{3+3-3+1}{2}\binom{\frac{3-1}{2}}{\frac{3+3-3+1}{4}}\binom{\frac{3+3+3-9}{4}}{\frac{3-3}{2}} = 2$$

pairs (A, B) of sets A and B such that $|\{1, t\} \cap A| = |\{1, t\} \cap B| = 0$. These pairs are

$$(\ A := \{\ ,\ 2,\ 3,\ \ ,\ \}, \quad B := \{\ ,\ \ ,\ \ ,\ 4,\ \}\ ), \quad \text{and} \quad (\ A := \{\ ,\ \ ,\ 3,\ 4,\ \}, \quad B := \{\ ,\ 2,\ \ ,\ \ ,\ \}\ ).$$

We have for instance

$$\boldsymbol{x}(\{2,3\}) = (1, -1, \ \ 0, \ \ 1, \ \ 0) ,$$
$$\boldsymbol{Q}({}_{-\{2,3\}}\mathrm{T}^{(+)}, \boldsymbol{R}) = \{R^0, R^3, R^6\} ,$$
$$\boldsymbol{x}(\{4\}) = (1, \ \ 0, \ \ 0, -1, \ \ 1) ,$$
$$\boldsymbol{Q}({}_{-4}\mathrm{T}^{(+)}, \boldsymbol{R}) = \{R^0, R^4, R^8\} ,$$

and

$$\boldsymbol{x}(\{2,3,4\}) = (1, -1, \ \ 0, \ \ 0, \ \ 1) ,$$
$$\boldsymbol{Q}({}_{-\{2,3,4\}}\mathrm{T}^{(+)}, \boldsymbol{R}) = \{R^0, R^4, R^6\} ;$$

see Figure 6.2.

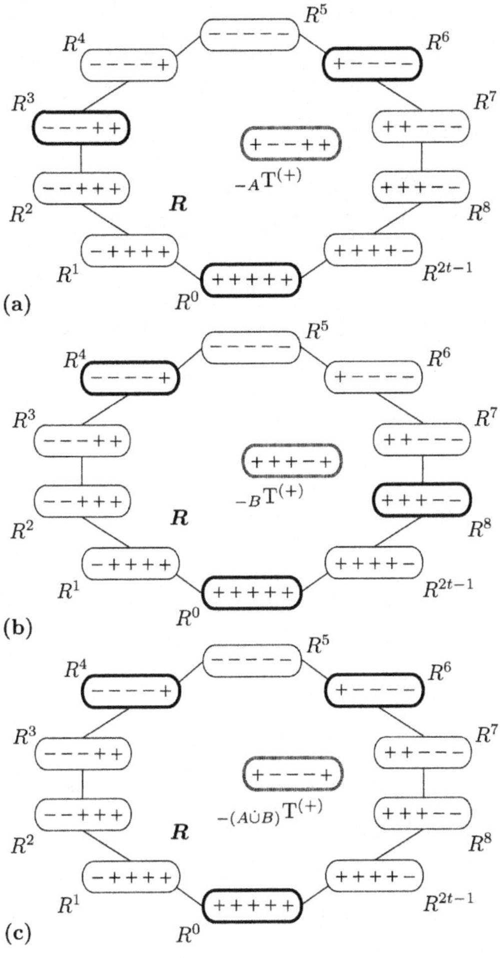

Figure 6.2 Three instances of the same distinguished symmetric cycle $\boldsymbol{R} := (R^0, R^1, \ldots, R^{2t-1}, R^0)$, defined by (6.1)(6.2), in the hypercube graph $\boldsymbol{H}((t := 5), 2)$.
Pick the subsets $A := \{2, 3\}$, $B := \{4\}$ and $A \dot{\cup} B = \{2, 3, 4\}$ of the set E_t. Note that $|\{1, t\} \cap A| = |\{1, t\} \cap B| = 0$.
(a) The vertex $_{-A}\mathrm{T}^{(+)}$ of the graph $\boldsymbol{H}(t, 2)$, with the corresponding decomposition set $\boldsymbol{Q}(_{-A}\mathrm{T}^{(+)}, \boldsymbol{R}) = \{R^0, R^3, R^6\}$.
(b) The vertex $_{-B}\mathrm{T}^{(+)}$, with its decomposition set $\boldsymbol{Q}(_{-B}\mathrm{T}^{(+)}, \boldsymbol{R}) = \{R^0, R^4, R^8\}$.
(c) The vertex $_{-(A\dot{\cup}B)}\mathrm{T}^{(+)}$, with its decomposition set $\boldsymbol{Q}(_{-(A\dot{\cup}B)}\mathrm{T}^{(+)}, \boldsymbol{R}) = \{R^0, R^4, R^6\}$.

(ii) Consider the family of pairs

$$\big\{(A,\,B)\in\mathbf{2}^{[t]}\times\mathbf{2}^{[t]}:\ |A\cap B|=0,\quad j':=|A|=2,\quad j'':=|B|=2,\\ \ell':=\mathfrak{q}(A)=3,\quad \ell'':=\mathfrak{q}(B)=5,\quad \ell:=\mathfrak{q}(A\dot{\cup}B)=3\big\}\,. \tag{6.19}$$

Since the quantity $\frac{\ell+\ell'+\ell''-1}{2}:=\frac{3+3+5-1}{2}=5$ is odd, Theorem 6.4(ii) asserts that in the family (6.19) there is

$$\binom{t-(j'+j'')-1}{\frac{\ell-3}{2}}\binom{j'-1}{\frac{\ell'-1}{2}}\binom{j''-1}{\frac{\ell''-3}{2}}\binom{\frac{\ell'-1}{2}}{\frac{\ell+\ell'-\ell''-1}{4}}\binom{\frac{\ell+\ell'+\ell''-7}{4}}{\frac{\ell'-3}{2}}$$

$$:=\binom{5-(2+2)-1}{\frac{3-3}{2}}\binom{2-1}{\frac{3-1}{2}}\binom{2-1}{\frac{5-3}{2}}\binom{\frac{3-1}{2}}{\frac{3+3-5-1}{4}}\binom{\frac{3+3+5-7}{4}}{\frac{3-3}{2}}=1$$

pair $(A,\,B)$ of sets A and B such that $\{1,\,t\}\cap A=\{1,\,t\}$, and $|\{1,\,t\}\cap B|=0$. This is the pair

$$\begin{aligned}(\ \ A &:= \{1,\ \ ,\ \ ,\ \ ,t\}\,,\\ B &:= \{\ \ ,2,\ \ ,4,\ \ \}\ )\,.\end{aligned}$$

We have

$$\begin{aligned}\boldsymbol{x}(\{1,\,t\}) &= (-1,\quad 1,\quad 0,\quad 0,-1)\,,\\ \boldsymbol{Q}({}_{-\{1,t\}}\mathrm{T}^{(+)},\,\boldsymbol{R}) &= \{R^1,\,R^5,\,R^{2t-1}\}\,,\\ \boldsymbol{x}(\{2,\,4\}) &= (\ \ 1,-1,\quad 1,-1,\quad 1)\,,\\ \boldsymbol{Q}({}_{-\{2,4\}}\mathrm{T}^{(+)},\,\boldsymbol{R}) &= \{R^0,\,R^2,\,R^4,\,R^6,\,R^8\}\,,\end{aligned}$$

and

$$\begin{aligned}\boldsymbol{x}(\{1,\,2,\,4,\,t\}) &= (-1,\quad 0,\quad 1,-1,\quad 0)\,,\\ \boldsymbol{Q}({}_{-\{1,2,4,t\}}\mathrm{T}^{(+)},\,\boldsymbol{R}) &= \{R^2,\,R^5,\,R^8\}\,;\end{aligned}$$

see Figure 6.3.

(iii) Consider the family of pairs

$$\big\{(A,\,B)\in\mathbf{2}^{[t]}\times\mathbf{2}^{[t]}:\ |A\cap B|=0,\quad j':=|A|=1,\quad j'':=|B|=3,\\ \ell':=\mathfrak{q}(A)=3,\quad \ell'':=\mathfrak{q}(B)=3,\quad \ell:=\mathfrak{q}(A\dot{\cup}B)=1\big\}\,. \tag{6.20}$$

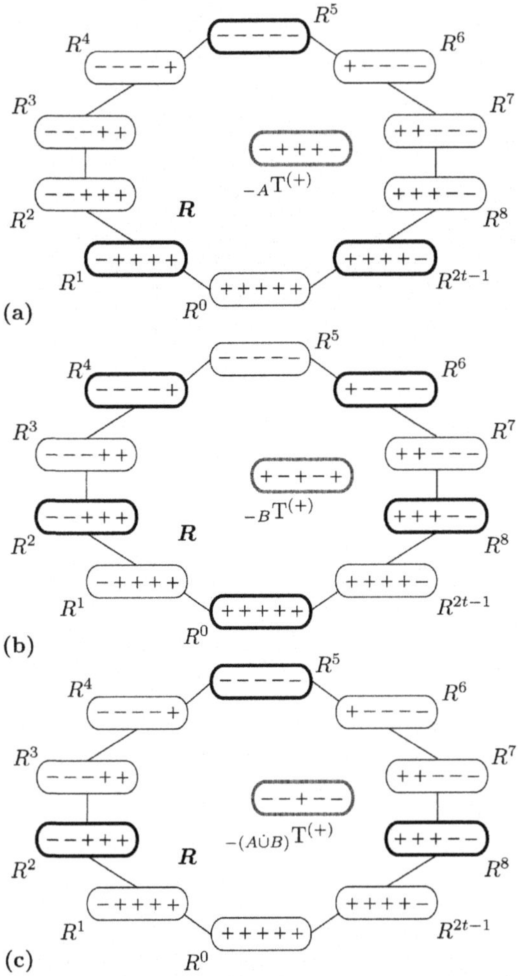

Figure 6.3 Three instances of the same distinguished symmetric cycle $\boldsymbol{R} := (R^0, R^1, \ldots, R^{2t-1}, R^0)$ in the hypercube graph $\boldsymbol{H}((t := 5), 2)$, defined by (6.1)(6.2).
Pick the subsets $A := \{1, t\}$, $B := \{2, 4\}$ and $A \dot{\cup} B = \{1, 2, 4, t\}$ of the set E_t. Note that $\{1, t\} \cap A = \{1, t\}$, and $|\{1, t\} \cap B| = 0$.
(a) The vertex ${}_{-A}\mathrm{T}^{(+)}$ of the graph $\boldsymbol{H}(t, 2)$, with the corresponding decomposition set $\boldsymbol{Q}({}_{-A}\mathrm{T}^{(+)}, \boldsymbol{R}) = \{R^1, R^5, R^{2t-1}\}$.
(b) The vertex ${}_{-B}\mathrm{T}^{(+)}$, with its decomposition set $\boldsymbol{Q}({}_{-B}\mathrm{T}^{(+)}, \boldsymbol{R}) = \{R^0, R^2, R^4, R^6, R^8\}$.
(c) The vertex ${}_{-(A\dot{\cup}B)}\mathrm{T}^{(+)}$, with its decomposition set $\boldsymbol{Q}({}_{-(A\dot{\cup}B)}\mathrm{T}^{(+)}, \boldsymbol{R}) = \{R^2, R^5, R^8\}$.

Since the quantity $\frac{\ell+\ell'+\ell''+1}{2} := \frac{1+3+3+1}{2} = 4$ is even, by Theorem 6.4(iii) we see that in the family (6.20) there are

$$\begin{aligned}
&\binom{t-(j'+j'')-1}{\frac{\ell-1}{2}}\binom{j'-1}{\frac{\ell'-3}{2}}\binom{j''-1}{\frac{\ell''-1}{2}}\Bigg(\binom{\frac{\ell-1}{2}}{\frac{\ell+\ell''-\ell'-1}{4}}\binom{\frac{\ell+\ell'+\ell''-3}{4}}{\frac{\ell-1}{2}}\\
&\qquad+\frac{\ell+\ell''-\ell'+3}{2}\binom{\frac{\ell-1}{2}}{\frac{\ell+\ell''-\ell'+3}{4}}\binom{\frac{\ell+\ell'+\ell''-7}{4}}{\frac{\ell-1}{2}}\Bigg)\\
&:=\binom{5-(1+3)-1}{\frac{1-1}{2}}\binom{1-1}{\frac{3-3}{2}}\binom{3-1}{\frac{3-1}{2}}\Bigg(\binom{\frac{1-1}{2}}{\frac{1+3-3-1}{4}}\binom{\frac{1+3+3-3}{4}}{\frac{1-1}{2}}\\
&\qquad+\frac{1+3-3+3}{2}\binom{\frac{1-1}{2}}{\frac{1+3-3+3}{4}}\binom{\frac{1+3+3-7}{4}}{\frac{1-1}{2}}\Bigg)=2
\end{aligned}$$

pairs (A, B) of sets A and B such that $|\{1, t\}\cap A| = 0$, and $\{1, t\}\cap B = \{t\}$. These pairs are

$$\begin{aligned}
&(\ A := \{\ ,\ ,3,\ ,\ \}\,, && (\ A := \{\ ,\ ,\ ,4,\ \}\,,\\
&\ \ B := \{\ ,2,\ ,4,t\}\ )\,, \quad\text{and} && \ \ B := \{\ ,2,3,\ ,t\}\ )\,.
\end{aligned}$$

We have for instance

$$\begin{aligned}
\boldsymbol{x}(\{4\}) &= (1,\ \ 0,\ \ 0, -1,\ \ 1)\,,\\
\boldsymbol{Q}({}_{-4}\mathrm{T}^{(+)}, \boldsymbol{R}) &= \{R^0, R^4, R^8\}\,,\\
\boldsymbol{x}(\{2,3,t\}) &= (0, -1,\ \ 0,\ \ 1, -1)\,,\\
\boldsymbol{Q}({}_{-\{2,3,t\}}\mathrm{T}^{(+)}, \boldsymbol{R}) &= \{R^3, R^6, R^{2t-1}\}\,,
\end{aligned}$$

and

$$\begin{aligned}
\boldsymbol{x}(\{2,3,4,t\}) &= (0, -1,\ \ 0,\ \ 0,\ \ 0)\,,\\
\boldsymbol{Q}({}_{-\{2,3,4,t\}}\mathrm{T}^{(+)}, \boldsymbol{R}) &= \{R^6\}\,;
\end{aligned}$$

see Figure 6.4.

(iv) Consider the family of pairs

$$\begin{aligned}
\big\{(A, B) \in \mathbf{2}^{[t]}\times\mathbf{2}^{[t]}:\ &|A\cap B| = 0,\ \ j' := |A| = 2,\ \ j'' := |B| = 1,\\
\ell' := \mathfrak{q}(A) = 3,\ \ &\ell'' := \mathfrak{q}(B) = 3,\ \ \ell := \mathfrak{q}(A\dot\cup B) = 3\big\}\,. \quad (6.21)
\end{aligned}$$

Since the quantity $\frac{\ell+\ell'+\ell''+1}{2} := \frac{3+3+3+1}{2} = 5$ is odd, Theorem 6.4(i) implies that in the family (6.21) there are

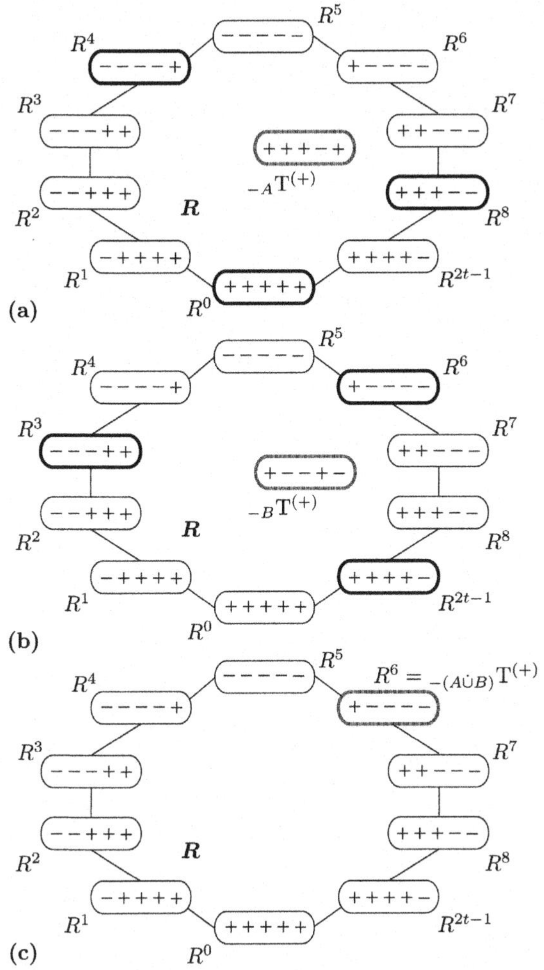

Figure 6.4 Three instances of the same distinguished symmetric cycle $\boldsymbol{R} := (R^0, R^1, \ldots, R^{2t-1}, R^0)$ in the hypercube graph $\boldsymbol{H}((t := 5), 2)$, defined by (6.1)(6.2).
Pick the subsets $A := \{4\}$, $B := \{2, 3, t\}$ and $A \dot{\cup} B = \{2, 3, 4, t\}$ of the set E_t. Note that $|\{1, t\} \cap A| = 0$, and $\{1, t\} \cap B = \{t\}$.
(a) The vertex $_{-A}\mathrm{T}^{(+)}$ of the graph $\boldsymbol{H}(t, 2)$, with the corresponding decomposition set $\boldsymbol{Q}({}_{-A}\mathrm{T}^{(+)}, \boldsymbol{R}) = \{R^0, R^4, R^8\}$.
(b) The vertex $_{-B}\mathrm{T}^{(+)}$, with its decomposition set $\boldsymbol{Q}({}_{-B}\mathrm{T}^{(+)}, \boldsymbol{R}) = \{R^3, R^6, R^{2t-1}\}$.
(c) The vertex $_{-(A\dot{\cup}B)}\mathrm{T}^{(+)} = R^6 \in \mathrm{V}(\boldsymbol{R})$.

$$
\begin{aligned}
&\binom{t-(j'+j'')-1}{\frac{\ell-1}{2}}\binom{j'-1}{\frac{\ell'-1}{2}}\binom{j''-1}{\frac{\ell''-3}{2}} \\
&\qquad\times\frac{\ell+\ell'-\ell''+3}{2}\binom{\frac{\ell'-1}{2}}{\frac{\ell+\ell'-\ell''+1}{4}}\binom{\frac{\ell+\ell'+\ell''-5}{4}}{\frac{\ell'-1}{2}} \\
&:=\binom{5-(2+1)-1}{\frac{3-1}{2}}\binom{2-1}{\frac{3-1}{2}}\binom{1-1}{\frac{3-3}{2}} \\
&\qquad\times\frac{3+3-3+3}{2}\binom{\frac{3-1}{2}}{\frac{3+3-3+1}{4}}\binom{\frac{3+3+3-5}{4}}{\frac{3-1}{2}}=3
\end{aligned}
$$

pairs (A, B) of sets A and B such that $\{1, t\} \cap A = \{1\}$, and $|\{1, t\} \cap B| = 0$. These are the pairs

$$
\begin{aligned}
&(\ A := \{1,\ ,3,\ ,\ \},\quad B := \{\ ,\ ,\ ,4,\ \}\ ), \\
&(\ A := \{1,\ ,\ ,4,\ \},\quad B := \{\ ,2,\ ,\ ,\ \}\ ),
\end{aligned}
$$

and

$$
(\ A := \{1,\ ,\ ,4,\ \},\quad B := \{\ ,\ ,3,\ ,\ \}\ ).
$$

We have for instance

$$
\begin{aligned}
\boldsymbol{x}(\{1, 4\}) &= (0,\quad 1,\quad 0, -1,\quad 1)\,, \\
\boldsymbol{Q}({}_{-\{1,4\}}\mathrm{T}^{(+)}, \boldsymbol{R}) &= \{R^1, R^4, R^8\}\,, \\
\boldsymbol{x}(\{2\}) &= (1, -1,\quad 1,\quad 0,\quad 0)\,, \\
\boldsymbol{Q}({}_{-2}\mathrm{T}^{(+)}, \boldsymbol{R}) &= \{R^0, R^2, R^6\}\,,
\end{aligned}
$$

and

$$
\begin{aligned}
\boldsymbol{x}(\{1, 2, 4\}) &= (0,\quad 0,\quad 1, -1,\quad 1)\,, \\
\boldsymbol{Q}({}_{-\{1,2,4\}}\mathrm{T}^{(+)}, \boldsymbol{R}) &= \{R^2, R^4, R^8\}\,;
\end{aligned}
$$

see Figure 6.5.

(v) Consider the family of pairs

$$
\begin{aligned}
\big\{(A, B) \in \mathbf{2}^{[t]}\times\mathbf{2}^{[t]}:\ &|A\cap B| = 0,\quad j' := |A| = 2,\quad j'' := |B| = 2, \\
&\ell' := \mathfrak{q}(A) = 3,\quad \ell'' := \mathfrak{q}(B) = 3,\quad \ell := \mathfrak{q}(A\dot{\cup}B) = 3\big\}\,. \qquad (6.22)
\end{aligned}
$$

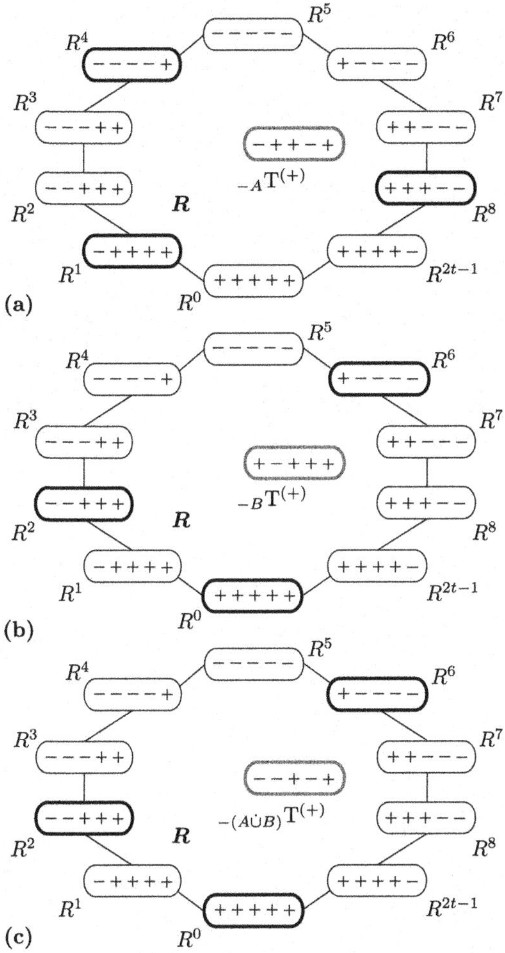

Figure 6.5 Three instances of the same distinguished symmetric cycle $\boldsymbol{R} := (R^0, R^1, \ldots, R^{2t-1}, R^0)$ in the hypercube graph $\boldsymbol{H}((t := 5), 2)$, defined by (6.1)(6.2).
Pick the subsets $A := \{1, 4\}$, $B := \{2\}$ and $A \dot{\cup} B = \{1, 2, 4\}$ of the set E_t. Note that $\{1, t\} \cap A = \{1\}$, and $|\{1, t\} \cap B| = 0$.
(a) The vertex ${}_{-A}\mathrm{T}^{(+)}$ of the graph $\boldsymbol{H}(t, 2)$, with the corresponding decomposition set $\boldsymbol{Q}({}_{-A}\mathrm{T}^{(+)}, \boldsymbol{R}) = \{R^1, R^4, R^8\}$.
(b) The vertex ${}_{-B}\mathrm{T}^{(+)}$, with its decomposition set $\boldsymbol{Q}({}_{-B}\mathrm{T}^{(+)}, \boldsymbol{R}) = \{R^0, R^2, R^6\}$.
(c) The vertex ${}_{-(A\dot{\cup}B)}\mathrm{T}^{(+)}$, with its decomposition set $\boldsymbol{Q}({}_{-B}\mathrm{T}^{(+)}, \boldsymbol{R}) = \{R^2, R^4, R^8\}$.

Since the quantity $\frac{\ell+\ell'+\ell''+1}{2} := \frac{3+3+3+1}{2} = 5$ is odd, Theorem 6.4(i) shows that in the family (6.22) there are

$$
\begin{aligned}
&\binom{t-(j'+j'')-1}{\frac{\ell-3}{2}}\binom{j'-1}{\frac{\ell'-1}{2}}\binom{j''-1}{\frac{\ell''-1}{2}} \\
&\qquad\qquad \times \frac{\ell'+\ell''-\ell+3}{2}\binom{\frac{\ell'-1}{2}}{\frac{\ell'+\ell''-\ell+1}{4}}\binom{\frac{\ell+\ell'+\ell''-5}{4}}{\frac{\ell'-1}{2}} \\
&\qquad\qquad := \binom{5-(2+2)-1}{\frac{3-3}{2}}\binom{2-1}{\frac{3-1}{2}}\binom{2-1}{\frac{3-1}{2}} \\
&\qquad\qquad\qquad \times \frac{3+3-3+3}{2}\binom{\frac{3-1}{2}}{\frac{3+3-3+1}{4}}\binom{\frac{3+3+3-5}{4}}{\frac{3-1}{2}} = 3
\end{aligned}
$$

pairs (A, B) of sets A and B such that $\{1, t\} \cap A = \{1\}$, and $\{1, t\} \cap B = \{t\}$. These pairs are

$$
\begin{aligned}
(\ A &:= \{1, \ \ , 3, \ \ , \ \ \}, \\
B &:= \{\ \ , 2, \ \ , \ \ , t\}\ ),
\end{aligned}
\qquad
\begin{aligned}
(\ A &:= \{1, \ \ , \ \ , 4, \ \ \}, \\
B &:= \{\ \ , 2, \ \ , \ \ , t\}\ ),
\end{aligned}
$$

and

$$
\begin{aligned}
(\ A &:= \{1, \ \ , \ \ , 4, \ \ \}, \\
B &:= \{\ \ , \ \ , 3, \ \ , t\}\ ).
\end{aligned}
$$

We have for instance

$$
\begin{aligned}
\boldsymbol{x}(\{1, 3\}) &= (\ \ 0, \ \ 1, -1, \ \ 1, \ \ 0)\,, \\
\boldsymbol{Q}({}_{-\{1,3\}}\mathrm{T}^{(+)}, \boldsymbol{R}) &= \{R^1, R^3, R^7\}\,, \\
\boldsymbol{x}(\{2, t\}) &= (\ \ 0, -1, \ \ 1, \ \ 0, -1)\,, \\
\boldsymbol{Q}({}_{-\{2,t\}}\mathrm{T}^{(+)}, \boldsymbol{R}) &= \{R^2, R^6, R^{2t-1}\}\,,
\end{aligned}
$$

and

$$
\begin{aligned}
\boldsymbol{x}(\{1, 2, 3, t\}) &= (-1, \ \ 0, \ \ 0, \ \ 1, -1)\,, \\
\boldsymbol{Q}({}_{-\{1,2,3,t\}}\mathrm{T}^{(+)}, \boldsymbol{R}) &= \{R^3, R^5, R^{2t-1}\}\,;
\end{aligned}
$$

see Figure 6.6.

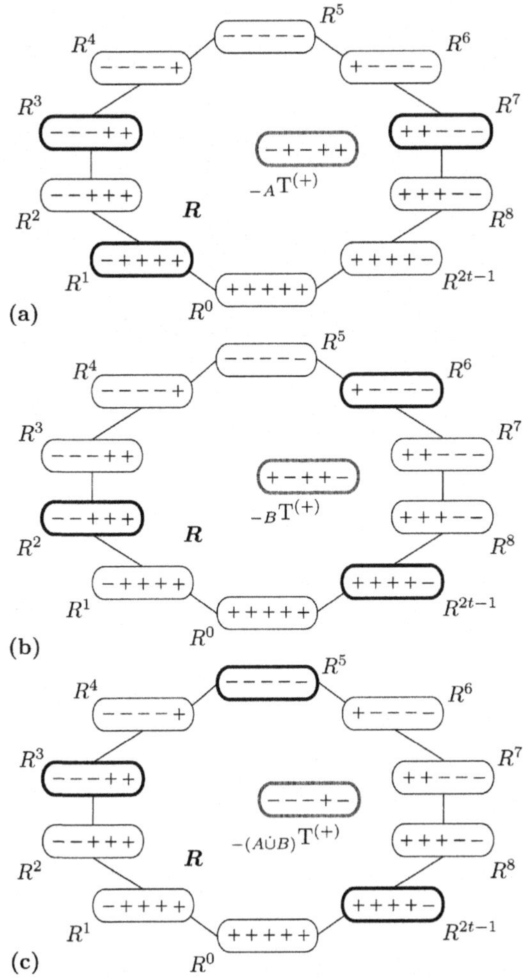

Figure 6.6 Three instances of the same distinguished symmetric cycle $\boldsymbol{R} := (R^0, R^1, \ldots, R^{2t-1}, R^0)$ in the hypercube graph $\boldsymbol{H}((t := 5), 2)$, defined by (6.1)(6.2).
Pick the subsets $A := \{1, 3\}$, $B := \{2, t\}$ and $A \dot{\cup} B = \{1, 2, 3, t\}$ of the set E_t. Note that $\{1, t\} \cap A = \{1\}$, and $\{1, t\} \cap B = \{t\}$.
(a) The vertex ${}_{-A}\mathrm{T}^{(+)}$ of the graph $\boldsymbol{H}(t, 2)$, with the corresponding decomposition set $\boldsymbol{Q}({}_{-A}\mathrm{T}^{(+)}, \boldsymbol{R}) = \{R^1, R^3, R^7\}$.
(b) The vertex ${}_{-B}\mathrm{T}^{(+)}$, with its decomposition set $\boldsymbol{Q}({}_{-B}\mathrm{T}^{(+)}, \boldsymbol{R}) = \{R^2, R^6, R^{2t-1}\}$.
(c) The vertex ${}_{-(A\dot{\cup}B)}\mathrm{T}^{(+)}$, with its decomposition set $\boldsymbol{Q}({}_{-B}\mathrm{T}^{(+)}, \boldsymbol{R}) = \{R^3, R^5, R^{2t-1}\}$.

Chapter 7

Distinguished Symmetric Cycles in Hypercube Graphs and Pairwise Decompositions of Vertices: Arbitrary Two-member Clutters

In this chapter, we present additional statistics on the decompositions of vertices $T \in \{1, -1\}^t$ of the hypercube graph $\boldsymbol{H}(t, 2)$ with respect to its distinguished symmetric cycle $\boldsymbol{R} := (R^0, R^1, \ldots, R^{2t-1}, R^0)$ defined, as earlier, by

$$R^0 := \mathrm{T}^{(+)} ,\qquad R^s := {}_{-[s]}R^0 , \quad 1 \leq s \leq t-1 , \tag{7.1}$$

and

$$R^{k+t} := -R^k , \quad 0 \leq k \leq t-1 . \tag{7.2}$$

Recall that the subsequence of vertices $(R^0, R^1, \ldots, R^{t-1})$ of the cycle $\boldsymbol{R}$ is an ordered *basis* of the space $\mathbb{R}^t$. For any vertex ${}_{-A}\mathrm{T}^{(+)}$ of the graph $\boldsymbol{H}(t, 2)$, there exists a unique row vector

$$\boldsymbol{x} := \boldsymbol{x}(A) := \boldsymbol{x}({}_{-A}\mathrm{T}^{(+)}) = (x_1, \ldots, x_t) \in \{-1, 0, 1\}^t ,$$

such that

$${}_{-A}\mathrm{T}^{(+)} = \sum_{i \in E_t} x_i \cdot R^{i-1} = \boldsymbol{x} \cdot \mathbf{M} ,$$

Symmetric Cycles
Andrey O. Matveev

ISBN 978-981-4968-81-2 (Hardcover), 978-1-003-43832-8 (eBook)
www.jennystanford.com

where

$$\mathbf{M} := \mathbf{M}(\boldsymbol{R}) := \begin{pmatrix} R^0 \\ R^1 \\ \vdots \\ R^{t-1} \end{pmatrix} .$$

Thus, the set

$$\boldsymbol{Q}({}_{-A}\mathrm{T}^{(+)}, \boldsymbol{R}) := \{x_i \cdot R^{i-1} \colon x_i \neq 0\}$$

is the *unique inclusion-minimal* and *linearly independent* subset (of *odd* cardinality) of the vertex sequence $\vec{\mathrm{V}}(\boldsymbol{R}) := (R^0, R^1, \ldots, R^{2t-1})$, such that

$$ {}_{-A}\mathrm{T}^{(+)} = \sum_{Q \in \boldsymbol{Q}({}_{-A}\mathrm{T}^{(+)}, \boldsymbol{R})} Q \; ; \tag{7.3}$$

see Subsection 2.1.4.

As shown in Chapter 5, the smart *interval* organization of the ordered *maximal positive basis* $\vec{\mathrm{V}}(\boldsymbol{R})$ of the space $\mathbb{R}^t$ allows us to get the *linear algebraic* decompositions of vertices of the graph $\boldsymbol{H}(t, 2)$ with respect to the cycle $\boldsymbol{R}$ in an *explicit* and *computation-free* way, no matter how large the dimension t of the discrete hypercube $\{1, -1\}^t$ is; see Proposition 5.9. The four assertions collected in Proposition 5.9 also concern the *interval* structure of subsets of the ground set E_t: any nonempty subset $A \subseteq E_t$ is regarded as a disjoint union

$$A = [i_1, j_1] \mathbin{\dot\cup} [i_2, j_2] \mathbin{\dot\cup} \cdots \mathbin{\dot\cup} [i_{\varrho(A)}, j_{\varrho(A)}]$$

of inclusion-maximal intervals of the set E_t, for some $\varrho(A)$, where

$$j_1 + 2 \leq i_2, \quad j_2 + 2 \leq i_3, \quad \ldots, \quad j_{\varrho(A)-1} + 2 \leq i_{\varrho(A)} .$$

Given subsets A and B of the ground set E_t, such that A and B are incomparable by inclusion, in this chapter we are interested in the decompositions $\boldsymbol{Q}({}_{-A}\mathrm{T}^{(+)}, \boldsymbol{R})$, $\boldsymbol{Q}({}_{-B}\mathrm{T}^{(+)}, \boldsymbol{R})$ and $\boldsymbol{Q}(T, \boldsymbol{R})$ of the vertices ${}_{-A}\mathrm{T}^{(+)}$, ${}_{-B}\mathrm{T}^{(+)}$ and T such that $\mathfrak{n}(T) = A \cup B$.

We consider ordered two-member *Sperner families* (A, B) on the ground set E_t, but the case of *strict inclusion* of subsets $F \subsetneqq G$ is implicitly taken into our consideration as well, if we set $A := F$ and $B := G - F$.

As shown in Section 6.3, *integer compositions* and *Smirnov words* (i.e., words in which adjacent letters always differ) over sufficiently large alphabets form a toolkit for a careful structural and enumerative analysis of single finite sequences of symbols and tuples of sequences. One could involve various *restricted integer compositions* in order to make such an analysis even more detailed. See, e.g., Ref. [22] on compositions and words, and Ref. [24] and Appendix A of the present monograph on Smirnov words.

A closely related family of research tools is that originating from "the first *Kaplansky's lemma*" (see Section 20 of Ref. [25]) that gives the number

$$\binom{n-k+1}{k} \tag{7.4}$$

of ways of selecting k objects, *no two consecutive*, from n objects arranged in a row; see, e.g., Section 1.8 of Ref. [26], Section 2.3.15 of Ref. [27], Ref. [28], Section 1.3 of Ref. [29], Section 3.2 of Ref. [30], and Ref. [31]. Thus, (7.4) is the number of words of length n, over the two-letter alphabet (θ, α), with k nonconsecutive letters α. Recall that the total number $\sum_{k=0}^{\lfloor (n+1)/2 \rfloor} \binom{n-k+1}{k}$ of such words of length n is the *Fibonacci number* F_{n+2}; see, e.g., Eq. (6.130) of Ref. [32], and Exercise 1.35a in Ref. [7].

Together with the four assertions of Proposition 5.9, three players in our statistics on the decompositions of vertices of the hypercube graph $\boldsymbol{H}(t, 2)$ with respect to its distinguished symmetric cycle $\boldsymbol{R}$ are as follows:

As earlier, we denote by $\mathsf{c}(m; n)$ the number of *compositions* of a positive integer n with m positive parts:

$$\mathsf{c}(m; n) := \binom{n-1}{m-1} .$$

Let (θ, α, β) be an ordered *three-letter* alphabet. If $\mathfrak{s}', \mathfrak{s}'' \in \{\theta, \alpha, \beta\}$, then we denote by

$$\mathfrak{T}(\mathfrak{s}', \mathfrak{s}''; k, i, j)$$

the number of ternary Smirnov words, over the alphabet (θ, α, β) and with the *Parikh vector* (k, i, j), that start with the letter $\mathfrak{s}'$ and end with the letter $\mathfrak{s}''$; see Remark A.1. Similarly, we denote by

$$\mathfrak{F}(\mathfrak{s}', \mathfrak{s}''; k, i, j, h)$$

the number of Smirnov words, over the ordered *four-letter* alphabet $(\theta, \alpha, \beta, \gamma)$ and with the Parikh vector (k, i, j, h), that start with a letter $\mathfrak{s}' \in \{\theta, \alpha, \beta, \gamma\}$ and end with a letter $\mathfrak{s}'' \in \{\theta, \alpha, \beta, \gamma\}$; see Remark A.2.

7.1 Statistics on Unions of the Negative Parts of Vertices of Hypercube Graphs and on Decompositions of Vertices

We deal exclusively with the distinguished symmetric cycle $\boldsymbol{R}$, defined by (7.1)(7.2), in the hypercube graph $\boldsymbol{H}(t, 2)$ on the vertex set $\{1, -1\}^t$.

We will present statistics on the decompositions $\boldsymbol{Q}({}_{-A}\mathrm{T}^{(+)}, \boldsymbol{R})$, $\boldsymbol{Q}({}_{-B}\mathrm{T}^{(+)}, \boldsymbol{R})$ and $\boldsymbol{Q}(T, \boldsymbol{R})$ of vertices ${}_{-A}\mathrm{T}^{(+)}$, ${}_{-B}\mathrm{T}^{(+)}$ and T of the graph $\boldsymbol{H}(t, 2)$, such that $\mathfrak{n}(T) = A \cup B$.

The following cases will be considered separately:

$|A \cap B| = 0$, and $|A \dot{\cup} B| = t$, in Subsection 7.1.2,

$|A \cap B| = 0$, and $|A \dot{\cup} B| < t$, in Subsection 7.1.3,

$|A \cap B| > 0$, and $|A \cup B| = t$, in Subsection 7.1.4, and

$|A \cap B| > 0$, and $|A \cup B| < t$, in Subsection 7.1.5.

Before proceeding to statistics on decompositions, we will recall well-known general information on vertex pairs of the discrete hypercube $\{1, -1\}^t$.

7.1.1 Vertex Pairs of the Discrete Hypercube $\{1, -1\}^t$

Remark 7.1. Pick two elements $j', j'' \in E_t$. We have

(i) (a)

$$\#\{(X, Y) \in \{1, -1\}^t \times \{1, -1\}^t : d(X, Y) = k\} = 2^t \binom{t}{k} .$$

(b) If t is even, then

$$\#\{(X, Y) \in \{1, -1\}^t \times \{1, -1\}^t : \langle X, Y \rangle = 0\} = 2^t \binom{t}{t/2} .$$

(c) The *intersection numbers* of the algebraic combinatorial *Hamming scheme* $\mathbf{H}(t, 2)$ (see Ref. [33], and Section 21.3 of Ref. [34]) suggest that for any vertices X and Y of the discrete hypercube $\{1, -1\}^t$ such that $\langle X, Y \rangle = 0$, we have

$$\left|\{Z \in \{1, -1\}^t : \langle Z, X \rangle = \langle Z, Y \rangle = 0\}\right| = \begin{cases} \binom{t/2}{t/4}^2 , & \text{if } 4|t, \\ 0 , & \text{otherwise.} \end{cases}$$

(ii) (a)

$$\#\{(X, Y) \in \{1, -1\}^t \times \{1, -1\}^t : \\ |\mathfrak{n}(X)| =: j',\ |\mathfrak{n}(Y)| =: j'',\ d(X, Y) = k\} \\ = \begin{cases} \binom{t}{k}\binom{t-k}{(j'+j''-k)/2}\binom{k}{(j'-j''+k)/2}, & \text{if } j'+j''+k \text{ even}, \\ 0, & \text{if } j'+j''+k \text{ odd}. \end{cases}$$

(b) If t is even, then

$$\#\{(X, Y) \in \{1, -1\}^t \times \{1, -1\}^t : \\ |\mathfrak{n}(X)| =: j',\ |\mathfrak{n}(Y)| =: j'',\ \langle X, Y\rangle = 0\} \\ = \begin{cases} \binom{t}{t/2}\binom{t/2}{(2j'+2j''-t)/4}\binom{k}{(2j'-2j''+t)/4}, \\ \qquad \text{if } j'+j''+\tfrac{t}{2} \text{ even}, \\ 0, \qquad \text{if } j'+j''+\tfrac{t}{2} \text{ odd}. \end{cases}$$

Pick an element $s \in E_t$. Let $\binom{E_t}{s}$ denote the family of all subsets $A \subseteq E_t$ with $|A| = s$. Recall that if $s \leq 2t$, then the family $\binom{E_t}{s}$ can be viewed as the set of elements of the algebraic combinatorial *Johnson scheme* $\mathbf{J}(t, s)$ (see Ref. [33]). Dealing with the sth layer $\binom{E_t}{s}$ of the Boolean lattice $\mathbb{B}(t)$ of subsets of the set E_t, one often uses a measure of (dis)similarity $\partial(A, B)$ between s-sets A and B, defined by

$$\partial(A, B) := s - |A \cap B| = |A \cup B| - s = \frac{1}{2}|A \triangle B| \\ = \frac{1}{2}d({}_{-A}\mathrm{T}^{(+)}, {}_{-B}\mathrm{T}^{(+)}) = \frac{1}{4}\big(t - \langle {}_{-A}\mathrm{T}^{(+)}, {}_{-B}\mathrm{T}^{(+)}\rangle\big).$$

Note that we have

$$A \in \tbinom{E_t}{s} \ni B,\ \langle {}_{-A}\mathrm{T}^{(+)}, {}_{-B}\mathrm{T}^{(+)}\rangle = 0 \iff \partial(A, B) = \frac{t}{4}.$$

Remark 7.2. For an element $s \in E_t$ we have

(a)

$$\#\{(X, Y) \in \{1, -1\}^t \times \{1, -1\}^t : \\ \mathfrak{n}(X) \in \tbinom{E_t}{s} \ni \mathfrak{n}(Y),\ \partial(\mathfrak{n}(X), \mathfrak{n}(Y)) = i\} \\ = \tbinom{t}{2i}\tbinom{t-2i}{s-i}\tbinom{2i}{i}.$$

(b)

$$\#\big\{(X, Y) \in \{1,-1\}^t \times \{1,-1\}^t:\ \mathfrak{n}(X) \in \tbinom{E_t}{s} \ni \mathfrak{n}(Y),\ \langle X, Y\rangle = 0\big\}$$
$$= \begin{cases} \binom{t}{t/2}\binom{t/2}{s-(t/4)}\binom{t/2}{t/4}, & \text{if } 4|t \text{ and } \frac{t}{4} \le s \le \frac{3t}{4}, \\ 0, & \text{otherwise}. \end{cases}$$

(c) Suppose that $4|t$, and $\frac{t}{4} \le s \le \frac{3t}{4}$. For any vertices X and Y of the discrete hypercube $\{1,-1\}^t$ such that $\mathfrak{n}(X) \in \binom{E_t}{s} \ni \mathfrak{n}(Y)$, and $\langle X, Y\rangle = 0$, we have

$$\Big|\big\{Z \in \{1,-1\}^t:\ \mathfrak{n}(Z) \in \tbinom{E_t}{s},\ \langle Z, X\rangle = \langle Z, Y\rangle = 0\big\}\Big|$$
$$= \sum_{c=\max\{0,s-(t/2)\}}^{\min\{t/4,s-(t/4)\}} \binom{s-(t/4)}{c}\binom{(3t/4)-s}{(t/4)-c}\binom{t/4}{s-(t/4)-c}^2. \quad (7.5)$$

The quantity (7.5) is suggested, in the case $2s \le t$, by the *intersection numbers* of the *Johnson scheme* $\mathbf{J}(t, s)$.

7.1.2 Case: $|A \cap B| = 0$, and $|A \dot{\cup} B| = t$

Let us consider the family

$$\big\{(A, B) \in \mathbf{2}^{[t]} \times \mathbf{2}^{[t]}:\ |A| = j' > 0,\ |B| = j'' > 0,$$
$$|A \cap B| = 0,\ |A \dot{\cup} B| = t,\ \mathfrak{q}(A) = \ell',\ \mathfrak{q}(B) = \ell''\big\} \quad (7.6)$$

of ordered two-member *partitions* of the set E_t; thus, $j' + j'' = t$. The integers $\ell', \ell'' \in [t]$ are *odd*. For any pair (A, B) in this family, we have

$$d({}_{-A}\mathrm{T}^{(+)}, {}_{-B}\mathrm{T}^{(+)}) = t, \quad \text{and} \quad \langle {}_{-A}\mathrm{T}^{(+)}, {}_{-B}\mathrm{T}^{(+)}\rangle = -t.$$

Note that if $\{1, t\} \cap A = \{1\}$, and $\{1, t\} \cap B = \{t\}$, with $\mathfrak{q}(A) = \ell'$ and $\mathfrak{q}(B) = \ell''$, then we have

$$\varrho(A) = \frac{\ell' + 1}{2}, \quad \varrho(B) = \frac{\ell'' + 1}{2},$$

by Proposition 5.9(i)(iv).

If $\{1, t\} \cap A = \{1, t\}$, and $|\{1, t\} \cap B| = 0$, then we have

$$\varrho(A) = \frac{\ell' + 1}{2}, \quad \varrho(B) = \frac{\ell'' - 1}{2},$$

by Proposition 5.9(ii)(iii).

Remark 7.3. Ordered pairs of sets (A, B) in the family (7.6) can be counted with the help of products of the form

$$\mathrm{c}(\varrho(A); j') \cdot \mathrm{c}(\varrho(B); j'') :$$

(i) In the family (7.6) there are

$$\mathrm{c}\left(\frac{\ell' + 1}{2}; j'\right) \cdot \mathrm{c}\left(\frac{\ell'' + 1}{2}; j''\right)$$

pairs (A, B) of sets A and B such that

$$\{1, t\} \cap A = \{1\}, \quad \text{and} \quad \{1, t\} \cap B = \{t\},$$

and if the subfamily of these pairs is nonempty, then $\ell' = \ell''$.

(ii) In the family (7.6) there are

$$\mathrm{c}\left(\frac{\ell' + 1}{2}; j'\right) \cdot \mathrm{c}\left(\frac{\ell'' - 1}{2}; j''\right)$$

pairs (A, B) such that

$$\{1, t\} \cap A = \{1, t\}, \quad \text{and} \quad |\{1, t\} \cap B| = 0,$$

and if the subfamily of these pairs is nonempty, then $\ell' = \ell''$.

7.1.3 Case: $|A \cap B| = 0$, and $|A \dot\cup B| < t$

Let us consider the family

$$\begin{aligned}\{(A, B) \in \mathbf{2}^{[t]} \times \mathbf{2}^{[t]} :\ & |A| = j' > 0, \ |B| = j'' > 0, \\ & |A \cap B| = 0, \ |A \dot\cup B| < t, \\ & \mathfrak{q}(A) = \ell', \ \mathfrak{q}(B) = \ell'', \ \mathfrak{q}(A \dot\cup B) = \ell\}\end{aligned} \quad (7.7)$$

of ordered pairs of *disjoint* subsets of the set E_t, with the *odd* integers $\ell', \ell'', \ell \in [t]$. Note that for pairs (A, B) in this family, we have

$$d({}_{-A}\mathrm{T}^{(+)}, {}_{-B}\mathrm{T}^{(+)}) = j' + j'',$$

and

$$\langle {}_{-A}\mathrm{T}^{(+)}, {}_{-B}\mathrm{T}^{(+)} \rangle = 0 \iff 2(j' + j'') = t.$$

Proposition 7.4 (Theorem 6.4, rephrased and abridged). *Ordered pairs of sets (A, B) in the family (7.7) can be counted with the help of products of the form*

$$\mathfrak{T}\left(\mathfrak{s}', \mathfrak{s}''; \varrho(E_t - (A \dot\cup B)), \varrho(A), \varrho(B)\right) \times \mathrm{c}\left(\varrho(E_t - (A \dot\cup B)); t - (j' + j'')\right) \cdot \mathrm{c}(\varrho(A); j') \cdot \mathrm{c}(\varrho(B); j'') :$$

(i) *In the family* (7.7) *there are*

$$\mathfrak{T}\left(\theta, \theta; \frac{\ell+1}{2}, \frac{\ell'-1}{2}, \frac{\ell''-1}{2}\right) \cdot \mathrm{c}\left(\frac{\ell+1}{2}; t - (j' + j'')\right) \times \mathrm{c}\left(\frac{\ell'-1}{2}; j'\right) \cdot \mathrm{c}\left(\frac{\ell''-1}{2}; j''\right)$$

pairs (A, B) *of sets* A *and* B *such that*

$$|\{1, t\} \cap A| = |\{1, t\} \cap B| = 0 .$$

(ii) *In the family* (7.7) *there are*

$$\mathfrak{T}\left(\alpha, \alpha; \frac{\ell-1}{2}, \frac{\ell'+1}{2}, \frac{\ell''-1}{2}\right) \cdot \mathrm{c}\left(\frac{\ell-1}{2}; t - (j' + j'')\right) \times \mathrm{c}\left(\frac{\ell'+1}{2}; j'\right) \cdot \mathrm{c}\left(\frac{\ell''-1}{2}; j''\right)$$

pairs (A, B) *such that*

$$\{1, t\} \cap A = \{1, t\}, \quad \textit{and} \quad |\{1, t\} \cap B| = 0 ;$$

(iii)

$$\mathfrak{T}\left(\theta, \beta; \frac{\ell+1}{2}, \frac{\ell'-1}{2}, \frac{\ell''+1}{2}\right) \cdot \mathrm{c}\left(\frac{\ell+1}{2}; t - (j' + j'')\right) \times \mathrm{c}\left(\frac{\ell'-1}{2}; j'\right) \cdot \mathrm{c}\left(\frac{\ell''+1}{2}; j''\right)$$

pairs (A, B) *such that*

$$|\{1, t\} \cap A| = 0 , \quad \textit{and} \quad \{1, t\} \cap B = \{t\} ;$$

(iv)

$$\mathfrak{T}\left(\alpha, \theta; \frac{\ell+1}{2}, \frac{\ell'+1}{2}, \frac{\ell''-1}{2}\right) \cdot \mathrm{c}\left(\frac{\ell+1}{2}; t - (j' + j'')\right) \times \mathrm{c}\left(\frac{\ell'+1}{2}; j'\right) \cdot \mathrm{c}\left(\frac{\ell''-1}{2}; j''\right)$$

pairs (A, B) *such that*

$$\{1, t\} \cap A = \{1\} , \quad \textit{and} \quad |\{1, t\} \cap B| = 0 ;$$

(v)

$$\mathfrak{T}\left(\alpha, \beta; \frac{\ell-1}{2}, \frac{\ell'+1}{2}, \frac{\ell''+1}{2}\right) \cdot \mathrm{c}\left(\frac{\ell-1}{2}; t-(j'+j'')\right) \times \mathrm{c}\left(\frac{\ell'+1}{2}; j'\right) \cdot \mathrm{c}\left(\frac{\ell''+1}{2}; j''\right)$$

pairs (A, B) *such that*

$$\{1, t\} \cap A = \{1\}, \quad \textit{and} \quad \{1, t\} \cap B = \{t\}.$$

7.1.4 Case: $|A \cap B| > 0$, and $|A \cup B| = t$

- Let us consider the family

$$\{(A, B) \in \mathbf{2}^{[t]} \times \mathbf{2}^{[t]} : \ |A| = j', \ |B| = j'', \ |A \cap B| > 0, \ \max\{j', j''\} < |A \cup B| = t, \ \mathfrak{q}(A \cap B) = \ell^{\cap}, \ \mathfrak{q}(A - B) = \ell', \ \mathfrak{q}(B - A) = \ell''\} \quad (7.8)$$

of ordered two-member *intersecting Sperner families* that *cover* the set E_t; the integers $\ell^{\cap}, \ell', \ell'' \in [t]$ are *odd*. Note that for pairs (A, B) in this family, we have

$$d({}_{-A}\mathrm{T}^{(+)}, {}_{-B}\mathrm{T}^{(+)}) = 2t - j' - j'', \quad (7.9)$$

and

$$\langle {}_{-A}\mathrm{T}^{(+)}, {}_{-B}\mathrm{T}^{(+)} \rangle = 0 \iff 2(j' + j'') = 3t. \quad (7.10)$$

Theorem 7.5. *Ordered pairs of sets* (A, B) *in the family* (7.8) *can be counted with the help of products of the form*

$$\mathfrak{T}(\mathfrak{s}', \mathfrak{s}''; \varrho(A \cap B), \varrho(A - B), \varrho(B - A)) \times \mathrm{c}(\varrho(A \cap B); (j' + j'') - t) \times \mathrm{c}(\varrho(A - B); t - j'') \cdot \mathrm{c}(\varrho(B - A); t - j') \ : \quad (7.11)$$

(i) *In the family* (7.8) *there are*

$$\mathfrak{T}\left(\theta, \beta; \frac{\ell^{\cap}+1}{2}, \frac{\ell'-1}{2}, \frac{\ell''+1}{2}\right) \cdot \mathrm{c}\left(\frac{\ell^{\cap}+1}{2}; (j'+j'')-t\right) \times \mathrm{c}\left(\frac{\ell'-1}{2}; t-j''\right) \cdot \mathrm{c}\left(\frac{\ell''+1}{2}; t-j'\right)$$

pairs (A, B) *of sets* A *and* B *such that*

$$\{1, t\} \cap A = \{1\}, \quad \textit{and} \quad \{1, t\} \cap B = \{1, t\}.$$

In the family (7.8) *there are*

(ii)

$$\mathfrak{T}\left(\alpha, \beta; \frac{\ell^{\cap}-1}{2}, \frac{\ell'+1}{2}, \frac{\ell''+1}{2}\right) \cdot \mathrm{c}\left(\frac{\ell^{\cap}-1}{2}; (j'+j'')-t\right)$$
$$\times\, \mathrm{c}\left(\frac{\ell'+1}{2}; t-j''\right) \cdot \mathrm{c}\left(\frac{\ell''+1}{2}; t-j'\right)$$

pairs (A, B) *such that*

$$\{1, t\} \cap A = \{1\}, \quad and \quad \{1, t\} \cap B = \{t\};$$

(iii)

$$\mathfrak{T}\left(\theta, \theta; \frac{\ell^{\cap}+1}{2}, \frac{\ell'-1}{2}, \frac{\ell''-1}{2}\right) \cdot \mathrm{c}\left(\frac{\ell^{\cap}+1}{2}; (j'+j'')-t\right)$$
$$\times\, \mathrm{c}\left(\frac{\ell'-1}{2}; t-j''\right) \cdot \mathrm{c}\left(\frac{\ell''-1}{2}; t-j'\right)$$

pairs (A, B) *such that*

$$\{1, t\} \cap A = \{1, t\} \cap B = \{1, t\};$$

(iv)

$$\mathfrak{T}\left(\alpha, \alpha; \frac{\ell^{\cap}-1}{2}, \frac{\ell'+1}{2}, \frac{\ell''-1}{2}\right) \cdot \mathrm{c}\left(\frac{\ell^{\cap}-1}{2}; (j'+j'')-t\right)$$
$$\times\, \mathrm{c}\left(\frac{\ell'+1}{2}; t-j''\right) \cdot \mathrm{c}\left(\frac{\ell''-1}{2}; t-j'\right)$$

pairs (A, B) *such that*

$$\{1, t\} \cap A = \{1, t\}, \quad and \quad |\{1, t\} \cap B| = 0;$$

(v)

$$\mathfrak{T}\left(\alpha, \theta; \frac{\ell^{\cap}+1}{2}, \frac{\ell'+1}{2}, \frac{\ell''-1}{2}\right) \cdot \mathrm{c}\left(\frac{\ell^{\cap}+1}{2}; (j'+j'')-t\right)$$
$$\times\, \mathrm{c}\left(\frac{\ell'+1}{2}; t-j''\right) \cdot \mathrm{c}\left(\frac{\ell''-1}{2}; t-j'\right)$$

pairs (A, B) *such that*

$$\{1, t\} \cap A = \{1, t\}, \quad and \quad \{1, t\} \cap B = \{t\}.$$

Proof. For each pair (A, B) in the family (7.8), pick an *ascending system of distinct representatives* $(e_1 < e_2 < \ldots < e_{\varrho(A\cap B)+\varrho(A-B)+\varrho(B-A)}) \subseteq E_t$ of the intervals composing the sets $(A \cap B)$, $(A - B)$ and $(B - A)$. By making the substitutions

$$e_i \mapsto \begin{cases} \theta\,, & \text{if } e_i \in A \cap B, \\ \alpha\,, & \text{if } e_i \in A - B, \\ \beta\,, & \text{if } e_i \in B - A, \end{cases} \quad 1 \le i \le \varrho(A\cap B)+\varrho(A-B)+\varrho(B-A)\,,$$

we obtain some ternary Smirnov words over the ordered alphabet (θ, α, β). Given any such Smirnov word, we find the number of pairs (A, B) in the corresponding subfamily of the family (7.8) by means of a product of the form (7.11).

(i) Since $\{1, t\} \cap A = \{1\}$, and $\{1, t\} \cap B = \{1, t\}$, we have

$$\varrho(A - B) = \frac{\ell' - 1}{2}\,, \quad \varrho(B - A) = \frac{\ell'' + 1}{2}\,,$$

by Proposition 5.9(iii)(iv), and

$$\varrho(A \cap B) = \frac{\ell^{\cap} + 1}{2}\,,$$

by Proposition 5.9(i).

(ii) Since $\{1, t\} \cap A = \{1\}$, and $\{1, t\} \cap B = \{t\}$, we have

$$\varrho(A - B) = \frac{\ell' + 1}{2}\,, \quad \varrho(B - A) = \frac{\ell'' + 1}{2}\,,$$

by Proposition 5.9(i)(iv), and

$$\varrho(A \cap B) = \frac{\ell^{\cap} - 1}{2}\,,$$

by Proposition 5.9(iii).

(iii) Since $\{1, t\} \cap A = \{1, t\} \cap B = \{1, t\}$, we have

$$\varrho(A - B) = \frac{\ell' - 1}{2}\,, \quad \varrho(B - A) = \frac{\ell'' - 1}{2}\,,$$

by Proposition 5.9(iii), and

$$\varrho(A \cap B) = \frac{\ell^{\cap} + 1}{2}\,,$$

by Proposition 5.9(ii).

(iv) Since $\{1, t\} \cap A = \{1, t\}$, and $|\{1, t\} \cap B| = 0$, we have

$$\varrho(A - B) = \frac{\ell' + 1}{2}, \quad \varrho(B - A) = \frac{\ell'' - 1}{2},$$

by Proposition 5.9(ii)(iii), and

$$\varrho(A \cap B) = \frac{\ell^{\cap} - 1}{2},$$

by Proposition 5.9(iii).

(v) Since $\{1, t\} \cap A = \{1, t\}$, and $\{1, t\} \cap B = \{t\}$, we have

$$\varrho(A - B) = \frac{\ell' + 1}{2}, \quad \varrho(B - A) = \frac{\ell'' - 1}{2},$$

by Proposition 5.9(i)(iii), and

$$\varrho(A \cap B) = \frac{\ell^{\cap} + 1}{2},$$

by Proposition 5.9(iv). □

Example 7.6. Suppose $t := 5$.

(i) In the family

$$\big\{(A, B) \in \mathbf{2}^{[t]} \times \mathbf{2}^{[t]} : \ j' := |A| = 3\,, \ \ j'' := |B| = 4\,, \\ |A \cap B| > 0\,, \ \ |A \cup B| = t\,, \\ \ell^{\cap} := \mathfrak{q}(A \cap B) = 3\,, \ \ \ell' := \mathfrak{q}(A - B) = 3\,, \ \ \ell'' := \mathfrak{q}(B - A) = 3\big\}$$

of ordered two-member *intersecting Sperner families* that cover the set E_t, there are

$$\mathfrak{T}\left(\theta, \beta; \frac{\ell^{\cap}+1}{2}, \frac{\ell'-1}{2}, \frac{\ell''+1}{2}\right) \cdot \mathrm{c}\left(\frac{\ell^{\cap}+1}{2}; (j'+j'') - t\right) \\ \times \mathrm{c}\left(\frac{\ell'-1}{2}; t - j''\right) \cdot \mathrm{c}\left(\frac{\ell''+1}{2}; t - j'\right) \\ := \mathfrak{T}\left(\theta, \beta; \frac{3+1}{2}, \frac{3-1}{2}, \frac{3+1}{2}\right) \cdot \mathrm{c}\left(\frac{3+1}{2}; (3+4) - 5\right) \\ \times \mathrm{c}\left(\frac{3-1}{2}; 5 - 4\right) \cdot \mathrm{c}\left(\frac{3+1}{2}; 5 - 3\right) \\ = \underbrace{\mathfrak{T}(\theta, \beta; 2, 1, 2)}_{\text{see (A.8)}} \cdot \mathrm{c}(2; 2) \cdot \mathrm{c}(1; 1) \cdot \mathrm{c}(2; 2) \\ := (2 + 2 - 1)\binom{2-1}{\frac{2+2-1-1}{2}}\binom{\frac{2+1+2-3}{2}}{2-1} \cdot \binom{2-1}{2-1} \cdot \binom{1-1}{1-1} \cdot \binom{2-1}{2-1} = 3$$

pairs $(A,\ B)$ of sets A and B such that

$$\{1,\,t\}\cap A=\{1\}\,,\quad\text{and}\quad\{1,\,t\}\cap B=\{1,\,t\}\,.$$

One such pair is

$$\begin{aligned}(\ A &:=\{1,\quad,\,3,\ 4,\quad\}\,,\\ B &:=\{1,\ 2,\ 3,\quad,\,t\}\ )\,,\end{aligned}$$

for which we have

$$\begin{aligned}A\cap B&=\{1,\quad,\,3,\quad,\quad\}\,,\\ A-B&=\{\quad,\quad,\quad,\,4,\quad\}\,,\\ B-A&=\{\quad,\,2,\quad,\quad,\,t\}\,,\end{aligned}$$

and

$$\begin{aligned}\boldsymbol{x}(A\cap B)&=(0,\quad1,-1,\quad1,\quad0)\,,\\ \boldsymbol{Q}({}_{-(A\cap B)}\mathrm{T}^{(+)},\,\boldsymbol{R})&=\{R^1,\,R^3,\,R^7\}\,,\\ \boldsymbol{x}(A-B)&=(1,\quad0,\quad0,-1,\quad1)\,,\\ \boldsymbol{Q}({}_{-(A-B)}\mathrm{T}^{(+)},\,\boldsymbol{R})&=\{R^0,\,R^4,\,R^8\}\,,\\ \boldsymbol{x}(B-A)&=(0,-1,\quad1,\quad0,-1)\,,\\ \boldsymbol{Q}({}_{-(B-A)}\mathrm{T}^{(+)},\,\boldsymbol{R})&=\{R^2,\,R^6,\,R^{2t-1}\}\,;\end{aligned}$$

see Figure 7.1.

(ii) The family

$$\begin{aligned}\big\{(A,\,B)\in\mathbf{2}^{[t]}\times\mathbf{2}^{[t]}:\ \ j':=|A|=3\,,\ \ j'':=|B|=4\,,\\ |A\cap B|>0\,,\ \ |A\cup B|=t\,,\\ \ell^{\cap}:=\mathfrak{q}(A\cap B)=5\,,\ \ \ell':=\mathfrak{q}(A-B)=1\,,\ \ \ell'':=\mathfrak{q}(B-A)=3\big\}\end{aligned} \tag{7.12}$$

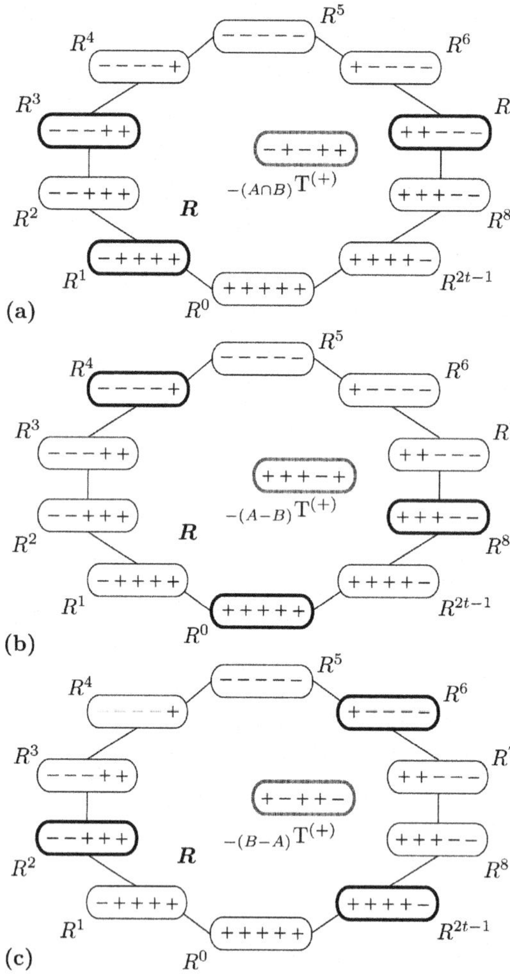

Figure 7.1 Three instances of the same distinguished symmetric cycle $\boldsymbol{R} := (R^0, R^1, \ldots, R^{2t-1}, R^0)$ in the hypercube graph $\boldsymbol{H}((t := 5), 2)$, defined by (7.1)(7.2).
Pick the subsets $A := \{1, 3, 4\}$ and $B := \{1, 2, 3, t\}$ of the set E_t. Note that $\{1, t\} \cap A = \{1\}$, and $\{1, t\} \cap B = \{1, t\}$.
(a) The vertex ${}_{-(A\cap B)}\mathrm{T}^{(+)}$ of the graph $\boldsymbol{H}(t, 2)$, with the corresponding decomposition set $\boldsymbol{Q}({}_{-(A\cap B)}\mathrm{T}^{(+)}, \boldsymbol{R}) = \{R^1, R^3, R^7\}$.
(b) The vertex ${}_{-(A-B)}\mathrm{T}^{(+)}$, with its decomposition set $\boldsymbol{Q}({}_{-(A-B)}\mathrm{T}^{(+)}, \boldsymbol{R}) = \{R^0, R^4, R^8\}$.
(c) The vertex ${}_{-(B-A)}\mathrm{T}^{(+)}$, with its decomposition set $\boldsymbol{Q}({}_{-(B-A)}\mathrm{T}^{(+)}, \boldsymbol{R}) = \{R^2, R^6, R^{2t-1}\}$.

consists of

$$
\begin{aligned}
&\mathfrak{T}\left(\alpha,\beta;\frac{\ell^{\cap}-1}{2},\frac{\ell'+1}{2},\frac{\ell''+1}{2}\right)\cdot \mathrm{c}\left(\frac{\ell^{\cap}-1}{2};(j'+j'')-t\right)\\
&\qquad\times \mathrm{c}\left(\frac{\ell'+1}{2};t-j''\right)\cdot \mathrm{c}\left(\frac{\ell''+1}{2};t-j'\right)\\
&\quad:=\mathfrak{T}\left(\alpha,\beta;\frac{5-1}{2},\frac{1+1}{2},\frac{3+1}{2}\right)\cdot \mathrm{c}\left(\frac{5-1}{2};(3+4)-5\right)\\
&\qquad\times \mathrm{c}\left(\frac{1+1}{2};5-4\right)\cdot \mathrm{c}\left(\frac{3+1}{2};5-3\right)\\
&\quad=\underbrace{\mathfrak{T}(\alpha,\beta;2,1,2)}_{\text{see (6.17)}}\cdot \mathrm{c}(2;2)\cdot \mathrm{c}(1;1)\cdot \mathrm{c}(2;2)\\
&\quad:=(1+2-2)\cdot\binom{1-1}{\frac{1+2-2-1}{2}}\binom{\frac{1+2+2-3}{2}}{1-1}\cdot\binom{2-1}{2-1}\cdot\binom{1-1}{1-1}\cdot\binom{2-1}{2-1}=1
\end{aligned}
$$

pair (A, B) such that

$$\{1, t\}\cap A=\{1\}\,,\quad\text{and}\quad\{1, t\}\cap B=\{t\}\,.$$

This pair is

$$
\begin{aligned}
(\ A &:= \{1,\ 2,\ \ ,\ 4,\ \ \}\,,\\
B &:= \{\ \ ,\ 2,\ 3,\ 4,\ t\}\ )\,,
\end{aligned}
$$

for which we have

$$
\begin{aligned}
A\cap B &= \{\ \ ,\ 2,\ \ ,\ 4,\ \ \}\,,\\
A-B &= \{1,\ \ ,\ \ ,\ \ ,\ \ \}\,,\\
B-A &= \{\ \ ,\ \ ,\ 3,\ \ ,\ t\}\,,
\end{aligned}
$$

and

$$
\begin{aligned}
\boldsymbol{x}(A\cap B) &= (1,-1,\ \ 1,-1,\ \ 1)\,,\\
\boldsymbol{Q}({}_{-(A\cap B)}\mathrm{T}^{(+)},\boldsymbol{R}) &= \{R^0, R^2, R^4, R^6, R^8\}\,,\\
\boldsymbol{x}(A-B) &= (0,\ \ 1,\ \ 0,\ \ 0,\ \ 0)\,,\\
\boldsymbol{Q}({}_{-(A-B)}\mathrm{T}^{(+)},\boldsymbol{R}) &= \{R^1\}\,,\\
\boldsymbol{x}(B-A) &= (0,\ \ 0,-1,\ \ 1,-1)\,,\\
\boldsymbol{Q}({}_{-(B-A)}\mathrm{T}^{(+)},\boldsymbol{R}) &= \{R^3, R^7, R^{2t-1}\}\,;
\end{aligned}
$$

see Figure 7.2.

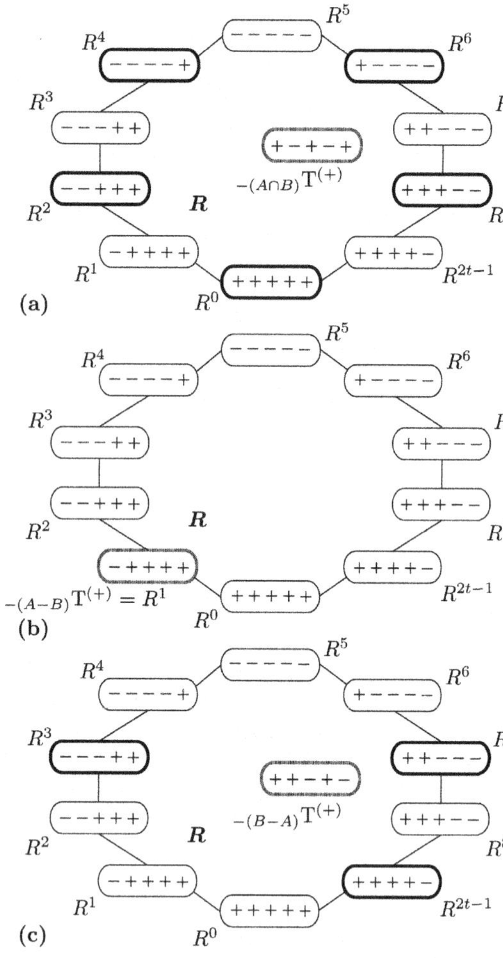

Figure 7.2 Three instances of the same distinguished symmetric cycle $\boldsymbol{R} := (R^0, R^1, \ldots, R^{2t-1}, R^0)$ in the hypercube graph $\boldsymbol{H}((t := 5), 2)$, defined by (7.1)(7.2).

Pick the subsets $A := \{1, 2, 4\}$ and $B := \{2, 3, 4, t\}$ of the set E_t. Note that $\{1, t\} \cap A = \{1\}$, and $\{1, t\} \cap B = \{t\}$.

(a) The vertex ${}_{-(A\cap B)}\mathrm{T}^{(+)}$ of the graph $\boldsymbol{H}(t, 2)$, with the corresponding decomposition set $\boldsymbol{Q}({}_{-(A\cap B)}\mathrm{T}^{(+)}, \boldsymbol{R}) = \{R^0, R^2, R^4, R^6, R^8\}$.

(b) The vertex ${}_{-(A-B)}\mathrm{T}^{(+)}$, with its decomposition set $\boldsymbol{Q}({}_{-(A-B)}\mathrm{T}^{(+)}, \boldsymbol{R}) = \{R^1\}$.

(c) The vertex ${}_{-(B-A)}\mathrm{T}^{(+)}$, with its decomposition set $\boldsymbol{Q}({}_{-(B-A)}\mathrm{T}^{(+)}, \boldsymbol{R}) = \{R^3, R^7, R^{2t-1}\}$.

(iii) In the family

$$\{(A,\,B)\in \mathbf{2}^{[t]}\times\mathbf{2}^{[t]}:\ \ j':=|A|=4\,,\ \ j'':=|B|=4\,,$$
$$|A\cap B|>0\,,\ \ |A\cup B|=t\,,$$
$$\ell^{\cap}:=\mathfrak{q}(A\cap B)=5\,,\ \ \ell':=\mathfrak{q}(A-B)=3\,,\ \ \ell'':=\mathfrak{q}(B-A)=3\} \tag{7.13}$$

there are

$$\begin{aligned}
&\mathfrak{T}\left(\theta,\theta;\frac{\ell^{\cap}+1}{2},\frac{\ell'-1}{2},\frac{\ell''-1}{2}\right)\cdot \mathrm{c}\left(\frac{\ell^{\cap}+1}{2};(j'+j'')-t\right)\\
&\qquad\times \mathrm{c}\left(\frac{\ell'-1}{2};t-j''\right)\cdot \mathrm{c}\left(\frac{\ell''-1}{2};t-j'\right)\\
&\quad:=\mathfrak{T}\left(\theta,\theta;\frac{5+1}{2},\frac{3-1}{2},\frac{3-1}{2}\right)\cdot \mathrm{c}\left(\frac{5+1}{2};(4+4)-5\right)\\
&\qquad\times \mathrm{c}\left(\frac{3-1}{2};5-4\right)\cdot \mathrm{c}\left(\frac{3-1}{2};5-4\right)\\
&\quad=\underbrace{\mathfrak{T}(\theta,\theta;3,1,1)}_{\text{see (A.7)}}\cdot \mathrm{c}(3;3)\cdot \mathrm{c}(1;1)\cdot \mathrm{c}(1;1)\\
&\quad=\binom{3-1}{\frac{3+1-1-1}{2}}\binom{\frac{3+1+1-3}{2}}{3-2}\cdot\binom{3-1}{3-1}\cdot\binom{1-1}{1-1}\cdot\binom{1-1}{1-1}=2
\end{aligned}$$

pairs $(A,\,B)$ such that

$$\{1,\,t\}\cap A=\{1,\,t\}\cap B=\{1,\,t\}\,.$$

One such pair is

$$(\ A:=\{1,\ \ ,\,3,\,4,\,t\}\,,$$
$$B:=\{1,\,2,\,3,\ \ ,\,t\}\ )\,,$$

for which we have

$$A\cap B=\{1,\ \ ,\,3,\ \ ,\,t\}\,,$$
$$A-B=\{\ ,\ \ ,\ \ ,\,4,\ \}\,,$$
$$B-A=\{\ ,\,2,\ \ ,\ \ ,\ \}\,,$$

and

$$\begin{aligned}
\boldsymbol{x}(A\cap B)&=(-1,\ \ 1,-1,\ \ 1,-1)\,,\\
\boldsymbol{Q}({}_{-(A\cap B)}\mathrm{T}^{(+)},\,\boldsymbol{R})&=\{R^1,\,R^3,\,R^5,\,R^7,\,R^{2t-1}\}\,,\\
\boldsymbol{x}(A-B)&=(\ \ 1,\ \ 0,\ \ 0,-1,\ \ 1)\,,\\
\boldsymbol{Q}({}_{-(A-B)}\mathrm{T}^{(+)},\,\boldsymbol{R})&=\{R^0,\,R^4,\,R^8\}\,,\\
\boldsymbol{x}(B-A)&=(\ \ 1,-1,\ \ 1,\ \ 0,\ \ 0)\,,\\
\boldsymbol{Q}({}_{-(B-A)}\mathrm{T}^{(+)},\,\boldsymbol{R})&=\{R^0,\,R^2,\,R^6\}\,;
\end{aligned}$$

see Figure 7.3.

(iv) In the family

$$\begin{aligned}\{(A,B)\in \mathbf{2}^{[t]}\times \mathbf{2}^{[t]}:\ \ j':=|A|=4\,,\ \ j'':=|B|=3\,,\\ |A\cap B|>0\,,\ \ |A\cup B|=t\,,\\ \ell^{\cap}:=\mathfrak{q}(A\cap B)=3\,,\ \ \ell':=\mathfrak{q}(A-B)=3\,,\ \ \ell'':=\mathfrak{q}(B-A)=3\}\end{aligned} \tag{7.14}$$

there are

$$\begin{aligned}&\mathfrak{T}\left(\alpha,\alpha;\frac{\ell^{\cap}-1}{2},\frac{\ell'+1}{2},\frac{\ell''-1}{2}\right)\cdot \mathrm{c}\left(\frac{\ell^{\cap}-1}{2};(j'+j'')-t\right)\\ &\quad\times \mathrm{c}\left(\frac{\ell'+1}{2};t-j''\right)\cdot \mathrm{c}\left(\frac{\ell''-1}{2};t-j'\right)\\ &:=\mathfrak{T}\left(\alpha,\alpha;\frac{3-1}{2},\frac{3+1}{2},\frac{3-1}{2}\right)\cdot \mathrm{c}\left(\frac{3-1}{2};(4+3)-5\right)\\ &\quad\times \mathrm{c}\left(\frac{3+1}{2};5-3\right)\cdot \mathrm{c}\left(\frac{3-1}{2};5-4\right)\\ &=\underbrace{\mathfrak{T}(\alpha,\alpha;1,2,1)}_{\text{see (6.15)}}\cdot \mathrm{c}(1;2)\cdot \mathrm{c}(2;2)\cdot \mathrm{c}(1;1)\\ &=(2+1-1)\cdot\binom{2-1}{\frac{2+1-1}{2}}\binom{\frac{2+1+1}{2}-2}{2-2}\cdot\binom{2-1}{1-1}\cdot\binom{2-1}{2-1}\cdot\binom{1-1}{1-1}=2\end{aligned}$$

pairs (A,B) such that

$$\{1,t\}\cap A=\{1,t\}\,,\quad \text{and}\quad |\{1,t\}\cap B|=0\,.$$

One such pair is

$$\begin{aligned}(\ A&:=\{1,\ 2,\ 3,\ \ ,\ t\}\,,\\ B&:=\{\ \ ,\ 2,\ 3,\ 4,\ \ \}\ )\,,\end{aligned}$$

for which we have

$$\begin{aligned}A\cap B&=\{\ \ ,\ 2,\ 3,\ \ ,\ \ \}\,,\\ A-B&=\{1,\ \ ,\ \ ,\ \ ,\ t\}\,,\\ B-A&=\{\ \ ,\ \ ,\ \ ,\ 4,\ \ \}\,,\end{aligned}$$

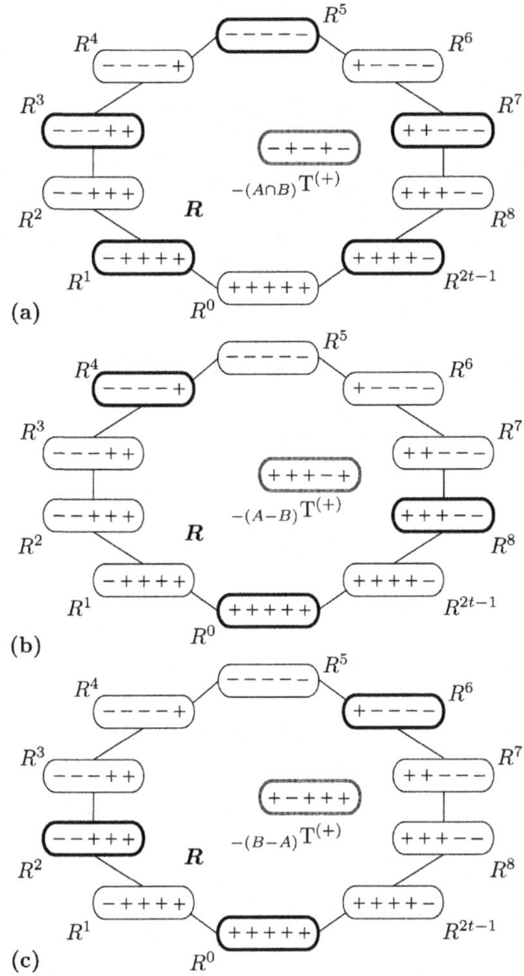

Figure 7.3 Three instances of the same distinguished symmetric cycle $\boldsymbol{R} := (R^0, R^1, \ldots, R^{2t-1}, R^0)$ in the hypercube graph $\boldsymbol{H}((t := 5), 2)$, defined by (7.1)(7.2).
Pick the subsets $A := \{1, 3, 4, t\}$ and $B := \{1, 2, 3, t\}$ of the set E_t. Note that $\{1, t\} \cap A = \{1, t\} \cap B = \{1, t\}$.
(a) The vertex ${}_{-(A \cap B)}\mathrm{T}^{(+)}$ of the graph $\boldsymbol{H}(t, 2)$, with the corresponding decomposition set $\boldsymbol{Q}({}_{-(A \cap B)}\mathrm{T}^{(+)}, \boldsymbol{R}) = \{R^1, R^3, R^5, R^7, R^{2t-1}\}$.
(b) The vertex ${}_{-(A-B)}\mathrm{T}^{(+)}$, with its decomposition set $\boldsymbol{Q}({}_{-(A-B)}\mathrm{T}^{(+)}, \boldsymbol{R}) = \{R^0, R^4, R^8\}$.
(c) The vertex ${}_{-(B-A)}\mathrm{T}^{(+)}$, with its decomposition set $\boldsymbol{Q}({}_{-(B-A)}\mathrm{T}^{(+)}, \boldsymbol{R}) = \{R^0, R^2, R^6\}$.

and

$$\begin{aligned}
\boldsymbol{x}(A\cap B) &= (\ \ 1, -1, \ \ 0, \ \ 1, \ \ 0)\,,\\
\boldsymbol{Q}({}_{-(A\cap B)}\mathrm{T}^{(+)}, \boldsymbol{R}) &= \{R^0, R^3, R^6\}\,,\\
\boldsymbol{x}(A-B) &= (-1, \ \ 1, \ \ 0, \ \ 0, -1)\,,\\
\boldsymbol{Q}({}_{-(A-B)}\mathrm{T}^{(+)}, \boldsymbol{R}) &= \{R^1, R^5, R^{2t-1}\}\,,\\
\boldsymbol{x}(B-A) &= (\ \ 1, \ \ 0, \ \ 0, -1, \ \ 1)\,,\\
\boldsymbol{Q}({}_{-(B-A)}\mathrm{T}^{(+)}, \boldsymbol{R}) &= \{R^0, R^4, R^8\}\,;
\end{aligned}$$

see Figure 7.4.

(v) The family

$$\begin{aligned}
\big\{(A, B) \in \mathbf{2}^{[t]} \times \mathbf{2}^{[t]} :\ \ j' := |A| = 3\,,\ \ j'' := |B| = 3\,,\\
|A\cap B| > 0\,,\ \ |A\cup B| = t\,,\\
\ell^{\cap} := \mathfrak{q}(A\cap B) = 1\,,\ \ \ell' := \mathfrak{q}(A-B) = 3\,,\ \ \ell'' := \mathfrak{q}(B-A) = 5\big\}
\end{aligned} \tag{7.15}$$

consists of

$$\begin{aligned}
&\mathfrak{T}\left(\alpha, \theta; \frac{\ell^{\cap}+1}{2}, \frac{\ell'+1}{2}, \frac{\ell''-1}{2}\right) \cdot \mathrm{c}\left(\frac{\ell^{\cap}+1}{2}; (j'+j'')-t\right)\\
&\quad \times \mathrm{c}\left(\frac{\ell'+1}{2}; t-j''\right) \cdot \mathrm{c}\left(\frac{\ell''-1}{2}; t-j'\right)\\
&:= \mathfrak{T}\left(\alpha, \theta; \frac{1+1}{2}, \frac{3+1}{2}, \frac{5-1}{2}\right) \cdot \mathrm{c}\left(\frac{1+1}{2}; (3+3)-5\right)\\
&\quad \times \mathrm{c}\left(\frac{3+1}{2}; 5-3\right) \cdot \mathrm{c}\left(\frac{3-1}{2}; 5-3\right)\\
&= \underbrace{\mathfrak{T}(\alpha, \theta; 1, 2, 2)}_{\text{see (6.16)}} \cdot \mathrm{c}(1; 1) \cdot \mathrm{c}(2; 2) \cdot \mathrm{c}(1; 2)\\
&:= (2+1-2)\binom{2-1}{\frac{2+1-2-1}{2}}\binom{\frac{2+2+1-3}{2}}{2-1} \cdot \binom{1-1}{1-1} \cdot \binom{2-1}{2-1} \cdot \binom{2-1}{1-1} = 1
\end{aligned}$$

pair (A, B) such that

$$\{1, t\} \cap A = \{1, t\}\,, \quad \text{and} \quad \{1, t\} \cap B = \{t\}\,.$$

This pair is

$$\begin{aligned}
(\ \ A &:= \{1, \ \ , 3, \ \ , t\}\,,\\
B &:= \{\ \ , 2, \ \ , 4, t\}\ \ )\,,
\end{aligned}$$

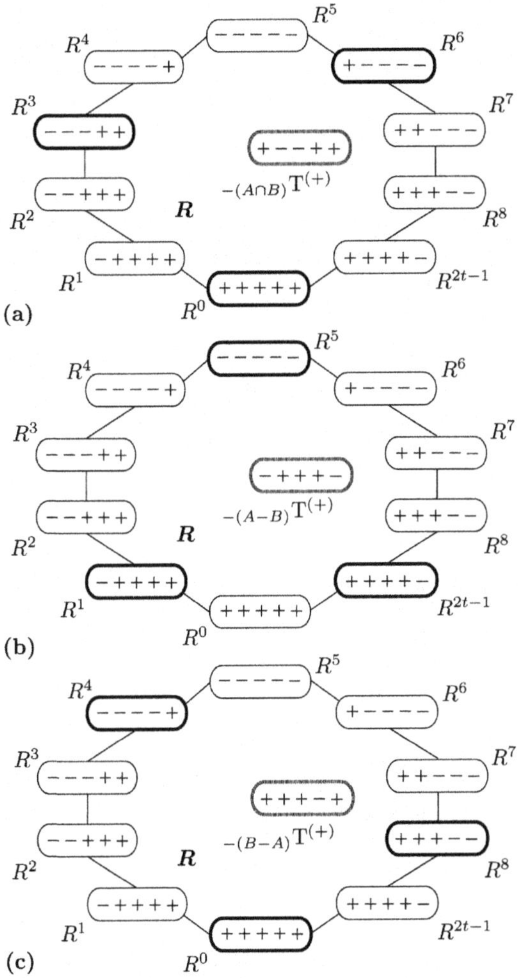

Figure 7.4 Three instances of the same distinguished symmetric cycle $\boldsymbol{R} := (R^0, R^1, \ldots, R^{2t-1}, R^0)$ in the hypercube graph $\boldsymbol{H}((t := 5), 2)$, defined by (7.1)(7.2).
Pick the subsets $A := \{1, 2, 3, t\}$ and $B := \{1, 3, 4\}$ of the set E_t. Note that $\{1, t\} \cap A = \{1, t\}$, and $|\{1, t\} \cap B| = 0$.
(a) The vertex ${}_{-(A\cap B)}\mathrm{T}^{(+)}$ of the graph $\boldsymbol{H}(t, 2)$, with the corresponding decomposition set $\boldsymbol{Q}({}_{-(A\cap B)}\mathrm{T}^{(+)}, \boldsymbol{R}) = \{R^0, R^3, R^6\}$.
(b) The vertex ${}_{-(A-B)}\mathrm{T}^{(+)}$, with its decomposition set $\boldsymbol{Q}({}_{-(A-B)}\mathrm{T}^{(+)}, \boldsymbol{R}) = \{R^1, R^5, R^{2t-1}\}$.
(c) The vertex ${}_{-(B-A)}\mathrm{T}^{(+)}$, with its decomposition set $\boldsymbol{Q}({}_{-(B-A)}\mathrm{T}^{(+)}, \boldsymbol{R}) = \{R^0, R^4, R^8\}$.

for which we have

$$A \cap B = \{\ ,\ \ ,\ \ ,\ \ , t\}\,,$$
$$A - B = \{1,\ \ , 3,\ \ ,\ \ \}\,,$$
$$B - A = \{\ , 2,\ \ , 4,\ \ \}\,,$$

and

$$\begin{aligned}
\boldsymbol{x}(A \cap B) &= (0,\quad 0,\quad 0,\quad 0, -1)\,,\\
\boldsymbol{Q}({}_{-(A\cap B)}\mathrm{T}^{(+)}, \boldsymbol{R}) &= \{R^{2t-1}\}\,,\\
\boldsymbol{x}(A - B) &= (0,\quad 1, -1,\quad 1,\quad 0)\,,\\
\boldsymbol{Q}({}_{-(A-B)}\mathrm{T}^{(+)}, \boldsymbol{R}) &= \{R^1, R^3, R^7\}\,,\\
\boldsymbol{x}(B - A) &= (1, -1,\quad 1, -1,\quad 1)\,,\\
\boldsymbol{Q}({}_{-(B-A)}\mathrm{T}^{(+)}, \boldsymbol{R}) &= \{R^0, R^2, R^4, R^6, R^8\}\,;
\end{aligned}$$

see Figure 7.5.

- Let us consider the family

$$\begin{aligned}
\big\{(A, B) \in \mathbf{2}^{[t]} \times \mathbf{2}^{[t]} :\ |A| = j'\,,\ |B| = j''\,,\\
|A \cap B| > 0\,,\ \max\{j', j''\} < |A \cup B| = t\,,\\
\mathfrak{q}(A \cap B) = \ell^{\cap}\,,\ \mathfrak{q}(A \triangle B) = \ell^{\triangle}\big\}
\end{aligned} \qquad (7.16)$$

of ordered two-member *intersecting Sperner families* that *cover* the set E_t, with the properties (7.9)(7.10).

Theorem 7.7. *Ordered pairs of sets (A, B) in the family (7.16) can be counted with the help of products that include subproducts of the form*

$$\mathrm{c}\left(\varrho(A \cap B); (j' + j'') - t\right) \cdot \mathrm{c}\left(\varrho(A \triangle B); 2t - (j' + j'')\right)\ :$$

(i) *In the family (7.16) there are*

$$\mathrm{c}\left(\frac{\ell^{\cap} + 1}{2}; (j' + j'') - t\right) \cdot \mathrm{c}\left(\frac{\ell^{\triangle} + 1}{2}; 2t - (j' + j'')\right) \cdot \tbinom{2t-(j'+j'')-1}{t-j''}$$

pairs (A, B) of sets A and B such that

$$\{1, t\} \cap A = \{1\}\,,\quad \textit{and}\quad \{1, t\} \cap B = \{1, t\}\,,$$

and if the subfamily of these pairs is nonempty, then $\ell^{\cap} = \ell^{\triangle}$.

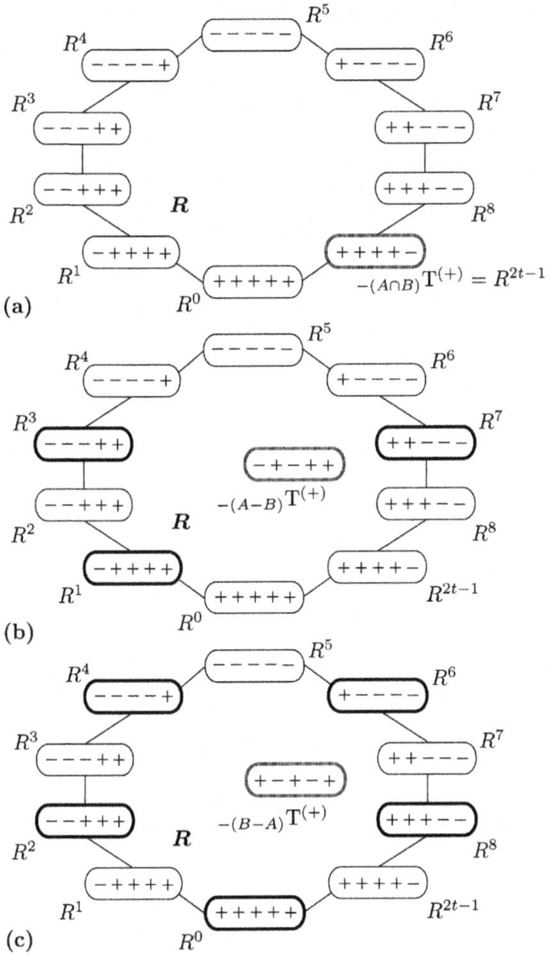

Figure 7.5 Three instances of the same distinguished symmetric cycle $\boldsymbol{R} := (R^0, R^1, \ldots, R^{2t-1}, R^0)$ in the hypercube graph $\boldsymbol{H}((t := 5), 2)$, defined by (7.1)(7.2).
Pick the subsets $A := \{1, 3, t\}$ and $B := \{2, 4, t\}$ of the set E_t. Note that $\{1, t\} \cap A = \{1, t\}$, and $\{1, t\} \cap B = \{t\}$.
(a) The vertex ${}_{-(A\cap B)}\mathrm{T}^{(+)}$ of the graph $\boldsymbol{H}(t, 2)$, with the corresponding decomposition set $\boldsymbol{Q}({}_{-(A\cap B)}\mathrm{T}^{(+)}, \boldsymbol{R}) = \{R^{2t-1}\}$.
(b) The vertex ${}_{-(A-B)}\mathrm{T}^{(+)}$, with its decomposition set $\boldsymbol{Q}({}_{-(A-B)}\mathrm{T}^{(+)}, \boldsymbol{R}) = \{R^1, R^3, R^7\}$.
(c) The vertex ${}_{-(B-A)}\mathrm{T}^{(+)}$, with its decomposition set $\boldsymbol{Q}({}_{-(B-A)}\mathrm{T}^{(+)}, \boldsymbol{R}) = \{R^0, R^2, R^4, R^6, R^8\}$.

In the family (7.16) *there are*

(ii)
$$\mathrm{c}\left(\frac{\ell^{\cap}-1}{2};(j'+j'')-t\right)\cdot\mathrm{c}\left(\frac{\ell^{\vartriangle}+1}{2};2t-(j'+j'')\right)\cdot\binom{2t-(j'+j'')-2}{t-j''-1}$$
pairs (A, B) *such that*
$$\{1, t\}\cap A=\{1\}\,,\quad \textit{and}\quad \{1, t\}\cap B=\{t\}\,,$$
and if the subfamily of these pairs is nonempty, then $\ell^{\cap}=\ell^{\vartriangle}$.

(iii)
$$\mathrm{c}\left(\frac{\ell^{\cap}+1}{2};(j'+j'')-t\right)\cdot\mathrm{c}\left(\frac{\ell^{\vartriangle}-1}{2};2t-(j'+j'')\right)\cdot\binom{2t-(j'+j'')}{t-j''}$$
pairs (A, B) *such that*
$$\{1, t\}\cap A=\{1, t\}\cap B=\{1, t\}\,,$$
and if the subfamily of these pairs is nonempty, then $\ell^{\cap}=\ell^{\vartriangle}$.

(iv)
$$\mathrm{c}\left(\frac{\ell^{\cap}-1}{2};(j'+j'')-t\right)\cdot\mathrm{c}\left(\frac{\ell^{\vartriangle}+1}{2};2t-(j'+j'')\right)\cdot\binom{2t-(j'+j'')-2}{t-j'}$$
pairs (A, B) *such that*
$$\{1, t\}\cap A=\{1, t\}\,,\quad \textit{and}\quad |\{1, t\}\cap B|=0\,,$$
and if the subfamily of these pairs is nonempty, then $\ell^{\cap}=\ell^{\vartriangle}$.

(v)
$$\mathrm{c}\left(\frac{\ell^{\cap}+1}{2};(j'+j'')-t\right)\cdot\mathrm{c}\left(\frac{\ell^{\vartriangle}+1}{2};2t-(j'+j'')\right)\cdot\binom{2t-(j'+j'')-1}{t-j'}$$
pairs (A, B) *such that*
$$\{1, t\}\cap A=\{1, t\}\,,\quad \textit{and}\quad \{1, t\}\cap B=\{t\}\,,$$
and if the subfamily of these pairs is nonempty, then $\ell^{\cap}=\ell^{\vartriangle}$.

Proof. (i) Since $\{1, t\}\cap A=\{1\}$, and $\{1, t\}\cap B=\{1, t\}$, we have
$$\varrho(A\cap B)=\frac{\ell^{\cap}+1}{2}\,,\quad \varrho(A\vartriangle B)=\frac{\ell^{\vartriangle}+1}{2}\,,$$
by Proposition 5.9(i)(iv).

(ii) Since $\{1, t\}\cap A=\{1\}$, and $\{1, t\}\cap B=\{t\}$, we have
$$\varrho(A\cap B)=\frac{\ell^{\cap}-1}{2}\,,\quad \varrho(A\vartriangle B)=\frac{\ell^{\vartriangle}+1}{2}\,,$$
by Proposition 5.9(iii)(ii).

(iii) Since $\{1, t\} \cap A = \{1, t\} \cap B = \{1, t\}$, we have

$$\varrho(A \cap B) = \frac{\ell^{\cap} + 1}{2}\,, \quad \varrho(A \triangle B) = \frac{\ell^{\triangle} - 1}{2}\,,$$

by Proposition 5.9(ii)(iii).

(iv) Since $\{1, t\} \cap A = \{1, t\}$, and $|\{1, t\} \cap B| = 0$, we have

$$\varrho(A \cap B) = \frac{\ell^{\cap} - 1}{2}\,, \quad \varrho(A \triangle B) = \frac{\ell^{\triangle} + 1}{2}\,,$$

by Proposition 5.9(iii)(ii).

(v) Since $\{1, t\} \cap A = \{1, t\}$, and $\{1, t\} \cap B = \{t\}$, we have

$$\varrho(A \cap B) = \frac{\ell^{\cap} + 1}{2}\,, \quad \varrho(A \triangle B) = \frac{\ell^{\triangle} + 1}{2}\,,$$

by Proposition 5.9(iv)(i). □

Example 7.8. Suppose $t := 5$.

(i) In the family

$$\big\{(A, B) \in \mathbf{2}^{[t]} \times \mathbf{2}^{[t]}:\ j' := |A| = 3\,,\ j'' := |B| = 4\,,\ |A \cap B| > 0\,,\ |A \cup B| = t\,,\ \ell^{\cap} := \mathfrak{q}(A \cap B) = 3\,,\ \ell^{\triangle} := \mathfrak{q}(A \triangle B) = 3\big\}$$

of ordered two-member *intersecting Sperner families* that *cover* the set E_t, there are

$$\begin{aligned} &\mathrm{c}\left(\frac{\ell^{\cap}+1}{2}; (j'+j'')-t\right) \cdot \mathrm{c}\left(\frac{\ell^{\triangle}+1}{2}; 2t-(j'+j'')\right) \cdot \tbinom{2t-(j'+j'')-1}{t-j''} \\ &:= \mathrm{c}\left(\frac{3+1}{2}; (3+4)-5\right) \cdot \mathrm{c}\left(\frac{3+1}{2}; 2\cdot 5-(3+4)\right) \cdot \tbinom{2\cdot 5-(3+4)-1}{5-4} \\ &= \tbinom{2-1}{2-1}\tbinom{3-1}{2-1}\tbinom{2}{1} = 4 \end{aligned}$$

pairs (A, B) of sets A and B such that

$$\{1, t\} \cap A = \{1\}\,, \quad \text{and} \quad \{1, t\} \cap B = \{1, t\}\,.$$

One such pair is

$$\begin{aligned} (\ A &:= \{1,\quad, 3,\ 4,\quad\}\,, \\ B &:= \{1,\ 2,\quad, 4,\ t\}\ )\,, \end{aligned}$$

for which we have

$$A \cap B = \{1, \quad , \quad , 4, \quad \} ,$$
$$A \triangle B = \{ \quad , 2, 3, \quad , t\} ,$$

and

$$\begin{aligned} \boldsymbol{x}(A \cap B) &= (0, \quad 1, \quad 0, -1, \quad 1) , \\ \boldsymbol{Q}({}_{-(A\cap B)}\mathrm{T}^{(+)}, \boldsymbol{R}) &= \{R^1, R^4, R^8\} , \\ \boldsymbol{x}(A \triangle B) &= (0, -1, \quad 0, \quad 1, -1) , \\ \boldsymbol{Q}({}_{-(A\triangle B)}\mathrm{T}^{(+)}, \boldsymbol{R}) &= \{R^3, R^6, R^{2t-1}\} ; \end{aligned}$$

see Figure 7.6.

(ii) In the family

$$\big\{(A, B) \in \mathbf{2}^{[t]} \times \mathbf{2}^{[t]} : \ j' := |A| = 3 , \ j'' := |B| = 3 , \ |A \cap B| > 0 , \ |A \cup B| = t , \ \ell^{\cap} := \mathfrak{q}(A \cap B) = 3 , \ \ell^{\triangle} := \mathfrak{q}(A \triangle B) = 3\big\}$$

there are

$$\begin{aligned} &\mathsf{c}\left(\frac{\ell^{\cap} - 1}{2}; (j' + j'') - t\right) \cdot \mathsf{c}\left(\frac{\ell^{\triangle} + 1}{2}; 2t - (j' + j'')\right) \cdot \binom{2t-(j'+j'')-2}{t-j''-1} \\ &:= \mathsf{c}\left(\frac{3-1}{2}; (3+3) - 5\right) \cdot \mathsf{c}\left(\frac{3+1}{2}; 2 \cdot 5 - (3+3)\right) \cdot \binom{2 \cdot 5-(3+3)-2}{5-3-1} \\ &= \binom{1-1}{1-1}\binom{4-1}{2-1}\binom{2}{1} = 6 \end{aligned}$$

pairs (A, B) such that

$$\{1, t\} \cap A = \{1\} , \quad \text{and} \quad \{1, t\} \cap B = \{t\} .$$

One such pair is

$$\big(\ A := \{1, 2, \quad , 4, \quad \} , \quad B := \{ \quad , 2, 3, \quad , t\} \ \big) ,$$

for which we have

$$A \cap B = \{ \quad , 2, \quad , \quad , \quad \} ,$$
$$A \triangle B = \{1, \quad , 3, 4, t\} ,$$

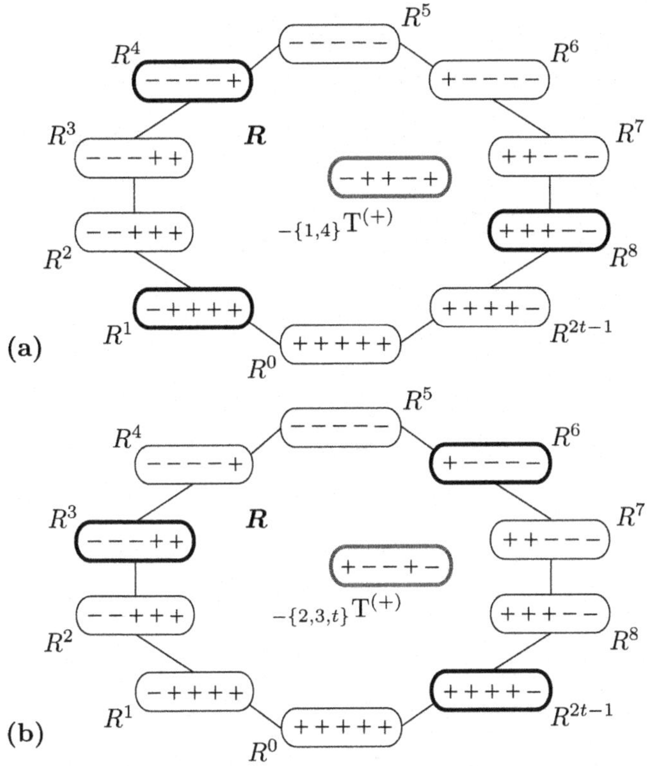

Figure 7.6 Two instances of the same distinguished symmetric cycle $\boldsymbol{R} := (R^0, R^1, \ldots, R^{2t-1}, R^0)$ in the hypercube graph $\boldsymbol{H}((t := 5), 2)$, defined by (7.1)(7.2).
Pick the subsets $A := \{1, 3, 4\}$ and $B := \{1, 2, 4, t\}$ of the set E_t.
Note that $\{1, t\} \cap A = \{1\}$, and $\{1, t\} \cap B = \{1, t\}$.
(a) The vertex $_{-(A\cap B)}\mathrm{T}^{(+)}$ of the graph $\boldsymbol{H}(t, 2)$, with the corresponding decomposition set $\boldsymbol{Q}(_{-(A\cap B)}\mathrm{T}^{(+)}, \boldsymbol{R}) = \{R^1, R^4, R^8\}$.
(b) The vertex $_{-(A\triangle B)}\mathrm{T}^{(+)}$, with its decomposition set $\boldsymbol{Q}(_{-(A\triangle B)}\mathrm{T}^{(+)}, \boldsymbol{R}) = \{R^3, R^6, R^{2t-1}\}$.

and

$$\boldsymbol{x}(A \cap B) = (\ \ 1, -1, \ \ 1, \ \ 0, \ \ 0)\,,$$
$$\boldsymbol{Q}(_{-(A\cap B)}\mathrm{T}^{(+)}, \boldsymbol{R}) = \{R^0, R^2, R^6\}\,,$$
$$\boldsymbol{x}(A \triangle B) = (-1, \ \ 1, -1, \ \ 0, \ \ 0)\,,$$
$$\boldsymbol{Q}(_{-(A\triangle B)}\mathrm{T}^{(+)}, \boldsymbol{R}) = \{R^1, R^5, R^7\}\,;$$

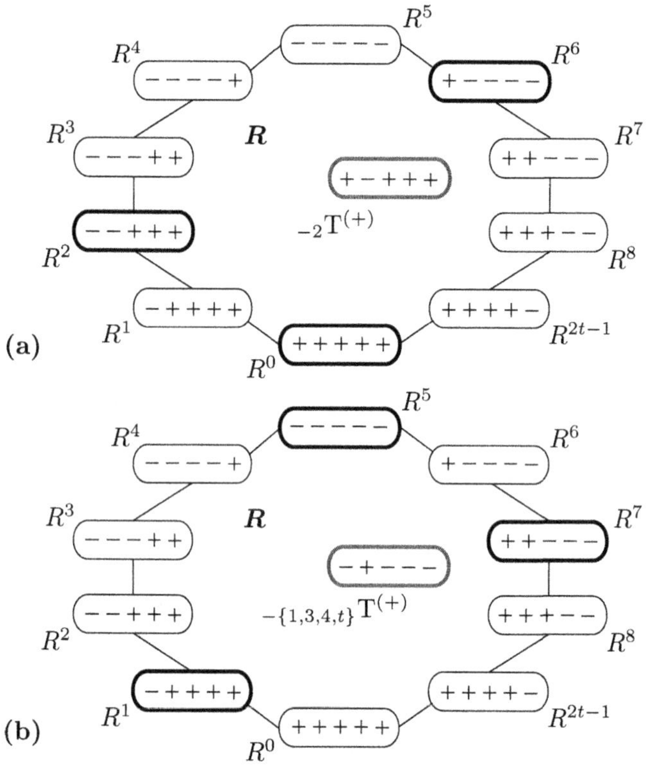

Figure 7.7 Two instances of the same distinguished symmetric cycle $\boldsymbol{R} := (R^0, R^1, \ldots, R^{2t-1}, R^0)$ in the hypercube graph $\boldsymbol{H}((t := 5), 2)$, defined by (7.1)(7.2).
Pick the subsets $A := \{1, 2, 4\}$ and $B := \{2, 3, t\}$ of the set E_t.
Note that $\{1, t\} \cap A = \{1\}$, and $\{1, t\} \cap B = \{t\}$.
(a) The vertex $_{-(A\cap B)}\mathrm{T}^{(+)}$ of the graph $\boldsymbol{H}(t, 2)$, with the corresponding decomposition set $\boldsymbol{Q}({}_{-(A\cap B)}\mathrm{T}^{(+)}, \boldsymbol{R}) = \{R^0, R^2, R^6\}$.
(b) The vertex $_{-(A\triangle B)}\mathrm{T}^{(+)}$, with its decomposition set $\boldsymbol{Q}({}_{-(A\triangle B)}\mathrm{T}^{(+)}, \boldsymbol{R}) = \{R^1, R^5, R^7\}$.

see Figure 7.7.

(iii) In the family

$$\bigl\{(A, B) \in \mathbf{2}^{[t]} \times \mathbf{2}^{[t]} :\ j' := |A| = 4\,,\ \ j'' := |B| = 4\,,\\ |A \cap B| > 0\,,\ \ |A \cup B| = t\,,\\ \ell^{\cap} := \mathfrak{q}(A \cap B) = 3\,,\ \ \ell^{\triangle} := \mathfrak{q}(A \triangle B) = 3\bigr\}$$

there are

$$\begin{aligned}
&\mathrm{c}\left(\frac{\ell^{\cap}+1}{2}; (j'+j'')-t\right)\cdot \mathrm{c}\left(\frac{\ell^{\vartriangle}-1}{2}; 2t-(j'+j'')\right)\cdot \tbinom{2t-(j'+j'')}{t-j''}\\
&\quad := \mathrm{c}\left(\frac{3+1}{2}; (4+4)-5\right)\cdot \mathrm{c}\left(\frac{3-1}{2}; 2\cdot 5-(4+4)\right)\cdot \tbinom{2\cdot 5-(4+4)}{5-4}\\
&\quad = \tbinom{3-1}{2-1}\tbinom{2-1}{1-1}\tbinom{2}{1} = 4
\end{aligned}$$

pairs (A, B) such that

$$\{1, t\}\cap A = \{1, t\}\cap B = \{1, t\}\ .$$

One such pair is

$$\begin{aligned}
(\ A &:= \{1, \quad , 3,\ 4,\ t\}\ ,\\
B &:= \{1,\ 2, \quad , 4,\ t\}\ \ )\ ,
\end{aligned}$$

for which we have

$$\begin{aligned}
A\cap B &= \{1, \quad , \quad , 4,\ t\}\ ,\\
A\vartriangle B &= \{\ \ , 2,\ 3, \quad , \quad \}\ ,
\end{aligned}$$

and

$$\begin{aligned}
\boldsymbol{x}(A\cap B) &= (\ \ 1, -1,\quad 0, -1,\quad 0)\ ,\\
\boldsymbol{Q}({}_{-(A\cap B)}\mathrm{T}^{(+)}, \boldsymbol{R}) &= \{R^1, R^5, R^8\}\ ,\\
\boldsymbol{x}(A\vartriangle B) &= (\ \ 1, -1,\quad 0,\quad 1,\quad 0)\ ,\\
\boldsymbol{Q}({}_{-(A\vartriangle B)}\mathrm{T}^{(+)}, \boldsymbol{R}) &= \{R^0, R^3, R^6\}\ ;
\end{aligned}$$

see Figure 7.8.

(iv) In the family

$$\begin{aligned}
\big\{(A, B)\in \mathbf{2}^{[t]}\times \mathbf{2}^{[t]}:\ \ & j' := |A| = 4\ ,\ \ j'' := |B| = 3\ ,\\
& |A\cap B| > 0\ ,\ \ |A\cup B| = t\ ,\\
& \ell^{\cap} := \mathfrak{q}(A\cap B) = 3\ ,\ \ \ell^{\vartriangle} := \mathfrak{q}(A\vartriangle B) = 3\big\}
\end{aligned}$$

there are

$$\begin{aligned}
&\mathrm{c}\left(\frac{\ell^{\cap}-1}{2}; (j'+j'')-t\right)\cdot \mathrm{c}\left(\frac{\ell^{\vartriangle}+1}{2}; 2t-(j'+j'')\right)\cdot \tbinom{2t-(j'+j'')-2}{t-j'}\\
&\quad := \mathrm{c}\left(\frac{3-1}{2}; (4+3)-5\right)\cdot \mathrm{c}\left(\frac{3+1}{2}; 2\cdot 5-(4+3)\right)\cdot \tbinom{2\cdot 5-(4+3)-2}{5-4}\\
&\quad = \tbinom{2-1}{1-1}\tbinom{3-1}{2-1}\tbinom{1}{1} = 2
\end{aligned}$$

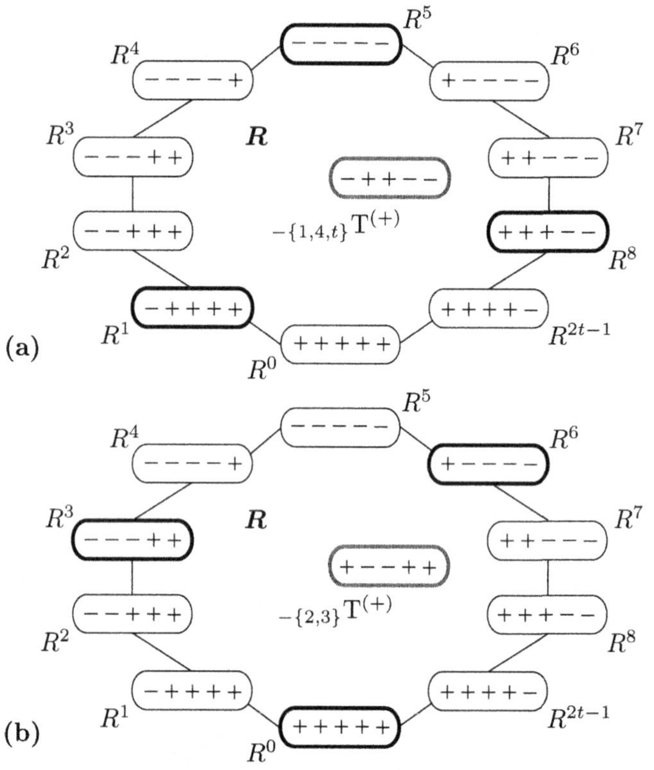

Figure 7.8 Two instances of the same distinguished symmetric cycle $\boldsymbol{R} := (R^0, R^1, \ldots, R^{2t-1}, R^0)$ in the hypercube graph $\boldsymbol{H}((t := 5), 2)$, defined by (7.1)(7.2).
Pick the subsets $A := \{1, 3, 4, t\}$ and $B := \{1, 2, 4, t\}$ of the set E_t.
Note that $\{1, t\} \cap A = \{1, t\} \cap B = \{1, t\}$.
(a) The vertex ${}_{-(A\cap B)}\mathrm{T}^{(+)}$ of the graph $\boldsymbol{H}(t, 2)$, with the corresponding decomposition set $\boldsymbol{Q}({}_{-(A\cap B)}\mathrm{T}^{(+)}, \boldsymbol{R}) = \{R^1, R^5, R^8\}$.
(b) The vertex ${}_{-(A\triangle B)}\mathrm{T}^{(+)}$, with its decomposition set $\boldsymbol{Q}({}_{-(A\triangle B)}\mathrm{T}^{(+)}, \boldsymbol{R}) = \{R^0, R^3, R^6\}$.

pairs (A, B) such that

$$\{1, t\} \cap A = \{1, t\}\,, \quad \text{and} \quad |\{1, t\} \cap B| = 0\,.$$

One such pair is

$$\begin{aligned}(\; A &:= \{1,\ 2,\ 3,\ \ ,\ t\}\,, \\ B &:= \{\ \ ,\ 2,\ 3,\ 4,\ \ \}\ )\,,\end{aligned}$$

for which we have

$$A \cap B = \{\ \ ,\ 2,\ 3,\ \ ,\ \ \}\,,$$

$$A \triangle B = \{1,\ \ ,\ \ ,\ 4,\ t\}\,,$$

and

$$\boldsymbol{x}(A \cap B) = (\ \ 1,\ -1,\ \ 0,\ \ 1,\ \ 0)\,,$$

$$\boldsymbol{Q}({}_{-(A\cap B)}\mathrm{T}^{(+)},\ \boldsymbol{R}) = \{R^0,\ R^3,\ R^6\}\,,$$

$$\boldsymbol{x}(A \triangle B) = (-1,\ \ 1,\ \ 0,\ -1,\ \ 0)\,,$$

$$\boldsymbol{Q}({}_{-(A\triangle B)}\mathrm{T}^{(+)},\ \boldsymbol{R}) = \{R^1,\ R^5,\ R^8\}\,;$$

see Figure 7.9.

(v) In the family

$$\big\{(A,\ B) \in \mathbf{2}^{[t]} \times \mathbf{2}^{[t]}:\ \ j' := |A| = 3\,,\ \ j'' := |B| = 4\,,$$
$$|A \cap B| > 0\,,\ \ |A \cup B| = t\,,$$
$$\ell^{\cap} := \mathrm{q}(A \cap B) = 3\,,\ \ \ell^{\triangle} := \mathrm{q}(A \triangle B) = 3\big\}$$

there are

$$\mathrm{c}\left(\frac{\ell^{\cap}+1}{2};(j'+j'')-t\right)\cdot\mathrm{c}\left(\frac{\ell^{\triangle}+1}{2};2t-(j'+j'')\right)\cdot\binom{2t-(j'+j'')-1}{t-j'}$$

$$:=\mathrm{c}\left(\frac{3+1}{2};(3+4)-5\right)\cdot\mathrm{c}\left(\frac{3+1}{2};2\cdot 5-(3+4)\right)\cdot\binom{2\cdot 5-(3+4)-1}{5-3}$$

$$=\binom{2-1}{2-1}\binom{3-1}{2-1}\binom{2}{2}=2$$

pairs $(A,\ B)$ such that

$$\{1,\ t\} \cap A = \{1,\ t\}\,,\quad \text{and}\quad \{1,\ t\} \cap B = \{t\}\,.$$

One such pair is

$$(\ \ A := \{1,\ 2,\ \ ,\ \ ,\ t\}\,,$$
$$B := \{\ \ ,\ 2,\ 3,\ 4,\ t\}\ \ )\,,$$

for which we have

$$A \cap B = \{\ \ ,\ 2,\ \ ,\ \ ,\ t\}\,,$$

$$A \triangle B = \{1,\ \ ,\ 3,\ 4,\ \ \}\,,$$

and

$$\boldsymbol{x}(A \cap B) = (0,\ -1,\ \ 1,\ \ 0,\ -1)\,,$$

$$\boldsymbol{Q}({}_{-(A\cap B)}\mathrm{T}^{(+)},\ \boldsymbol{R}) = \{R^2,\ R^6,\ R^{2t-1}\}\,,$$

$$\boldsymbol{x}(A \triangle B) = (0,\ \ 1,\ -1,\ \ 0,\ \ 1)\,,$$

$$\boldsymbol{Q}({}_{-(A\triangle B)}\mathrm{T}^{(+)},\ \boldsymbol{R}) = \{R^1,\ R^4,\ R^7\}\,;$$

see Figure 7.10.

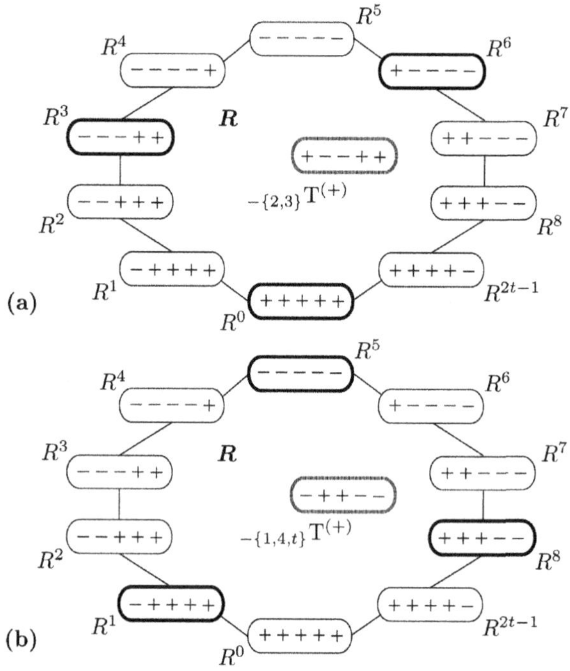

Figure 7.9 Two instances of the same distinguished symmetric cycle $\boldsymbol{R} := (R^0, R^1, \ldots, R^{2t-1}, R^0)$ in the hypercube graph $\boldsymbol{H}((t := 5), 2)$, defined by (7.1)(7.2).
Pick the subsets $A := \{1, 2, 3, t\}$ and $B := \{2, 3, 4\}$ of the set E_t.
Note that $\{1, t\} \cap A = \{1, t\}$, and $|\{1, t\} \cap B| = 0$.
(a) The vertex $_{-(A\cap B)}\mathrm{T}^{(+)}$ of the graph $\boldsymbol{H}(t, 2)$, with the corresponding decomposition set $\boldsymbol{Q}({}_{-(A\cap B)}\mathrm{T}^{(+)}, \boldsymbol{R}) = \{R^0, R^3, R^6\}$.
(b) The vertex $_{-(A\triangle B)}\mathrm{T}^{(+)}$, with its decomposition set $\boldsymbol{Q}({}_{-(A\triangle B)}\mathrm{T}^{(+)}, \boldsymbol{R}) = \{R^1, R^5, R^8\}$.

7.1.5 Case: $|A \cap B| > 0$, and $|A \cup B| < t$

- Here we consider the family

$$\begin{gathered}\big\{(A, B) \in \mathbf{2}^{[t]} \times \mathbf{2}^{[t]} :\ |A| = j' < t\,,\ |B| = j'' < t\,,\\ |A \cap B| = j > 0\,,\ \max\{j', j''\} < |A \cup B| < t\,,\\ \mathfrak{q}(A - B) = \ell'\,,\ \mathfrak{q}(B - A) = \ell''\,,\ \mathfrak{q}(A \cap B) = \ell^{\cap}\,,\ \mathfrak{q}(A \cup B) = \ell\big\}\end{gathered} \tag{7.17}$$

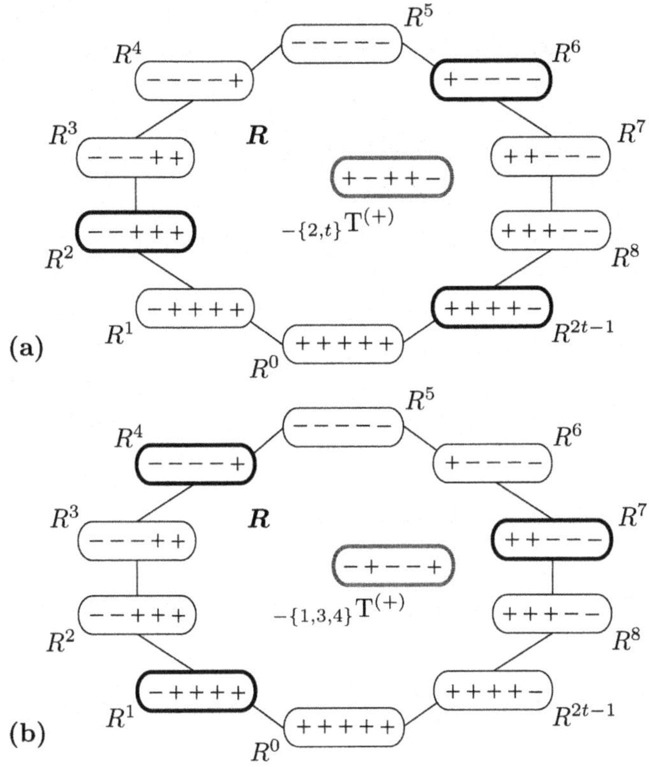

Figure 7.10 Two instances of the same distinguished symmetric cycle $\boldsymbol{R} := (R^0, R^1, \ldots, R^{2t-1}, R^0)$ in the hypercube graph $\boldsymbol{H}((t := 5), 2)$, defined by (7.1)(7.2).
Pick the subsets $A := \{1, 2, t\}$ and $B := \{2, 3, 4, t\}$ of the set E_t.
Note that $\{1, t\} \cap A = \{1, t\}$, and $\{1, t\} \cap B = \{t\}$.
(a) The vertex ${}_{-(A\cap B)}\mathrm{T}^{(+)}$ of the graph $\boldsymbol{H}(t, 2)$, with the corresponding decomposition set $\boldsymbol{Q}({}_{-(A\cap B)}\mathrm{T}^{(+)}, \boldsymbol{R}) = \{R^2, R^6, R^{2t-1}\}$.
(b) The vertex ${}_{-(A\triangle B)}\mathrm{T}^{(+)}$, with its decomposition set $\boldsymbol{Q}({}_{-(A\triangle B)}\mathrm{T}^{(+)}, \boldsymbol{R}) = \{R^1, R^4, R^7\}$.

of ordered two-member *intersecting Sperner families* that *do not* cover the set E_t. For pairs in this family, we have

$$d({}_{-A}\mathrm{T}^{(+)}, {}_{-B}\mathrm{T}^{(+)}) = j' + j'' - 2j\,, \tag{7.18}$$

and

$$\langle {}_{-A}\mathrm{T}^{(+)}, {}_{-B}\mathrm{T}^{(+)} \rangle = 0 \iff 2(j' + j'' - 2j) = t\,. \tag{7.19}$$

Theorem 7.9. *Ordered pairs of sets* $(A,\ B)$ *in the family* (7.17) *can be counted with the help of products of the form*

$$\begin{aligned}\mathfrak{F}\left(\mathfrak{s}',\ \mathfrak{s}'';\ \varrho(E_t-(A\cup B)),\ \varrho(A-B),\ \varrho(B-A),\ \varrho(A\cap B)\right)\\ \times\,\mathsf{c}\left(\varrho(E_t-(A\cup B));t-(j'+j''-j)\right)\\ \times\,\mathsf{c}\left(\varrho(A-B);j'-j\right)\cdot\mathsf{c}\left(\varrho(B-A);j''-j\right)\cdot\mathsf{c}\left(\varrho(A\cap B);j\right)\ :\end{aligned} \tag{7.20}$$

(i) *In the family* (7.17) *there are*

$$\begin{aligned}&\mathfrak{F}\left(\gamma,\theta;\frac{\ell+1}{2},\frac{\ell'-1}{2},\frac{\ell''-1}{2},\frac{\ell^{\cap}+1}{2}\right)\\ &\times\mathsf{c}\left(\frac{\ell+1}{2};t-(j'+j''-j)\right)\\ &\times\mathsf{c}\left(\frac{\ell'-1}{2};j'-j\right)\cdot\mathsf{c}\left(\frac{\ell''-1}{2};j''-j\right)\cdot\mathsf{c}\left(\frac{\ell^{\cap}+1}{2};j\right)\end{aligned}$$

pairs $(A,\ B)$ *of sets* A *and* B *such that*

$$\{1,\ t\}\cap A=\{1,\ t\}\cap B=\{1\}\ .$$

In the family (7.17) *there are*

(ii)

$$\begin{aligned}&\mathfrak{F}\left(\gamma,\beta;\frac{\ell-1}{2},\frac{\ell'-1}{2},\frac{\ell''+1}{2},\frac{\ell^{\cap}+1}{2}\right)\\ &\times\mathsf{c}\left(\frac{\ell-1}{2};t-(j'+j''-j)\right)\\ &\times\mathsf{c}\left(\frac{\ell'-1}{2};j'-j\right)\cdot\mathsf{c}\left(\frac{\ell''+1}{2};j''-j\right)\cdot\mathsf{c}\left(\frac{\ell^{\cap}+1}{2};j\right)\end{aligned}$$

pairs $(A,\ B)$ *such that*

$$\{1,\ t\}\cap A=\{1\}\ ,\quad \textit{and}\quad \{1,\ t\}\cap B=\{1,\ t\}\ ;$$

(iii)

$$\begin{aligned}&\mathfrak{F}\left(\alpha,\theta;\frac{\ell+1}{2},\frac{\ell'+1}{2},\frac{\ell''-1}{2},\frac{\ell^{\cap}-1}{2}\right)\\ &\times\mathsf{c}\left(\frac{\ell+1}{2};t-(j'+j''-j)\right)\\ &\times\mathsf{c}\left(\frac{\ell'+1}{2};j'-j\right)\cdot\mathsf{c}\left(\frac{\ell''-1}{2};j''-j\right)\cdot\mathsf{c}\left(\frac{\ell^{\cap}-1}{2};j\right)\end{aligned}$$

pairs $(A,\ B)$ *such that*

$$\{1,\ t\}\cap A=\{1\}\ ,\quad \textit{and}\quad |\{1,\ t\}\cap B|=0\ ;$$

(iv)

$$\mathfrak{F}\left(\alpha, \beta; \frac{\ell-1}{2}, \frac{\ell'+1}{2}, \frac{\ell''+1}{2}, \frac{\ell^{\cap}-1}{2}\right)$$
$$\times \mathrm{c}\left(\frac{\ell-1}{2}; t-(j'+j''-j)\right)$$
$$\times \mathrm{c}\left(\frac{\ell'+1}{2}; j'-j\right) \cdot \mathrm{c}\left(\frac{\ell''+1}{2}; j''-j\right) \cdot \mathrm{c}\left(\frac{\ell^{\cap}-1}{2}; j\right)$$

pairs (A, B) *such that*

$$\{1, t\} \cap A = \{1\}, \quad \textit{and} \quad \{1, t\} \cap B = \{t\};$$

(v)

$$\mathfrak{F}\left(\gamma, \gamma; \frac{\ell-1}{2}, \frac{\ell'-1}{2}, \frac{\ell''-1}{2}, \frac{\ell^{\cap}+1}{2}\right)$$
$$\times \mathrm{c}\left(\frac{\ell-1}{2}; t-(j'+j''-j)\right)$$
$$\times \mathrm{c}\left(\frac{\ell'-1}{2}; j'-j\right) \cdot \mathrm{c}\left(\frac{\ell''-1}{2}; j''-j\right) \cdot \mathrm{c}\left(\frac{\ell^{\cap}+1}{2}; j\right)$$

pairs (A, B) *such that*

$$\{1, t\} \cap A = \{1, t\} \cap B = \{1, t\};$$

(vi)

$$\mathfrak{F}\left(\alpha, \alpha; \frac{\ell-1}{2}, \frac{\ell'+1}{2}, \frac{\ell''-1}{2}, \frac{\ell^{\cap}-1}{2}\right)$$
$$\times \mathrm{c}\left(\frac{\ell-1}{2}; t-(j'+j''-j)\right)$$
$$\times \mathrm{c}\left(\frac{\ell'+1}{2}; j'-j\right) \cdot \mathrm{c}\left(\frac{\ell''-1}{2}; j''-j\right) \cdot \mathrm{c}\left(\frac{\ell^{\cap}-1}{2}; j\right)$$

pairs (A, B) *such that*

$$\{1, t\} \cap A = \{1, t\}, \quad \textit{and} \quad |\{1, t\} \cap B| = 0;$$

(vii)

$$\mathfrak{F}\left(\alpha, \gamma; \frac{\ell-1}{2}, \frac{\ell'+1}{2}, \frac{\ell''-1}{2}, \frac{\ell^{\cap}+1}{2}\right)$$
$$\times \mathrm{c}\left(\frac{\ell-1}{2}; t-(j'+j''-j)\right)$$
$$\times \mathrm{c}\left(\frac{\ell'+1}{2}; j'-j\right) \cdot \mathrm{c}\left(\frac{\ell''-1}{2}; j''-j\right) \cdot \mathrm{c}\left(\frac{\ell^{\cap}+1}{2}; j\right)$$

pairs (A, B) *such that*

$$\{1, t\} \cap A = \{1, t\}, \quad \textit{and} \quad \{1, t\} \cap B = \{t\};$$

(viii)

$$\mathfrak{F}\left(\theta, \theta; \frac{\ell+1}{2}, \frac{\ell'-1}{2}, \frac{\ell''-1}{2}, \frac{\ell^{\cap}-1}{2}\right)$$
$$\times \mathrm{c}\left(\frac{\ell+1}{2}; t-(j'+j''-j)\right)$$
$$\times \mathrm{c}\left(\frac{\ell'-1}{2}; j'-j\right) \cdot \mathrm{c}\left(\frac{\ell''-1}{2}; j''-j\right) \cdot \mathrm{c}\left(\frac{\ell^{\cap}-1}{2}; j\right)$$

pairs (A, B) *such that*

$$|\{1, t\} \cap A| = |\{1, t\} \cap B| = 0\,;$$

(ix)

$$\mathfrak{F}\left(\theta, \beta; \frac{\ell+1}{2}, \frac{\ell'-1}{2}, \frac{\ell''+1}{2}, \frac{\ell^{\cap}-1}{2}\right)$$
$$\times \mathrm{c}\left(\frac{\ell+1}{2}; t-(j'+j''-j)\right)$$
$$\times \mathrm{c}\left(\frac{\ell'-1}{2}; j'-j\right) \cdot \mathrm{c}\left(\frac{\ell''+1}{2}; j''-j\right) \cdot \mathrm{c}\left(\frac{\ell^{\cap}-1}{2}; j\right)$$

pairs (A, B) *such that*

$$|\{1, t\} \cap A| = 0\,, \quad \textit{and} \quad \{1, t\} \cap B = \{t\}\,;$$

(x)

$$\mathfrak{F}\left(\theta, \gamma; \frac{\ell+1}{2}, \frac{\ell'-1}{2}, \frac{\ell''-1}{2}, \frac{\ell^{\cap}+1}{2}\right)$$
$$\times \mathrm{c}\left(\frac{\ell+1}{2}; t-(j'+j''-j)\right)$$
$$\times \mathrm{c}\left(\frac{\ell'-1}{2}; j'-j\right) \cdot \mathrm{c}\left(\frac{\ell''-1}{2}; j''-j\right) \cdot \mathrm{c}\left(\frac{\ell^{\cap}+1}{2}; j\right)$$

pairs (A, B) *such that*

$$\{1, t\} \cap A = \{1, t\} \cap B = \{t\}\,.$$

Proof. For each pair (A, B) in the family (7.17), pick an *ascending system* of *distinct representatives*

$$(e_1 < e_2 < \ldots < e_{\varrho(E_t-(A\cup B))+\varrho(A-B)+\varrho(B-A)+\varrho(A\cap B)}) \subseteq E_t$$

of the intervals composing the sets $(E_t - (A \cup B))$, $(A - B)$, $(B - A)$ and $(A \cap B)$. By making the substitutions

$$e_i \mapsto \begin{cases} \theta\,, & \text{if } e_i \in E_t - (A \cup B), \\ \alpha\,, & \text{if } e_i \in A - B, \\ \beta\,, & \text{if } e_i \in B - A, \\ \gamma\,, & \text{if } e_i \in A \cap B, \end{cases}$$

$$1 \le i \le \varrho(E_t - (A \cup B)) + \varrho(A - B) + \varrho(B - A) + \varrho(A \cap B)\,,$$

we obtain some Smirnov words over the ordered four-letter alphabet $(\theta, \alpha, \beta, \gamma)$. Given any such Smirnov word, we find the number of pairs (A, B) in the corresponding subfamily of the family (7.17) by means of a product of the form (7.20).

(i) Since $\{1, t\} \cap A = \{1, t\} \cap B = \{1\}$, we have

$$\varrho(A - B) = \frac{\ell' - 1}{2}\,, \quad \varrho(B - A) = \frac{\ell'' - 1}{2}\,,$$

by Proposition 5.9(iii), and

$$\varrho(A \cap B) = \frac{\ell^{\cap} + 1}{2}\,, \quad \varrho(A \cup B) = \frac{\ell + 1}{2}\,,$$

by Proposition 5.9(i). Note that $\varrho(E_t - (A \cup B)) = \varrho(A \cup B)$, that is,

$$\varrho\,(E_t - (A \cup B)) = \frac{\ell + 1}{2}\,.$$

(ii) Since $\{1, t\} \cap A = \{1\}$, and $\{1, t\} \cap B = \{1, t\}$, we have

$$\varrho(A - B) = \frac{\ell' - 1}{2}\,, \quad \varrho(B - A) = \frac{\ell'' + 1}{2}\,,$$

by Proposition 5.9(iii)(iv), and

$$\varrho(A \cap B) = \frac{\ell^{\cap} + 1}{2}\,, \quad \varrho(A \cup B) = \frac{\ell + 1}{2}\,,$$

by Proposition 5.9(i)(ii). Note that $\varrho(E_t - (A \cup B)) = \varrho(A \cup B) - 1$, that is,

$$\varrho\,(E_t - (A \cup B)) = \frac{\ell - 1}{2}\,.$$

(iii) Since $\{1, t\} \cap A = \{1\}$, and $|\{1, t\} \cap B| = 0$, we have

$$\varrho(A - B) = \frac{\ell' + 1}{2}\,, \quad \varrho(B - A) = \frac{\ell'' - 1}{2}\,,$$

by Proposition 5.9(i)(iii), and

$$\varrho(A \cap B) = \frac{\ell^{\cap} - 1}{2}, \quad \varrho(A \cup B) = \frac{\ell + 1}{2},$$

by Proposition 5.9(iii)(i). Note that

$$\varrho\left(E_t - (A \cup B)\right) = \frac{\ell + 1}{2}.$$

(iv) Since $\{1, t\} \cap A = \{1\}$, and $\{1, t\} \cap B = \{t\}$, we have

$$\varrho(A - B) = \frac{\ell' + 1}{2}, \quad \varrho(B - A) = \frac{\ell'' + 1}{2},$$

by Proposition 5.9(i)(iv), and

$$\varrho(A \cap B) = \frac{\ell^{\cap} - 1}{2}, \quad \varrho(A \cup B) = \frac{\ell + 1}{2},$$

by Proposition 5.9(iii)(ii). Note that

$$\varrho\left(E_t - (A \cup B)\right) = \frac{\ell - 1}{2}.$$

(v) Since $\{1, t\} \cap A = \{1, t\} \cap B = \{1, t\}$, we have

$$\varrho(A - B) = \frac{\ell' - 1}{2}, \quad \varrho(B - A) = \frac{\ell'' - 1}{2},$$

by Proposition 5.9(iii), and

$$\varrho(A \cap B) = \frac{\ell^{\cap} + 1}{2}, \quad \varrho(A \cup B) = \frac{\ell + 1}{2},$$

by Proposition 5.9(ii). Note that

$$\varrho\left(E_t - (A \cup B)\right) = \frac{\ell - 1}{2}.$$

(vi) Since $\{1, t\} \cap A = \{1, t\}$, and $|\{1, t\} \cap B| = 0$, we have

$$\varrho(A - B) = \frac{\ell' + 1}{2}, \quad \varrho(B - A) = \frac{\ell'' - 1}{2}.$$

by Proposition 5.9(ii)(iii), and

$$\varrho(A \cap B) = \frac{\ell^{\cap} - 1}{2}, \quad \varrho(A \cup B) = \frac{\ell + 1}{2},$$

by Proposition 5.9(iii)(ii). Note that

$$\varrho\left(E_t - (A \cup B)\right) = \frac{\ell - 1}{2}.$$

(vii) Since $\{1, t\} \cap A = \{1, t\}$, and $\{1, t\} \cap B = \{t\}$, we have

$$\varrho(A - B) = \frac{\ell' + 1}{2}, \quad \varrho(B - A) = \frac{\ell'' - 1}{2},$$

by Proposition 5.9(i)(iii), and

$$\varrho(A \cap B) = \frac{\ell^{\cap} + 1}{2}, \quad \varrho(A \cup B) = \frac{\ell + 1}{2},$$

by Proposition 5.9(iv)(ii). Note that

$$\varrho\,(E_t - (A \cup B)) = \frac{\ell - 1}{2}.$$

(viii) Since $|\{1, t\} \cap A| = |\{1, t\} \cap B| = 0$, we have

$$\varrho(A - B) = \frac{\ell' - 1}{2}, \quad \varrho(B - A) = \frac{\ell'' - 1}{2}.$$

and

$$\varrho(A \cap B) = \frac{\ell^{\cap} - 1}{2}, \quad \varrho(A \cup B) = \frac{\ell - 1}{2},$$

by Proposition 5.9(iii). Note that $\varrho(E_t - (A \cup B)) = \varrho(A \cup B) + 1$, that is,

$$\varrho\,(E_t - (A \cup B)) = \frac{\ell + 1}{2}.$$

(ix) Since $|\{1, t\} \cap A| = 0$, and $\{1, t\} \cap B = \{t\}$, we have

$$\varrho(A - B) = \frac{\ell' - 1}{2}, \quad \varrho(B - A) = \frac{\ell'' + 1}{2},$$

and

$$\varrho(A \cap B) = \frac{\ell^{\cap} - 1}{2}, \quad \varrho(A \cup B) = \frac{\ell + 1}{2},$$

by Proposition 5.9(iii)(iv). Note that $\varrho(E_t - (A \cup B)) = \varrho(A \cup B)$, that is,

$$\varrho\,(E_t - (A \cup B)) = \frac{\ell + 1}{2}.$$

(x) Since $\{1, t\} \cap A = \{1, t\} \cap B = \{t\}$, we have

$$\varrho(A - B) = \frac{\ell' - 1}{2}, \quad \varrho(B - A) = \frac{\ell'' - 1}{2},$$

by Proposition 5.9(iii), and

$$\varrho(A \cap B) = \frac{\ell^{\cap} + 1}{2}, \quad \varrho(A \cup B) = \frac{\ell + 1}{2},$$

by Proposition 5.9(iv). Note that

$$\varrho\,(E_t - (A \cup B)) = \frac{\ell + 1}{2}. \qquad \square$$

Example 7.10. (We will only present explicit calculations for the case (viii) on page 171). Suppose $t := 5$.
(i) In the family

$$\begin{aligned}\big\{(A,\, B) \in \mathbf{2}^{[t]} \times \mathbf{2}^{[t]}:\ \ & j' := |A| = 3\,,\ \ j'' := |B| = 3\,,\\ j := |A \cap B| = 2\,,\ \ & \ell' := \mathfrak{q}(A - B) = 3\,,\ \ \ell'' := \mathfrak{q}(B - A) = 3\,,\\ & \ell^{\cap} := \mathfrak{q}(A \cap B) = 3\,,\ \ \ell := \mathfrak{q}(A \cup B) = 1\big\}\end{aligned}$$

there are

$$\begin{aligned}&\mathfrak{F}\left(\gamma, \theta; \frac{\ell+1}{2}, \frac{\ell'-1}{2}, \frac{\ell''-1}{2}, \frac{\ell^{\cap}+1}{2}\right)\\ &\quad \times \mathrm{c}\left(\frac{\ell+1}{2}; t - (j' + j'' - j)\right)\\ &\quad \times \mathrm{c}\left(\frac{\ell'-1}{2}; j' - j\right) \cdot \mathrm{c}\left(\frac{\ell''-1}{2}; j'' - j\right) \cdot \mathrm{c}\left(\frac{\ell^{\cap}+1}{2}; j\right)\\ &:= \mathfrak{F}\left(\gamma, \theta; \frac{1+1}{2}, \frac{3-1}{2}, \frac{3-1}{2}, \frac{3+1}{2}\right)\\ &\quad \times \mathrm{c}\left(\frac{1+1}{2}; 5 - (3 + 3 - 2)\right)\\ &\quad \times \mathrm{c}\left(\frac{3-1}{2}; 3 - 2\right) \cdot \mathrm{c}\left(\frac{3-1}{2}; 3 - 2\right) \cdot \mathrm{c}\left(\frac{3+1}{2}; 2\right)\\ &= \underbrace{\mathfrak{F}(\gamma, \theta; 1, 1, 1, 2)}_{\substack{\text{certainly, } 2 \cdot 2! = 4;\\ \text{but cf. (A.17)}}} \cdot \mathrm{c}(1; 1) \cdot \mathrm{c}(1; 1) \cdot \mathrm{c}(1; 1) \cdot \mathrm{c}(2; 2)\\ &= 4 \cdot 1 \cdot 1 \cdot 1 \cdot 1 = 4\end{aligned}$$

pairs (A, B) of sets A and B such that

$$\{1, t\} \cap A = \{1, t\} \cap B = \{1\}\,.$$

One such pair is

$$\begin{aligned}(\ A &:= \{1,\ \ ,\ 3,\ 4,\ \ \}\,,\\ B &:= \{1,\ 2,\ 3,\ \ ,\ \ \}\ )\,,\end{aligned}$$

for which we have

$$\begin{aligned}A - B &= \{\ \ ,\ \ ,\ 4,\ \ ,\ \ \}\,,\\ B - A &= \{\ \ ,\ 2,\ \ ,\ \ ,\ \ \}\,,\\ A \cap B &= \{1,\ \ ,\ 3,\ \ ,\ \ \}\,,\\ A \cup B &= \{1,\ 2,\ 3,\ 4,\ \ \}\,,\end{aligned}$$

and

$$\begin{aligned}
\boldsymbol{x}(A-B) &= (1,\quad 0,\quad 0, -1,\quad 1)\,,\\
\boldsymbol{Q}({}_{-(A-B)}\mathrm{T}^{(+)}, \boldsymbol{R}) &= \{R^0, R^4, R^8\}\,,\\
\boldsymbol{x}(B-A) &= (1, -1,\quad 1,\quad 0,\quad 0)\,,\\
\boldsymbol{Q}({}_{-(B-A)}\mathrm{T}^{(+)}, \boldsymbol{R}) &= \{R^0, R^2, R^6\}\,,\\
\boldsymbol{x}(A\cap B) &= (0,\quad 1, -1,\quad 1,\quad 0)\,,\\
\boldsymbol{Q}({}_{-(A\cap B)}\mathrm{T}^{(+)}, \boldsymbol{R}) &= \{R^1, R^3, R^7\}\,,\\
\boldsymbol{x}(A\cup B) &= (0,\quad 0,\quad 0,\quad 0,\quad 1)\,,\\
\boldsymbol{Q}({}_{-(A\cup B)}\mathrm{T}^{(+)}, \boldsymbol{R}) &= \{R^4\}\,;
\end{aligned}$$

see Figure 7.11.

(ii) In the family

$$\begin{aligned}
\big\{(A, B) \in \mathbf{2}^{[t]} \times \mathbf{2}^{[t]}:\ & j' := |A| = 3\,,\ j'' := |B| = 3\,,\\
j := |A\cap B| = 2\,,\ \ell' &:= \mathfrak{q}(A-B) = 3\,,\ \ell'' := \mathfrak{q}(B-A) = 1\,,\\
\ell^{\cap} &:= \mathfrak{q}(A\cap B) = 3\,,\ \ell := \mathfrak{q}(A\cup B) = 3\big\}
\end{aligned}$$

there are

$$\begin{aligned}
&\mathfrak{F}\left(\gamma, \beta; \frac{\ell-1}{2}, \frac{\ell'-1}{2}, \frac{\ell''+1}{2}, \frac{\ell^{\cap}+1}{2}\right)\\
&\qquad \times \mathrm{c}\left(\frac{\ell-1}{2}; t-(j'+j''-j)\right)\\
&\qquad \times \mathrm{c}\left(\frac{\ell'-1}{2}; j'-j\right)\cdot \mathrm{c}\left(\frac{\ell''+1}{2}; j''-j\right)\cdot \mathrm{c}\left(\frac{\ell^{\cap}+1}{2}; j\right)\\
&\quad := \mathfrak{F}\left(\gamma, \beta; \frac{3-1}{2}, \frac{3-1}{2}, \frac{1+1}{2}, \frac{3+1}{2}\right)\\
&\qquad \times \mathrm{c}\left(\frac{3-1}{2}; 5-(3+3-2)\right)\\
&\qquad \times \mathrm{c}\left(\frac{3-1}{2}; 3-2\right)\cdot \mathrm{c}\left(\frac{1+1}{2}; 3-2\right)\cdot \mathrm{c}\left(\frac{3+1}{2}; 2\right)\\
&\quad = \underbrace{\mathfrak{F}(\gamma, \beta; 1, 1, 1, 2)}_{\substack{\text{certainly, } 2\cdot 2! = 4;\\ \text{but cf. (A.17)}}}\cdot \mathrm{c}(1;1)\cdot \mathrm{c}(1;1)\cdot \mathrm{c}(1;1)\cdot \mathrm{c}(2;2)\\
&\quad = 4\cdot 1\cdot 1\cdot 1\cdot 1 = 4
\end{aligned}$$

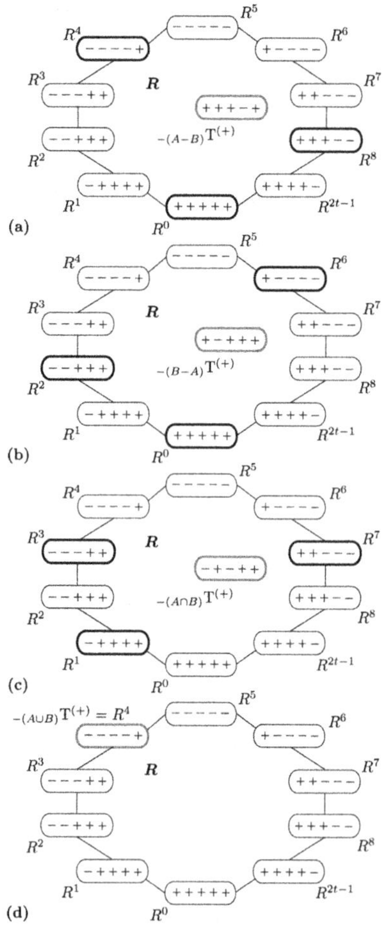

Figure 7.11 Four instances of the same distinguished symmetric cycle $\boldsymbol{R} := (R^0, R^1, \ldots, R^{2t-1}, R^0)$ in the hypercube graph $\boldsymbol{H}((t := 5), 2)$, defined by (7.1)(7.2).
Pick the subsets $A := \{1, 3, 4\}$ and $B := \{1, 2, 3\}$ of the set E_t.
Note that $\{1, t\} \cap A = \{1, t\} \cap B = \{1\}$.
(a) The vertex ${}_{-(A-B)}\mathrm{T}^{(+)}$ of the graph $\boldsymbol{H}(t, 2)$, with the corresponding decomposition set $\boldsymbol{Q}({}_{-(A-B)}\mathrm{T}^{(+)}, \boldsymbol{R}) = \{R^0, R^4, R^8\}$.
(b) The vertex ${}_{-(B-A)}\mathrm{T}^{(+)}$, with its decomposition set $\boldsymbol{Q}({}_{-(B-A)}\mathrm{T}^{(+)}, \boldsymbol{R}) = \{R^0, R^2, R^6\}$.
(c) The vertex ${}_{-(A\cap B)}\mathrm{T}^{(+)}$, with its decomposition set $\boldsymbol{Q}({}_{-(A\cap B)}\mathrm{T}^{(+)}, \boldsymbol{R}) = \{R^1, R^3, R^7\}$.
(d) The vertex ${}_{-(A\cup B)}\mathrm{T}^{(+)} = R^4 \in \mathrm{V}(\boldsymbol{R})$.

pairs (A, B) such that

$$\{1, t\} \cap A = \{1\}\,, \quad \text{and} \quad \{1, t\} \cap B = \{1, t\}\,.$$

One such pair is

$$\begin{aligned} (\; A &:= \{1, \quad, 3,\ 4, \quad\}\,, \\ B &:= \{1, \quad, 3, \quad, t\} \;)\,, \end{aligned}$$

for which we have

$$\begin{aligned} A - B &= \{\ ,\quad ,\quad ,4, \quad\}\,, \\ B - A &= \{\ ,\quad ,\quad ,\quad , t\}\,, \\ A \cap B &= \{1, \quad, 3, \quad, \quad\}\,, \\ A \cup B &= \{1, \quad, 3,\ 4,\ t\}\,, \end{aligned}$$

and

$$\begin{aligned} \boldsymbol{x}(A - B) &= (\ \ 1, \quad 0, \quad 0, -1, \quad 1)\,, \\ \boldsymbol{Q}({}_{-(A-B)}\mathrm{T}^{(+)}, \boldsymbol{R}) &= \{R^0, R^4, R^8\}\,, \\ \boldsymbol{x}(B - A) &= (\ \ 0, \quad 0, \quad 0, \quad 0, -1)\,, \\ \boldsymbol{Q}({}_{-(B-A)}\mathrm{T}^{(+)}, \boldsymbol{R}) &= \{R^{2t-1}\}\,, \\ \boldsymbol{x}(A \cap B) &= (\ \ 0, \quad 1, -1, \quad 1, \quad 0)\,, \\ \boldsymbol{Q}({}_{-(A\cap B)}\mathrm{T}^{(+)}, \boldsymbol{R}) &= \{R^1, R^3, R^7\}\,, \\ \boldsymbol{x}(A \cup B) &= (-1, \quad 1, -1, \quad 0, \quad 0)\,, \\ \boldsymbol{Q}({}_{-(A\cup B)}\mathrm{T}^{(+)}, \boldsymbol{R}) &= \{R^1, R^5, R^7\}\,; \end{aligned}$$

see Figure 7.12.

(iii) In the family

$$\begin{aligned} &\big\{(A, B) \in \mathbf{2}^{[t]} \times \mathbf{2}^{[t]} \colon \; j' := |A| = 2\,, \;\; j'' := |B| = 2;\,, \\ &j := |A \cap B| = 1\,, \;\; \ell' := \mathfrak{q}(A - B) = 1\,, \;\; \ell'' := \mathfrak{q}(B - A) = 3\,, \\ &\qquad\qquad\qquad \ell^{\cap} := \mathfrak{q}(A \cap B) = 3\,, \;\; \ell := \mathfrak{q}(A \cup B) = 1\big\} \end{aligned}$$

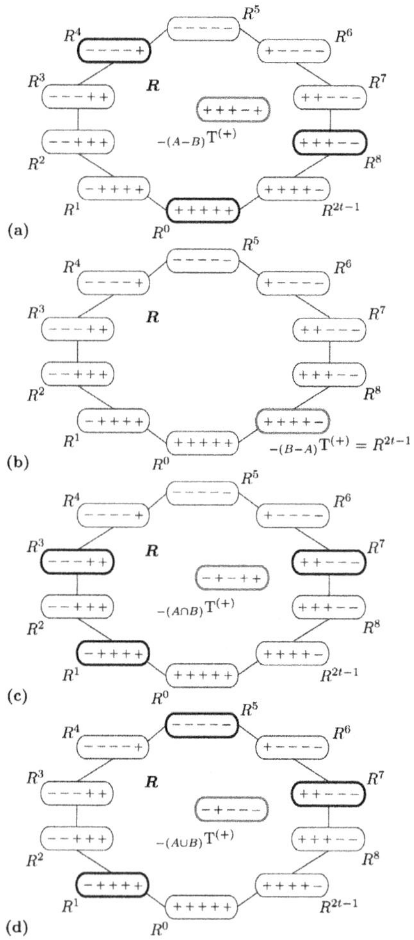

Figure 7.12 Four instances of the same distinguished symmetric cycle $\boldsymbol{R} := (R^0, R^1, \ldots, R^{2t-1}, R^0)$ in the hypercube graph $\boldsymbol{H}((t := 5), 2)$, defined by (7.1)(7.2).
Pick the subsets $A := \{1, 3, 4\}$ and $B := \{1, 3, t\}$ of the set E_t.
Note that $\{1, t\} \cap A = \{1\}$, and $\{1, t\} \cap B = \{1, t\}$.
(a) The vertex ${}_{-(A-B)}\mathrm{T}^{(+)}$ of the graph $\boldsymbol{H}(t, 2)$, with the corresponding decomposition set $\boldsymbol{Q}({}_{-(A-B)}\mathrm{T}^{(+)}, \boldsymbol{R}) = \{R^0, R^4, R^8\}$.
(b) The vertex ${}_{-(B-A)}\mathrm{T}^{(+)} = R^{2t-1} \in \mathrm{V}(\boldsymbol{R})$.
(c) The vertex ${}_{-(A\cap B)}\mathrm{T}^{(+)}$, with its decomposition set $\boldsymbol{Q}({}_{-(A\cap B)}\mathrm{T}^{(+)}, \boldsymbol{R}) = \{R^1, R^3, R^7\}$.
(d) The vertex ${}_{-(A\cup B)}\mathrm{T}^{(+)}$, with its decomposition set $\boldsymbol{Q}({}_{-(A\cup B)}\mathrm{T}^{(+)}, \boldsymbol{R}) = \{R^1, R^5, R^7\}$.

there are

$$\mathfrak{F}\left(\alpha, \theta; \frac{\ell+1}{2}, \frac{\ell'+1}{2}, \frac{\ell''-1}{2}, \frac{\ell^{\cap}-1}{2}\right)$$

$$\times \mathrm{c}\left(\frac{\ell+1}{2}; t-(j'+j''-j)\right)$$

$$\times \mathrm{c}\left(\frac{\ell'+1}{2}; j'-j\right) \cdot \mathrm{c}\left(\frac{\ell''-1}{2}; j''-j\right) \cdot \mathrm{c}\left(\frac{\ell^{\cap}-1}{2}; j\right)$$

$$:= \mathfrak{F}\left(\alpha, \theta; \frac{1+1}{2}, \frac{1+1}{2}, \frac{3-1}{2}, \frac{3-1}{2}\right)$$

$$\times \mathrm{c}\left(\frac{1+1}{2}; 5-(2+2-1)\right) \cdot \mathrm{c}\left(\frac{1+1}{2}; 2-1\right)$$

$$\times \mathrm{c}\left(\frac{3-1}{2}; 2-1\right) \cdot \mathrm{c}\left(\frac{3-1}{2}; 1\right)$$

$$= \underbrace{\mathfrak{F}(\alpha, \theta; 1, 1, 1, 1)}_{\substack{\text{certainly, } 2! = 2; \\ \text{but cf. (A.17)}}} \cdot \mathrm{c}(1; 2) \cdot \mathrm{c}(1; 1) \cdot \mathrm{c}(1; 1) \cdot \mathrm{c}(1; 1)$$

$$= 2 \cdot 1 \cdot 1 \cdot 1 \cdot 1 = 2$$

pairs (A, B) such that

$$\{1, t\} \cap A = \{1\}, \quad \text{and} \quad |\{1, t\} \cap B| = 0 .$$

One such pair is

$$\begin{aligned}(\ A &:= \{1, \ \ , 3, \ \ , \ \}, \\ B &:= \{ \ , 2, 3, \ \ , \ \} \ ),\end{aligned}$$

for which we have

$$\begin{aligned} A - B &= \{1, \ , \ , \ , \ \}, \\ B - A &= \{ \ , 2, \ , \ , \ \}, \\ A \cap B &= \{ \ , \ , 3, \ , \ \}, \\ A \cup B &= \{1, 2, 3, \ , \ \}, \end{aligned}$$

and

$$
\begin{aligned}
\boldsymbol{x}(A-B) &= (0,\quad 1,\quad 0,\quad 0,\quad 0)\,,\\
\boldsymbol{Q}({}_{-(A-B)}\mathrm{T}^{(+)},\ \boldsymbol{R}) &= \{R^1\}\,,\\
\boldsymbol{x}(B-A) &= (1,\ -1,\quad 1,\quad 0,\quad 0)\,,\\
\boldsymbol{Q}({}_{-(B-A)}\mathrm{T}^{(+)},\ \boldsymbol{R}) &= \{R^0,\ R^2,\ R^6\}\,,\\
\boldsymbol{x}(A\cap B) &= (1,\quad 0,\ -1,\quad 1,\quad 0)\,,\\
\boldsymbol{Q}({}_{-(A\cap B)}\mathrm{T}^{(+)},\ \boldsymbol{R}) &= \{R^0,\ R^3,\ R^7\}\,,\\
\boldsymbol{x}(A\cup B) &= (0,\quad 0,\quad 0,\quad 1,\quad 0)\,,\\
\boldsymbol{Q}({}_{-(A\cup B)}\mathrm{T}^{(+)},\ \boldsymbol{R}) &= \{R^3\}\,;
\end{aligned}
$$

see Figure 7.13.

(iv) In the family

$$
\begin{aligned}
\big\{(A,\,B)\in \mathbf{2}^{[t]}\times\mathbf{2}^{[t]}:\ \ & j' := |A| = 2\,,\ \ j'' := |B| = 3\,,\\
j := |A\cap B| = 1\,,\ \ \ell' := \mathfrak{q}(A-B) = 1\,,\ \ & \ell'' := \mathfrak{q}(B-A) = 3\,,\\
& \ell^{\cap} := \mathfrak{q}(A\cap B) = 3\,,\ \ \ell := \mathfrak{q}(A\cup B) = 3\big\}
\end{aligned}
$$

there are

$$
\begin{aligned}
&\mathfrak{F}\left(\alpha,\,\beta;\,\frac{\ell-1}{2},\,\frac{\ell'+1}{2},\,\frac{\ell''+1}{2},\,\frac{\ell^{\cap}-1}{2}\right)\cdot \mathrm{c}\left(\frac{\ell-1}{2};t-(j'+j''-j)\right)\\
&\qquad\times \mathrm{c}\left(\frac{\ell'+1}{2};j'-j\right)\cdot \mathrm{c}\left(\frac{\ell''+1}{2};j''-j\right)\cdot \mathrm{c}\left(\frac{\ell^{\cap}-1}{2};j\right)\\
&:= \mathfrak{F}\left(\alpha,\,\beta;\,\frac{3-1}{2},\,\frac{1+1}{2},\,\frac{3+1}{2},\,\frac{3-1}{2}\right)\cdot \mathrm{c}\left(\frac{3-1}{2};5-(2+3-1)\right)\\
&\qquad\times \mathrm{c}\left(\frac{1+1}{2};2-1\right)\cdot \mathrm{c}\left(\frac{3+1}{2};3-1\right)\cdot \mathrm{c}\left(\frac{3-1}{2};j\right)\\
&= \underbrace{\mathfrak{F}(\alpha,\beta;1,\,1,\,2,\,1)}_{\substack{\text{certainly, } 2\cdot 2! = 4;\\ \text{but cf. (A.17)}}}\cdot \mathrm{c}(1;1)\cdot \mathrm{c}(1;1)\cdot \mathrm{c}(2;2)\cdot \mathrm{c}(1;1)\\
&= 4\cdot 1\cdot 1\cdot 1\cdot 1 = 4
\end{aligned}
$$

pairs $(A,\,B)$ such that

$$
\{1,\,t\}\cap A = \{1\}\,,\quad \text{and}\quad \{1,\,t\}\cap B = \{t\}\,.
$$

One such pair is

$$
\begin{aligned}
(\ \ A &:= \{1,\ 2,\ \ ,\ \ ,\ \ \}\,,\\
B &:= \{\ \ ,\ 2,\ 3,\ \ ,\ t\}\ \ )\,,
\end{aligned}
$$

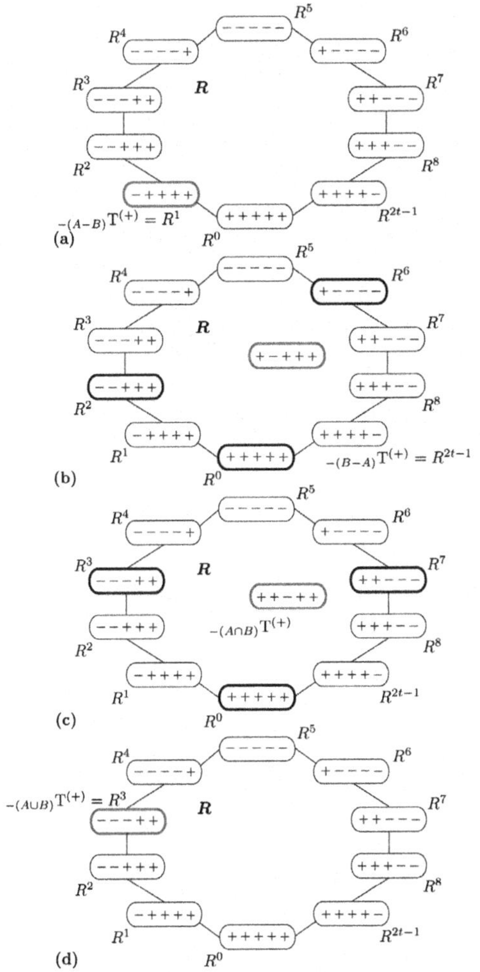

Figure 7.13 Four instances of the same distinguished symmetric cycle $\boldsymbol{R} := (R^0, R^1, \ldots, R^{2t-1}, R^0)$ in the hypercube graph $\boldsymbol{H}((t := 5), 2)$, defined by (7.1)(7.2).
Pick the subsets $A := \{1, 3\}$ and $B := \{2, 3\}$ of the set E_t.
Note that $\{1, t\} \cap A = \{1\}$, and $|\{1, t\} \cap B| = 0$.
(a) The vertex ${}_{-(A-B)}\mathrm{T}^{(+)} = R^1 \in \mathrm{V}(\boldsymbol{R})$.
(b) The vertex ${}_{-(B-A)}\mathrm{T}^{(+)}$ of the graph $\boldsymbol{H}(t, 2)$, with the corresponding decomposition set $\boldsymbol{Q}({}_{-(B-A)}\mathrm{T}^{(+)}, \boldsymbol{R}) = \{R^0, R^2, R^6\}$.
(c) The vertex ${}_{-(A\cap B)}\mathrm{T}^{(+)}$, with its decomposition set $\boldsymbol{Q}({}_{-(A\cap B)}\mathrm{T}^{(+)}, \boldsymbol{R}) = \{R^0, R^3, R^7\}$.
(d) The vertex ${}_{-(A\cup B)}\mathrm{T}^{(+)} = R^3 \in \mathrm{V}(\boldsymbol{R})$.

for which we have

$$\begin{aligned}
A - B &= \{1,\ \ ,\ \ ,\ \ ,\ \ \}\,,\\
B - A &= \{\ \ ,\ \ ,3,\ \ ,t\}\,,\\
A \cap B &= \{\ \ ,2,\ \ ,\ \ ,\ \ \}\,,\\
A \cup B &= \{1,2,3,\ \ ,t\}\,,
\end{aligned}$$

and

$$\begin{aligned}
\boldsymbol{x}(A-B) &= (\ \ 0,\ \ \ 1,\ \ \ 0,\ \ \ 0,\ \ \ 0)\,,\\
\boldsymbol{Q}({}_{-(A-B)}\mathrm{T}^{(+)}, \boldsymbol{R}) &= \{R^1\}\,,\\
\boldsymbol{x}(B-A) &= (\ \ 0,\ \ \ 0,-1,\ \ \ 1,-1)\,,\\
\boldsymbol{Q}({}_{-(B-A)}\mathrm{T}^{(+)}, \boldsymbol{R}) &= \{R^3, R^7, R^{2t-1}\}\,,\\
\boldsymbol{x}(A\cap B) &= (\ \ 1,-1,\ \ \ 1,\ \ \ 0,\ \ \ 0)\,,\\
\boldsymbol{Q}({}_{-(A\cap B)}\mathrm{T}^{(+)}, \boldsymbol{R}) &= \{R^0, R^2, R^6\}\,,\\
\boldsymbol{x}(A\cup B) &= (-1,\ \ \ 0,\ \ \ 0,\ \ \ 1,-1)\,,\\
\boldsymbol{Q}({}_{-(A\cup B)}\mathrm{T}^{(+)}, \boldsymbol{R}) &= \{R^3, R^5, R^{2t-1}\}\,;
\end{aligned}$$

see Figure 7.14.

(v) In the family

$$\begin{aligned}
\big\{(A,B) \in \mathbf{2}^{[t]} \times \mathbf{2}^{[t]}:\ \ j' := |A| = 3\,,\ \ j'' := |B| = 3\,,\\
j := |A\cap B| = 2\,,\ \ \ell' := \mathfrak{q}(A-B) = 3\,,\ \ \ell'' := \mathfrak{q}(B-A) = 3\,,\\
\ell^{\cap} := \mathfrak{q}(A\cap B) = 3\,,\ \ \ell := \mathfrak{q}(A\cup B) = 3\big\}
\end{aligned}$$

there are

$$\begin{aligned}
&\mathfrak{F}\left(\gamma,\gamma;\frac{\ell-1}{2},\frac{\ell'-1}{2},\frac{\ell''-1}{2},\frac{\ell^{\cap}+1}{2}\right)\cdot \mathrm{c}\left(\frac{\ell-1}{2};t-(j'+j''-j)\right)\\
&\quad\times \mathrm{c}\left(\frac{\ell'-1}{2};j'-j\right)\cdot \mathrm{c}\left(\frac{\ell''-1}{2};j''-j\right)\cdot \mathrm{c}\left(\frac{\ell^{\cap}+1}{2};j\right)\\
&:= \mathfrak{F}\left(\gamma,\gamma;\frac{3-1}{2},\frac{3-1}{2},\frac{3-1}{2},\frac{3+1}{2}\right)\cdot \mathrm{c}\left(\frac{3-1}{2};5-(3+3-2)\right)\\
&\quad\times \mathrm{c}\left(\frac{3-1}{2};3-2\right)\cdot \mathrm{c}\left(\frac{3-1}{2};3-2\right)\cdot \mathrm{c}\left(\frac{3+1}{2};2\right)\\
&= \underbrace{\mathfrak{F}(\gamma,\gamma;1,1,1,2)}_{\substack{\text{certainly, } 3! = 6;\\ \text{but cf. (A.16)}}}\cdot \mathrm{c}(1;1)\cdot \mathrm{c}(1;1)\cdot \mathrm{c}(1;1)\cdot \mathrm{c}(2;2)\\
&= 6\cdot 1\cdot 1\cdot 1\cdot 1 = 6
\end{aligned}$$

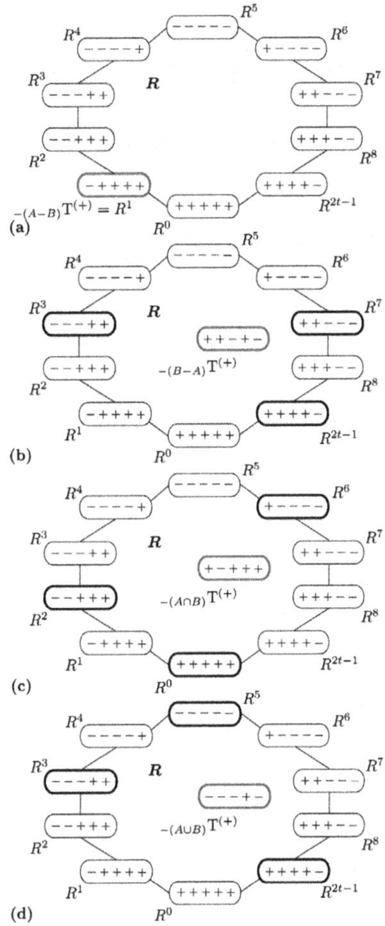

Figure 7.14 Four instances of the same distinguished symmetric cycle $\boldsymbol{R} := (R^0, R^1, \ldots, R^{2t-1}, R^0)$ in the hypercube graph $\boldsymbol{H}((t := 5), 2)$, defined by (7.1)(7.2).
Pick the subsets $A := \{1, 2\}$ and $B := \{2, 3, t\}$ of the set E_t.
Note that $\{1, t\} \cap A = \{1\}$, and $\{1, t\} \cap B = \{t\}$.
(a) The vertex ${}_{-(A-B)}\mathrm{T}^{(+)} = R^1 \in \mathrm{V}(\boldsymbol{R})$.
(b) The vertex ${}_{-(B-A)}\mathrm{T}^{(+)}$ of the graph $\boldsymbol{H}(t, 2)$, with the corresponding decomposition set $\boldsymbol{Q}({}_{-(B-A)}\mathrm{T}^{(+)}, \boldsymbol{R}) = \{R^3, R^7, R^{2t-1}\}$.
(c) The vertex ${}_{-(A\cap B)}\mathrm{T}^{(+)}$, with its decomposition set $\boldsymbol{Q}({}_{-(A\cap B)}\mathrm{T}^{(+)}, \boldsymbol{R}) = \{R^0, R^2, R^6\}$.
(d) The vertex ${}_{-(A\cup B)}\mathrm{T}^{(+)}$, with its decomposition set $\boldsymbol{Q}({}_{-(A\cup B)}\mathrm{T}^{(+)}, \boldsymbol{R}) = \{R^3, R^5, R^{2t-1}\}$.

pairs (A, B) such that

$$\{1, t\} \cap A = \{1, t\} \cap B = \{1, t\} .$$

One such pair is

$$\begin{aligned} (\; A &:= \{1, \;\;, \;\;, 4, t\} , \\ B &:= \{1, 2, \;\;, \;\;, t\} \;) , \end{aligned}$$

for which we have

$$\begin{aligned} A - B &= \{ \;\;, \;\;, \;\;, 4, \;\; \} , \\ B - A &= \{ \;\;, 2, \;\;, \;\;, \;\; \} , \\ A \cap B &= \{1, \;\;, \;\;, \;\;, t\} , \\ A \cup B &= \{1, 2, \;\;, 4, t\} , \end{aligned}$$

and

$$\begin{aligned} \boldsymbol{x}(A - B) &= (\;\; 1, \;\; 0, \;\; 0, -1, \;\; 1) , \\ \boldsymbol{Q}({}_{-(A-B)}\mathrm{T}^{(+)}, \boldsymbol{R}) &= \{R^0, R^4, R^8\} , \\ \boldsymbol{x}(B - A) &= (\;\; 1, -1, \;\; 1, \;\; 0, \;\; 0) , \\ \boldsymbol{Q}({}_{-(B-A)}\mathrm{T}^{(+)}, \boldsymbol{R}) &= \{R^0, R^2, R^6\} , \\ \boldsymbol{x}(A \cap B) &= (-1, \;\; 1, \;\; 0, \;\; 0, -1) , \\ \boldsymbol{Q}({}_{-(A\cap B)}\mathrm{T}^{(+)}, \boldsymbol{R}) &= \{R^1, R^5, R^{2t-1}\} , \\ \boldsymbol{x}(A \cup B) &= (-1, \;\; 0, \;\; 1, -1, \;\; 0) , \\ \boldsymbol{Q}({}_{-(A\cup B)}\mathrm{T}^{(+)}, \boldsymbol{R}) &= \{R^2, R^5, R^8\} ; \end{aligned}$$

see Figure 7.15.

(vi) In the family

$$\Big\{(A, B) \in \mathbf{2}^{[t]} \times \mathbf{2}^{[t]} : \;\; j' := |A| = 3 \,, \;\; j'' := |B| = 2 \,, \\ j := |A \cap B| = 1 \,, \;\; \ell' := \mathfrak{q}(A - B) = 3 \,, \;\; \ell'' := \mathfrak{q}(B - A) = 3 \,, \\ \ell^{\cap} := \mathfrak{q}(A \cap B) = 3 \,, \;\; \ell := \mathfrak{q}(A \cup B) = 3\Big\}$$

there are

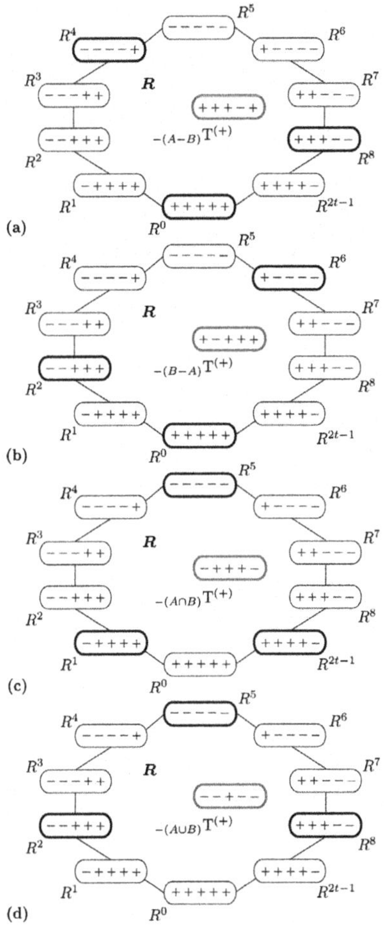

Figure 7.15 Four instances of the same distinguished symmetric cycle $\boldsymbol{R} := (R^0, R^1, \ldots, R^{2t-1}, R^0)$ in the hypercube graph $\boldsymbol{H}((t := 5), 2)$, defined by (7.1)(7.2).
Pick the subsets $A := \{1, 4, t\}$ and $B := \{1, 2, t\}$ of the set E_t.
Note that $\{1, t\} \cap A = \{1, t\} \cap B = \{1, t\}$.
(a) The vertex ${}_{-(A-B)}\mathrm{T}^{(+)}$ of the graph $\boldsymbol{H}(t, 2)$, with the corresponding decomposition set $\boldsymbol{Q}({}_{-(A-B)}\mathrm{T}^{(+)}, \boldsymbol{R}) = \{R^0, R^4, R^8\}$.
(b) The vertex ${}_{-(B-A)}\mathrm{T}^{(+)}$, with its decomposition set $\boldsymbol{Q}({}_{-(B-A)}\mathrm{T}^{(+)}, \boldsymbol{R}) = \{R^0, R^2, R^6\}$.
(c) The vertex ${}_{-(A\cap B)}\mathrm{T}^{(+)}$, with its decomposition set $\boldsymbol{Q}({}_{-(A\cap B)}\mathrm{T}^{(+)}, \boldsymbol{R}) = \{R^1, R^5, R^{2t-1}\}$.
(d) The vertex ${}_{-(A\cup B)}\mathrm{T}^{(+)}$, with its decomposition set $\boldsymbol{Q}({}_{-(A\cup B)}\mathrm{T}^{(+)}, \boldsymbol{R}) = \{R^2, R^5, R^8\}$.

$$\begin{aligned}
&\mathfrak{F}\left(\alpha, \alpha; \frac{\ell-1}{2}, \frac{\ell'+1}{2}, \frac{\ell''-1}{2}, \frac{\ell^{\cap}-1}{2}\right) \cdot \mathrm{c}\left(\frac{\ell-1}{2}; t-(j'+j''-j)\right) \\
&\quad \times \mathrm{c}\left(\frac{\ell'+1}{2}; j'-j\right) \cdot \mathrm{c}\left(\frac{\ell''-1}{2}; j''-j\right) \cdot \mathrm{c}\left(\frac{\ell^{\cap}-1}{2}; j\right) \\
&:= \mathfrak{F}\left(\alpha, \alpha; \frac{3-1}{2}, \frac{3+1}{2}, \frac{3-1}{2}, \frac{3-1}{2}\right) \cdot \mathrm{c}\left(\frac{3-1}{2}; 5-(3+2-1)\right) \\
&\quad \times \mathrm{c}\left(\frac{3+1}{2}; 3-1\right) \cdot \mathrm{c}\left(\frac{3-1}{2}; 2-1\right) \cdot \mathrm{c}\left(\frac{3-1}{2}; 1\right) \\
&= \underbrace{\mathfrak{F}\left(\alpha, \alpha; 1, 2, 1, 1\right)}_{\substack{\text{certainly, } 3! = 6; \\ \text{but cf. (A.16)}}} \cdot \mathrm{c}\left(1; 1\right) \cdot \mathrm{c}\left(2; 2\right) \cdot \mathrm{c}\left(1; 1\right) \cdot \mathrm{c}\left(1; 1\right) \\
&= 6 \cdot 1 \cdot 1 \cdot 1 \cdot 1 = 6
\end{aligned}$$

pairs (A, B) such that

$$\{1, t\} \cap A = \{1, t\}, \quad \text{and} \quad |\{1, t\} \cap B| = 0 .$$

One such pair is

$$\begin{aligned}
(\; A &:= \{1, \;\;, 3, \;\;, t\}, \\
B &:= \{\;\;, \;\;, 3, 4, \;\;\}\;),
\end{aligned}$$

for which we have

$$\begin{aligned}
A - B &= \{1, \;\;, \;\;, \;\;, t\}, \\
B - A &= \{\;\;, \;\;, \;\;, 4, \;\;\}, \\
A \cap B &= \{\;\;, \;\;, 3, \;\;, \;\;\}, \\
A \cup B &= \{1, \;\;, 3, 4, t\},
\end{aligned}$$

and

$$\begin{aligned}
\boldsymbol{x}(A - B) &= (-1, \;\; 1, \;\; 0, \;\; 0, -1), \\
\boldsymbol{Q}({}_{-(A-B)}\mathrm{T}^{(+)}, \boldsymbol{R}) &= \{R^1, R^5, R^{2t-1}\}, \\
\boldsymbol{x}(B - A) &= (\;\; 1, \;\; 0, \;\; 0, -1, \;\; 1), \\
\boldsymbol{Q}({}_{-(B-A)}\mathrm{T}^{(+)}, \boldsymbol{R}) &= \{R^0, R^4, R^8\}, \\
\boldsymbol{x}(A \cap B) &= (\;\; 1, \;\; 0, -1, \;\; 1, \;\; 0), \\
\boldsymbol{Q}({}_{-(A\cap B)}\mathrm{T}^{(+)}, \boldsymbol{R}) &= \{R^0, R^3, R^7\}, \\
\boldsymbol{x}(A \cup B) &= (-1, \;\; 1, -1, \;\; 0, \;\; 0), \\
\boldsymbol{Q}({}_{-(A\cup B)}\mathrm{T}^{(+)}, \boldsymbol{R}) &= \{R^1, R^5, R^7\};
\end{aligned}$$

see Figure 7.16.

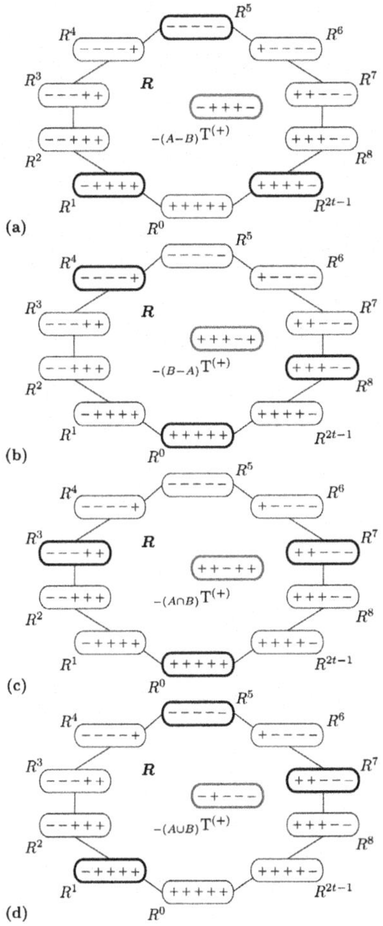

Figure 7.16 Four instances of the same distinguished symmetric cycle $\boldsymbol{R} := (R^0, R^1, \ldots, R^{2t-1}, R^0)$ in the hypercube graph $\boldsymbol{H}((t := 5), 2)$, defined by (7.1)(7.2).
Pick the subsets $A := \{1, 3, t\}$ and $B := \{3, 4\}$ of the set E_t.
Note that $\{1, t\} \cap A = \{1, t\}$, and $|\{1, t\} \cap B| = 0$.
(a) The vertex $_{-(A-B)}\mathrm{T}^{(+)}$ of the graph $\boldsymbol{H}(t, 2)$, with the corresponding decomposition set $\boldsymbol{Q}(_{-(A-B)}\mathrm{T}^{(+)}, \boldsymbol{R}) = \{R^1, R^5, R^{2t-1}\}$.
(b) The vertex $_{-(B-A)}\mathrm{T}^{(+)}$, with its decomposition set $\boldsymbol{Q}(_{-(B-A)}\mathrm{T}^{(+)}, \boldsymbol{R}) = \{R^0, R^4, R^8\}$.
(c) The vertex $_{-(A\cap B)}\mathrm{T}^{(+)}$, with its decomposition set $\boldsymbol{Q}(_{-(A\cap B)}\mathrm{T}^{(+)}, \boldsymbol{R}) = \{R^0, R^3, R^7\}$.
(d) The vertex $_{-(A\cup B)}\mathrm{T}^{(+)}$, with its decomposition set $\boldsymbol{Q}(_{-(A\cup B)}\mathrm{T}^{(+)}, \boldsymbol{R}) = \{R^1, R^5, R^7\}$.

(vii) The family

$$\big\{(A,B)\in \mathbf{2}^{[t]}\times\mathbf{2}^{[t]}:\ \ j':=|A|=2\,,\ \ j'':=|B|=2\,,\\ j:=|A\cap B|=1\,,\ \ \ell':=\mathfrak{q}(A-B)=1\,,\ \ \ell'':=\mathfrak{q}(B-A)=3\,,\\ \ell^{\cap}:=\mathfrak{q}(A\cap B)=1\,,\ \ \ell:=\mathfrak{q}(A\cup B)=5\big\}$$

consists of

$$\begin{aligned}
&\mathfrak{F}\left(\alpha,\gamma;\frac{\ell-1}{2},\frac{\ell'+1}{2},\frac{\ell''-1}{2},\frac{\ell^{\cap}+1}{2}\right)\\
&\qquad\times\mathrm{c}\left(\frac{\ell-1}{2};t-(j'+j''-j)\right)\\
&\qquad\times\mathrm{c}\left(\frac{\ell'+1}{2};j'-j\right)\cdot\mathrm{c}\left(\frac{\ell''-1}{2};j''-j\right)\cdot\mathrm{c}\left(\frac{\ell^{\cap}+1}{2};j\right)\\
&\quad:=\mathfrak{F}\left(\alpha,\gamma;\frac{5-1}{2},\frac{1+1}{2},\frac{3-1}{2},\frac{1+1}{2}\right)\\
&\qquad\times\mathrm{c}\left(\frac{5-1}{2};5-(2+2-1)\right)\\
&\qquad\times\mathrm{c}\left(\frac{1+1}{2};2-1\right)\cdot\mathrm{c}\left(\frac{3-1}{2};2-1\right)\cdot\mathrm{c}\left(\frac{1+1}{2};1\right)\\
&\quad=\underbrace{\mathfrak{F}(\alpha,\gamma;2,1,1,1)}_{\substack{\text{certainly, }1!=1;\\ \text{but cf. (A.17)}}}\cdot\mathrm{c}(2;2)\cdot\mathrm{c}(1;1)\cdot\mathrm{c}(1;1)\cdot\mathrm{c}(1;1)\\
&\quad=1\cdot1\cdot1\cdot1\cdot1=1
\end{aligned}$$

pair (A, B) such that

$$\{1,t\}\cap A=\{1,t\}\,,\quad\text{and}\quad\{1,t\}\cap B=\{t\}\,.$$

This is the pair

$$\begin{aligned}(\ A&:=\{1,\ \ ,\ \ ,\ \ ,t\}\,,\\ B&:=\{\ \ ,\ \ ,3,\ \ ,t\}\ )\,,\end{aligned}$$

for which we have

$$\begin{aligned}A-B&=\{1,\ \ ,\ \ ,\ \ ,\ \ \}\,,\\ B-A&=\{\ \ ,\ \ ,3,\ \ ,\ \ \}\,,\\ A\cap B&=\{\ \ ,\ \ ,\ \ ,\ \ ,t\}\,,\\ A\cup B&=\{1,\ \ ,3,\ \ ,t\}\,,\end{aligned}$$

and

$$
\begin{aligned}
\boldsymbol{x}(A-B) &= (\ \ 0,\ \ 1,\ \ 0,\ \ 0,\ \ 0)\,,\\
\boldsymbol{Q}({}_{-(A-B)}\mathrm{T}^{(+)},\,\boldsymbol{R}) &= \{R^1\}\,,\\
\boldsymbol{x}(B-A) &= (\ \ 1,\ \ 0,-1,\ \ 1,\ \ 0)\,,\\
\boldsymbol{Q}({}_{-(B-A)}\mathrm{T}^{(+)},\,\boldsymbol{R}) &= \{R^0,\,R^3,\,R^7\}\,,\\
\boldsymbol{x}(A\cap B) &= (\ \ 0,\ \ 0,\ \ 0,\ \ 0,-1)\,,\\
\boldsymbol{Q}({}_{-(A\cap B)}\mathrm{T}^{(+)},\,\boldsymbol{R}) &= \{R^{2t-1}\}\,,\\
\boldsymbol{x}(A\cup B) &= (-1,\ \ 1,-1,\ \ 1,-1)\,,\\
\boldsymbol{Q}({}_{-(A\cup B)}\mathrm{T}^{(+)},\,\boldsymbol{R}) &= \{R^1,\,R^3,\,R^5,\,R^7,\,R^{2t-1}\}\,;
\end{aligned}
$$

see Figure 7.17.

(viii) In the family

$$
\begin{aligned}
&\big\{(A,\,B)\in\mathbf{2}^{[t]}\times\mathbf{2}^{[t]}:\ \ j':=|A|=2\,,\ \ j'':=|B|=2\,,\\
&j:=|A\cap B|=1\,,\ \ \ell':=\mathfrak{q}(A-B)=3\,,\ \ \ell'':=\mathfrak{q}(B-A)=3\,,\\
&\qquad\qquad\qquad \ell^{\cap}:=\mathfrak{q}(A\cap B)=3\,,\ \ \ell:=\mathfrak{q}(A\cup B)=3\big\}
\end{aligned}
$$

there are

$$
\begin{aligned}
&\mathfrak{F}\left(\theta,\theta;\frac{\ell+1}{2},\frac{\ell'-1}{2},\frac{\ell''-1}{2},\frac{\ell^{\cap}-1}{2}\right)\\
&\qquad\times\mathrm{c}\left(\frac{\ell+1}{2};t-(j'+j''-j)\right)\\
&\qquad\times\mathrm{c}\left(\frac{\ell'-1}{2};j'-j\right)\cdot\mathrm{c}\left(\frac{\ell''-1}{2};j''-j\right)\cdot\mathrm{c}\left(\frac{\ell^{\cap}-1}{2};j\right)\\
&:=\mathfrak{F}\left(\theta,\theta;\frac{3+1}{2},\frac{3-1}{2},\frac{3-1}{2},\frac{3-1}{2}\right)\\
&\qquad\times\mathrm{c}\left(\frac{3+1}{2};5-(2+2-1)\right)\\
&\qquad\times\mathrm{c}\left(\frac{3-1}{2};2-1\right)\cdot\mathrm{c}\left(\frac{3-1}{2};2-1\right)\cdot\mathrm{c}\left(\frac{3-1}{2};1\right)\\
&=\underbrace{\mathfrak{F}(\theta,\theta;2,1,1,1)}_{\substack{\text{certainly, } 3!=6;\\ \text{but we will apply (A.16)}}}\cdot\mathrm{c}(2;2)\cdot\mathrm{c}(1;1)\cdot\mathrm{c}(1;1)\cdot\mathrm{c}(1;1)
\end{aligned}
$$

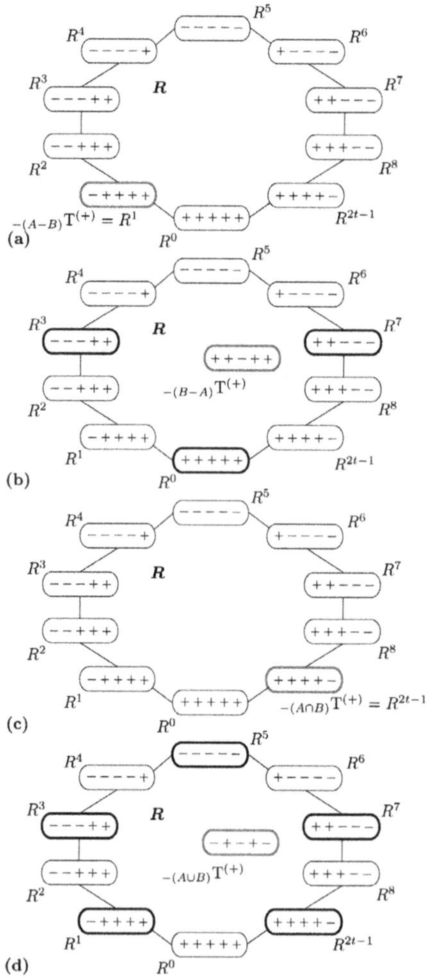

Figure 7.17 Four instances of the same distinguished symmetric cycle $\boldsymbol{R} := (R^0, R^1, \ldots, R^{2t-1}, R^0)$ in the hypercube graph $\boldsymbol{H}((t := 5), 2)$, defined by (7.1)(7.2).
Pick the subsets $A := \{1, t\}$ and $B := \{3, t\}$ of the set E_t.
Note that $\{1, t\} \cap A = \{1, t\}$, and $\{1, t\} \cap B = \{t\}$.
(a) The vertex ${}_{-(A-B)}\mathrm{T}^{(+)} = R^1 \in \mathrm{V}(\boldsymbol{R})$.
(b) The vertex ${}_{-(B-A)}\mathrm{T}^{(+)}$ of the graph $\boldsymbol{H}(t, 2)$, with the corresponding decomposition set $\boldsymbol{Q}({}_{-(B-A)}\mathrm{T}^{(+)}, \boldsymbol{R}) = \{R^0, R^3, R^7\}$.
(c) The vertex ${}_{-(A\cap B)}\mathrm{T}^{(+)} = R^{2t-1} \in \mathrm{V}(\boldsymbol{R})$.
(d) The vertex ${}_{-(A\cup B)}\mathrm{T}^{(+)}$, with its decomposition set $\boldsymbol{Q}({}_{-(A\cup B)}\mathrm{T}^{(+)}, \boldsymbol{R}) = \{R^1, R^3, R^5, R^7, R^{2t-1}\}$.

$$= \Bigg(\sum_{\substack{0 \le p \le 2-1, \\ 0 \le r \le \lfloor \frac{1}{2}(1+1+1-2+1) \rfloor}} \sum_{\substack{p \le s \le 2-1, \\ r \le t \le \lfloor \frac{1}{2}(1+1+1-2+1) \rfloor}} \binom{2+t-1}{p,\ r,\ s-p,\ t-r,\ 2-s-1}$$

$$\times 2^{-1-1-1+3\cdot 2-2s-2r+4t-3} \cdot 3^{1+1+1-2\cdot 2+s+r-3t+2}$$

$$\times \Bigg(\binom{2-s-1}{-1-1-1+3\cdot 2-2s-r+3t-3} \binom{-1-1-1+3\cdot 2-2s+3t-3}{-1+2+p-s+t-1}$$

$$\times \binom{-1-1+2\cdot 2-2p+2t-2}{-1+2-p+t-1} - \frac{4}{9} \cdot \binom{2-s-1}{-1-1-1+3\cdot 2-2s-r+3t-1}$$

$$\times \binom{-1-1-1+3\cdot 2-2s+3t}{-1+2+p-s+t} \binom{-1-1+2\cdot 2-2p+2t}{-1+2-p+t} \Bigg) \Bigg)$$

$$\times \binom{2-1}{2-1} \cdot \binom{1-1}{1-1} \cdot \binom{1-1}{1-1} \cdot \binom{1-1}{1-1}$$

$$= \Bigg(\sum_{\substack{0 \le p \le 1, \\ 0 \le r \le 1}} \sum_{\substack{p \le s \le 1, \\ r \le t \le 1}} \binom{1+t}{p,\ r,\ s-p,\ t-r,\ 1-s}$$

$$\times 2^{-2s-2r+4t} \cdot 3^{1+s+r-3t} \cdot \Bigg(\binom{1-s}{-2s-r+3t} \binom{-2s+3t}{p-s+t} \binom{-2p+2t}{-p+t}$$

$$- \frac{4}{9} \cdot \binom{1-s}{2-2s-r+3t} \binom{3-2s+3t}{1+p-s+t} \binom{2-2p+2t}{1-p+t} \Bigg) \Bigg)$$

$$\times 1 \cdot 1 \cdot 1 \cdot 1 = 6 \cdot 1 \cdot 1 \cdot 1 \cdot 1 = 6$$

pairs (A, B) such that

$$|\{1, t\} \cap A| = |\{1, t\} \cap B| = 0 .$$

One such pair is

$$(\ A := \{ \ , 2, \ \ , 4, \ \} ,$$
$$B := \{ \ , 2, 3, \ \ , \ \} \ ) ,$$

for which we have

$$A - B = \{ \ , \ \ , \ \ , 4, \ \} ,$$
$$B - A = \{ \ , \ \ , 3, \ \ , \ \} ,$$
$$A \cap B = \{ \ , 2, \ \ , \ \ , \ \} ,$$
$$A \cup B = \{ \ , 2, 3, 4, \ \} ,$$

and

$$\begin{aligned}\boldsymbol{x}(A-B) &= (1,\quad 0,\quad 0,-1,\quad 1)\,,\\ \boldsymbol{Q}({}_{-(A-B)}\mathrm{T}^{(+)},\,\boldsymbol{R}) &= \{R^0,\,R^4,\,R^8\}\,,\\ \boldsymbol{x}(B-A) &= (1,\quad 0,-1,\quad 1,\quad 0)\,,\\ \boldsymbol{Q}({}_{-(B-A)}\mathrm{T}^{(+)},\,\boldsymbol{R}) &= \{R^0,\,R^3,\,R^7\}\,,\\ \boldsymbol{x}(A\cap B) &= (1,-1,\quad 1,\quad 0,\quad 0)\,,\\ \boldsymbol{Q}({}_{-(A\cap B)}\mathrm{T}^{(+)},\,\boldsymbol{R}) &= \{R^0,\,R^2,\,R^6\}\,,\\ \boldsymbol{x}(A\cup B) &= (1,-1,\quad 0,\quad 0,\quad 1)\,,\\ \boldsymbol{Q}({}_{-(A\cup B)}\mathrm{T}^{(+)},\,\boldsymbol{R}) &= \{R^0,\,R^4,\,R^6\}\,;\end{aligned}$$

see Figure 7.18.

(ix) In the family

$$\begin{aligned}\big\{(A,\,B)\in\mathbf{2}^{[t]}\times\mathbf{2}^{[t]}:\ \ j' := |A| = 3\,,\ \ j'' := |B| = 2\,,\\ j := |A\cap B| = 1\,,\ \ \ell' := \mathfrak{q}(A-B) = 3\,,\ \ \ell'' := \mathfrak{q}(B-A) = 1\,,\\ \ell^{\cap} := \mathfrak{q}(A\cap B) = 3\,,\ \ \ell := \mathfrak{q}(A\cup B) = 1\big\}\end{aligned}$$

there are

$$\begin{aligned}&\mathfrak{F}\left(\theta,\,\beta;\,\frac{\ell+1}{2},\,\frac{\ell'-1}{2},\,\frac{\ell''+1}{2},\,\frac{\ell^{\cap}-1}{2}\right)\cdot\mathrm{c}\left(\frac{\ell+1}{2};t-(j'+j''-j)\right)\\ &\quad\times\mathrm{c}\left(\frac{\ell'-1}{2};j'-j\right)\cdot\mathrm{c}\left(\frac{\ell''+1}{2};j''-j\right)\cdot\mathrm{c}\left(\frac{\ell^{\cap}-1}{2};j\right)\\ &:=\mathfrak{F}\left(\theta,\,\beta;\,\frac{1+1}{2},\,\frac{3-1}{2},\,\frac{1+1}{2},\,\frac{3-1}{2}\right)\cdot\mathrm{c}\left(\frac{1+1}{2};5-(3+2-1)\right)\\ &\quad\times\mathrm{c}\left(\frac{3-1}{2};3-1\right)\cdot\mathrm{c}\left(\frac{1+1}{2};2-1\right)\cdot\mathrm{c}\left(\frac{3-1}{2};1\right)\\ &=\underbrace{\mathfrak{F}(\theta,\beta;1,\,1,\,1,\,1)}_{\substack{\text{certainly, }2!\,=\,2;\\ \text{but cf. (A.17)}}}\cdot\mathrm{c}(1;1)\cdot\mathrm{c}(1;2)\cdot\mathrm{c}(1;1)\cdot\mathrm{c}(1;1)\\ &=2\cdot1\cdot1\cdot1\cdot1=2\end{aligned}$$

pairs $(A,\,B)$ such that

$$|\{1,\,t\}\cap A| = 0\,,\quad\text{and}\quad\{1,\,t\}\cap B = \{t\}\,.$$

One such pair is

$$\begin{aligned}(\ A &:= \{\ \ ,\,2,\,3,\,4,\ \ \}\,,\\ B &:= \{\ \ ,\,2,\ \ ,\ \ ,\,t\}\ )\,,\end{aligned}$$

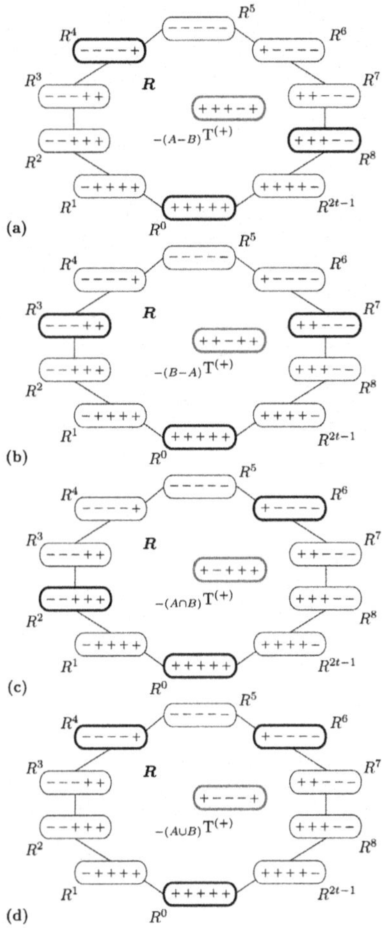

Figure 7.18 Four instances of the same distinguished symmetric cycle $\boldsymbol{R} := (R^0, R^1, \ldots, R^{2t-1}, R^0)$ in the hypercube graph $\boldsymbol{H}((t := 5), 2)$, defined by (7.1)(7.2).
Pick the subsets $A := \{2, 4\}$ and $B := \{2, 3\}$ of the set E_t.
Note that $|\{1, t\} \cap A| = |\{1, t\} \cap B| = 0$.
(a) The vertex $_{-(A-B)}\mathrm{T}^{(+)}$ of the graph $\boldsymbol{H}(t, 2)$, with the corresponding decomposition set $\boldsymbol{Q}(_{-(A-B)}\mathrm{T}^{(+)}, \boldsymbol{R}) = \{R^0, R^4, R^8\}$.
(b) The vertex $_{-(B-A)}\mathrm{T}^{(+)}$, with its decomposition set $\boldsymbol{Q}(_{-(B-A)}\mathrm{T}^{(+)}, \boldsymbol{R}) = \{R^0, R^3, R^7\}$.
(c) The vertex $_{-(A\cap B)}\mathrm{T}^{(+)}$, with its decomposition set $\boldsymbol{Q}(_{-(A\cap B)}\mathrm{T}^{(+)}, \boldsymbol{R}) = \{R^0, R^2, R^6\}$.
(d) The vertex $_{-(A\cup B)}\mathrm{T}^{(+)}$, with its decomposition set $\boldsymbol{Q}(_{-(A\cup B)}\mathrm{T}^{(+)}, \boldsymbol{R}) = \{R^0, R^4, R^6\}$.

for which we have

$$\begin{aligned} A - B &= \{\ ,\ \ ,3,\ 4,\ \ \}\,, \\ B - A &= \{\ ,\ \ ,\ \ ,\ \ ,t\}\,, \\ A \cap B &= \{\ ,2,\ \ ,\ \ ,\ \ \}\,, \\ A \cup B &= \{\ ,2,\ 3,\ 4,\ t\}\,, \end{aligned}$$

and

$$\begin{aligned} \boldsymbol{x}(A - B) &= (1, \quad 0, -1, \quad 0, \quad 1)\,, \\ \boldsymbol{Q}({}_{-(A-B)}\mathrm{T}^{(+)}, \boldsymbol{R}) &= \{R^0, R^4, R^7\}\,, \\ \boldsymbol{x}(B - A) &= (0, \quad 0, \quad 0, \quad 0, -1)\,, \\ \boldsymbol{Q}({}_{-(B-A)}\mathrm{T}^{(+)}, \boldsymbol{R}) &= \{R^{2t-1}\}\,, \\ \boldsymbol{x}(A \cap B) &= (1, -1, \quad 1, \quad 0, \quad 0)\,, \\ \boldsymbol{Q}({}_{-(A\cap B)}\mathrm{T}^{(+)}, \boldsymbol{R}) &= \{R^0, R^2, R^6\}\,, \\ \boldsymbol{x}(A \cup B) &= (0, -1, \quad 0, \quad 0, \quad 0)\,, \\ \boldsymbol{Q}({}_{-(A\cup B)}\mathrm{T}^{(+)}, \boldsymbol{R}) &= \{R^6\}\,; \end{aligned}$$

see Figure 7.19.

(x) In the family

$$\begin{aligned} \big\{(A, B) \in \mathbf{2}^{[t]} \times \mathbf{2}^{[t]}:\ & j' := |A| = 2\,,\ j'' := |B| = 2\,, \\ j := |A \cap B| = 1\,,\ \ell' &:= \mathfrak{q}(A - B) = 3\,,\ \ell'' := \mathfrak{q}(B - A) = 3\,, \\ & \ell^{\cap} := \mathfrak{q}(A \cap B) = 1\,,\ \ell := \mathfrak{q}(A \cup B) = 3\big\} \end{aligned}$$

there are

$$\begin{aligned} &\mathfrak{F}\left(\theta, \gamma; \frac{\ell+1}{2}, \frac{\ell'-1}{2}, \frac{\ell''-1}{2}, \frac{\ell^{\cap}+1}{2}\right) \cdot \mathrm{c}\left(\frac{\ell+1}{2}; t - (j' + j'' - j)\right) \\ &\quad \times \mathrm{c}\left(\frac{\ell'-1}{2}; j' - j\right) \cdot \mathrm{c}\left(\frac{\ell''-1}{2}; j'' - j\right) \cdot \mathrm{c}\left(\frac{\ell^{\cap}+1}{2}; j\right) \\ &:= \mathfrak{F}\left(\theta, \gamma; \frac{3+1}{2}, \frac{3-1}{2}, \frac{3-1}{2}, \frac{1+1}{2}\right) \cdot \mathrm{c}\left(\frac{3+1}{2}; 5 - (2 + 2 - j)\right) \\ &\quad \times \mathrm{c}\left(\frac{3-1}{2}; 2 - 1\right) \cdot \mathrm{c}\left(\frac{3-1}{2}; 2 - 1\right) \cdot \mathrm{c}\left(\frac{1+1}{2}; 1\right) \\ &= \underbrace{\mathfrak{F}(\theta, \gamma; 2, 1, 1, 1)}_{\substack{\text{certainly, } 2 \cdot 2! = 4; \\ \text{but cf. (A.17)}}} \cdot \mathrm{c}(2; 2) \cdot \mathrm{c}(1; 1) \cdot \mathrm{c}(1; 1) \cdot \mathrm{c}(1; 1) \\ &= 4 \cdot 1 \cdot 1 \cdot 1 \cdot 1 = 4 \end{aligned}$$

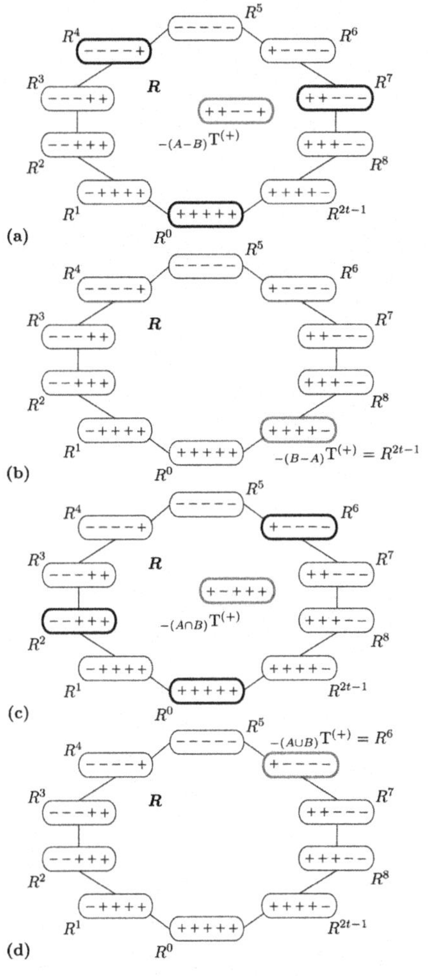

Figure 7.19 Four instances of the same distinguished symmetric cycle $\boldsymbol{R} := (R^0, R^1, \ldots, R^{2t-1}, R^0)$ in the hypercube graph $\boldsymbol{H}((t := 5), 2)$, defined by (7.1)(7.2).
Pick the subsets $A := \{2, 3, 4\}$ and $B := \{2, t\}$ of the set E_t.
Note that $|\{1, t\} \cap A| = 0$, and $\{1, t\} \cap B = \{t\}$.
(a) The vertex $_{-(A-B)}\mathrm{T}^{(+)}$ of the graph $\boldsymbol{H}(t, 2)$, with the corresponding decomposition set $\boldsymbol{Q}(_{-(A-B)}\mathrm{T}^{(+)}, \boldsymbol{R}) = \{R^0, R^4, R^7\}$.
(b) The vertex $_{-(B-A)}\mathrm{T}^{(+)} = R^{2t-1} \in \mathrm{V}(\boldsymbol{R})$.
(c) The vertex $_{-(A\cap B)}\mathrm{T}^{(+)}$, with its decomposition set $\boldsymbol{Q}(_{-(A\cap B)}\mathrm{T}^{(+)}, \boldsymbol{R}) = \{R^0, R^2, R^6\}$.
(d) The vertex $_{-(A\cup B)}\mathrm{T}^{(+)} = R^6 \in \mathrm{V}(\boldsymbol{R})$.

pairs $(A,\ B)$ such that

$$\{1,\,t\}\cap A=\{1,\,t\}\cap B=\{t\}\,.$$

One such pair is

$$\begin{aligned}(\ A &:=\{\ ,\ \ ,3,\ \ ,t\}\,,\\ B &:=\{\ ,2,\ \ ,\ \ ,t\}\ )\,,\end{aligned}$$

for which we have

$$\begin{aligned}A-B &=\{\ ,\ \ ,3,\ \ ,\ \}\,,\\ B-A &=\{\ ,2,\ \ ,\ \ ,\ \}\,,\\ A\cap B &=\{\ ,\ \ ,\ \ ,\ \ ,t\}\,,\\ A\cup B &=\{\ ,2,3,\ \ ,t\}\,,\end{aligned}$$

and

$$\begin{aligned}\mathbf{x}(A-B) &=(1,\ \ 0,-1,\ \ 1,\ \ 0)\,,\\ \boldsymbol{Q}({}_{-(A-B)}\mathrm{T}^{(+)},\,\boldsymbol{R}) &=\{R^0,\,R^3,\,R^7\}\,,\\ \mathbf{x}(B-A) &=(1,-1,\ \ 1,\ \ 0,\ \ 0)\,,\\ \boldsymbol{Q}({}_{-(B-A)}\mathrm{T}^{(+)},\,\boldsymbol{R}) &=\{R^0,\,R^2,\,R^6\}\,,\\ \mathbf{x}(A\cap B) &=(0,\ \ 0,\ \ 0,\ \ 0,-1)\,,\\ \boldsymbol{Q}({}_{-(A\cap B)}\mathrm{T}^{(+)},\,\boldsymbol{R}) &=\{R^{2t-1}\}\,,\\ \mathbf{x}(A\cup B) &=(0,-1,\ \ 0,\ \ 1,-1)\,,\\ \boldsymbol{Q}({}_{-(A\cup B)}\mathrm{T}^{(+)},\,\boldsymbol{R}) &=\{R^3,\,R^6,\,R^{2t-1}\}\,;\end{aligned}$$

see Figure 7.20.

- Let us consider the family

$$\begin{aligned}\big\{(A,\,B)\in\mathbf{2}^{[t]}\times\mathbf{2}^{[t]}:\ &|A|=j'<t\,,\ |B|=j''<t\,,\\ &|A\cap B|=j>0\,,\ \max\{j',\,j''\}<|A\cup B|<t\,,\\ &\mathfrak{q}(A\triangle B)=\ell^{\triangle}\,,\ \mathfrak{q}(A\cap B)=\ell^{\cap}\,,\ \mathfrak{q}(A\cup B)=\ell\big\}\end{aligned}\qquad(7.21)$$

of ordered two-member *intersecting Sperner families* that *do not* cover the set E_t, with the properties (7.18)(7.19).

Theorem 7.11. *Ordered pairs of sets $(A,\ B)$ in the family (7.21) can be counted with the help of products that include subproducts of the form*

$$\begin{aligned}&\mathfrak{T}\left(\mathfrak{s}',\,\mathfrak{s}'';\,\varrho(E_t-(A\cup B)),\,\varrho(A\triangle B),\,\varrho(A\cap B)\right)\\ &\qquad\times\mathrm{c}\left(\varrho(E_t-(A\cup B));t-(j'+j''-j)\right)\\ &\qquad\times\mathrm{c}\left(\varrho(A\triangle B);j'+j''-2j\right)\cdot\mathrm{c}\left(\varrho(A\cap B);j\right)\ :\end{aligned}\qquad(7.22)$$

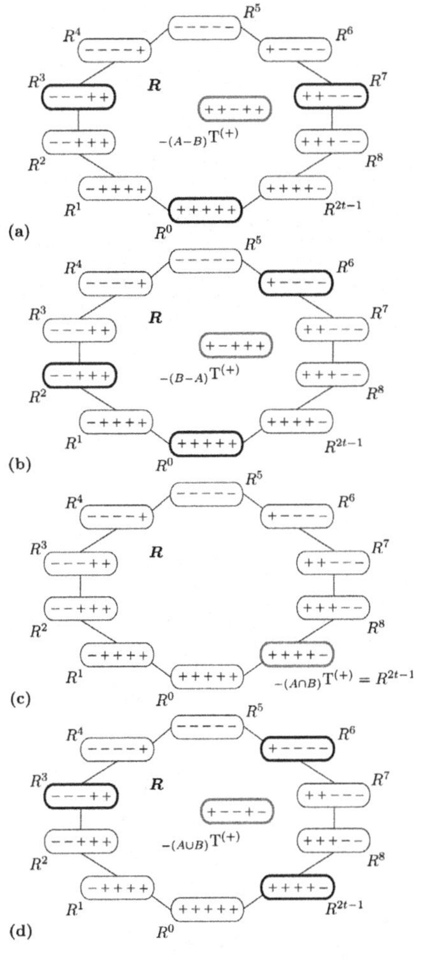

Figure 7.20 Four instances of the same distinguished symmetric cycle $\boldsymbol{R} := (R^0, R^1, \ldots, R^{2t-1}, R^0)$ in the hypercube graph $\boldsymbol{H}((t := 5), 2)$, defined by (7.1)(7.2).
Pick the subsets $A := \{3, t\}$ and $B := \{2, t\}$ of the set E_t.
Note that $\{1, t\} \cap A = \{1, t\} \cap B = \{t\}$.
(a) The vertex ${}_{-(A-B)}\mathrm{T}^{(+)}$ of the graph $\boldsymbol{H}(t, 2)$, with the corresponding decomposition set $\boldsymbol{Q}({}_{-(A-B)}\mathrm{T}^{(+)}, \boldsymbol{R}) = \{R^0, R^3, R^7\}$.
(b) The vertex ${}_{-(B-A)}\mathrm{T}^{(+)}$, with its decomposition set $\boldsymbol{Q}({}_{-(B-A)}\mathrm{T}^{(+)}, \boldsymbol{R}) = \{R^0, R^2, R^6\}$.
(c) The vertex ${}_{-(A\cap B)}\mathrm{T}^{(+)} = R^{2t-1} \in \mathrm{V}(\boldsymbol{R})$.
(d) The vertex ${}_{-(A\cup B)}\mathrm{T}^{(+)}$, with its decomposition set $\boldsymbol{Q}({}_{-(A\cup B)}\mathrm{T}^{(+)}, \boldsymbol{R}) = \{R^3, R^6, R^{2t-1}\}$.

(i) *In the family* (7.21) *there are*

$$\mathfrak{T}\left(\beta, \theta; \frac{\ell+1}{2}, \frac{\ell^{\triangle}-1}{2}, \frac{\ell^{\cap}+1}{2}\right) \cdot \mathrm{c}\left(\frac{\ell+1}{2}; t-(j'+j''-j)\right)$$
$$\times \mathrm{c}\left(\frac{\ell^{\triangle}-1}{2}; j'+j''-2j\right) \cdot \mathrm{c}\left(\frac{\ell^{\cap}+1}{2}; j\right) \cdot \binom{j'+j''-2j}{j'-j}$$

pairs (A, B) *of sets* A *and* B *such that*

$$\{1, t\} \cap A = \{1, t\} \cap B = \{1\}\,.$$

In the family (7.21) *there are*

(ii)

$$\mathfrak{T}\left(\beta, \alpha; \frac{\ell-1}{2}, \frac{\ell^{\triangle}+1}{2}, \frac{\ell^{\cap}+1}{2}\right) \cdot \mathrm{c}\left(\frac{\ell-1}{2}; t-(j'+j''-j)\right)$$
$$\times \mathrm{c}\left(\frac{\ell^{\triangle}+1}{2}; j'+j''-2j\right) \cdot \mathrm{c}\left(\frac{\ell^{\cap}+1}{2}; j\right) \cdot \binom{j'+j''-2j-1}{j'-j}$$

pairs (A, B) *such that*

$$\{1, t\} \cap A = \{1\}\,, \quad \textit{and} \quad \{1, t\} \cap B = \{1, t\}\,;$$

(iii)

$$\mathfrak{T}\left(\alpha, \theta; \frac{\ell+1}{2}, \frac{\ell^{\triangle}+1}{2}, \frac{\ell^{\cap}-1}{2}\right) \cdot \mathrm{c}\left(\frac{\ell+1}{2}; t-(j'+j''-j)\right)$$
$$\times \mathrm{c}\left(\frac{\ell^{\triangle}+1}{2}; j'+j''-2j\right) \cdot \mathrm{c}\left(\frac{\ell^{\cap}-1}{2}; j\right) \cdot \binom{j'+j''-2j-1}{j''-j}$$

pairs (A, B) *such that*

$$\{1, t\} \cap A = \{1\}\,, \quad \textit{and} \quad |\{1, t\} \cap B| = 0\,;$$

(iv)

$$\mathfrak{T}\left(\alpha, \alpha; \frac{\ell-1}{2}, \frac{\ell^{\triangle}+1}{2}, \frac{\ell^{\cap}-1}{2}\right) \cdot \mathrm{c}\left(\frac{\ell-1}{2}; t-(j'+j''-j)\right)$$
$$\times \mathrm{c}\left(\frac{\ell^{\triangle}+1}{2}; j'+j''-2j\right) \cdot \mathrm{c}\left(\frac{\ell^{\cap}-1}{2}; j\right) \cdot \binom{j'+j''-2j-2}{j'-j-1}$$

pairs (A, B) *such that*

$$\{1, t\} \cap A = \{1\}\,, \quad \textit{and} \quad \{1, t\} \cap B = \{t\}\,;$$

(v)

$$\mathfrak{T}\left(\beta, \beta; \frac{\ell-1}{2}, \frac{\ell^{\Delta}-1}{2}, \frac{\ell^{\cap}+1}{2}\right) \cdot \mathrm{c}\left(\frac{\ell-1}{2}; t-(j'+j''-j)\right)$$
$$\times \mathrm{c}\left(\frac{\ell^{\Delta}-1}{2}; j'+j''-2j\right) \cdot \mathrm{c}\left(\frac{\ell^{\cap}+1}{2}; j\right) \cdot \binom{j'+j''-2j}{j'-j}$$

pairs (A, B) *such that*

$$\{1, t\} \cap A = \{1, t\} \cap B = \{1, t\} ;$$

(vi)

$$\mathfrak{T}\left(\alpha, \alpha; \frac{\ell-1}{2}, \frac{\ell^{\Delta}+1}{2}, \frac{\ell^{\cap}-1}{2}\right) \cdot \mathrm{c}\left(\frac{\ell-1}{2}; t-(j'+j''-j)\right)$$
$$\times \mathrm{c}\left(\frac{\ell^{\Delta}+1}{2}; j'+j''-2j\right) \cdot \mathrm{c}\left(\frac{\ell^{\cap}-1}{2}; j\right) \cdot \binom{j'+j''-2j-2}{j''-j}$$

pairs (A, B) *such that*

$$\{1, t\} \cap A = \{1, t\}, \quad \textit{and} \quad |\{1, t\} \cap B| = 0 ;$$

(vii)

$$\mathfrak{T}\left(\alpha, \beta; \frac{\ell-1}{2}, \frac{\ell^{\Delta}+1}{2}, \frac{\ell^{\cap}+1}{2}\right) \cdot \mathrm{c}\left(\frac{\ell-1}{2}; t-(j'+j''-j)\right)$$
$$\times \mathrm{c}\left(\frac{\ell^{\Delta}+1}{2}; j'+j''-2j\right) \cdot \mathrm{c}\left(\frac{\ell^{\cap}+1}{2}; j\right) \cdot \binom{j'+j''-2j-1}{j''-j}$$

pairs (A, B) *such that*

$$\{1, t\} \cap A = \{1, t\}, \quad \textit{and} \quad \{1, t\} \cap B = \{t\} ;$$

(viii)

$$\mathfrak{T}\left(\theta, \theta; \frac{\ell+1}{2}, \frac{\ell^{\Delta}-1}{2}, \frac{\ell^{\cap}-1}{2}\right) \cdot \mathrm{c}\left(\frac{\ell+1}{2}; t-(j'+j''-j)\right)$$
$$\times \mathrm{c}\left(\frac{\ell^{\Delta}-1}{2}; j'+j''-2j\right) \cdot \mathrm{c}\left(\frac{\ell^{\cap}-1}{2}; j\right) \cdot \binom{j'+j''-2j}{j'-j}$$

pairs (A, B) *such that*

$$|\{1, t\} \cap A| = |\{1, t\} \cap B| = 0 ;$$

(ix)

$$\mathfrak{T}\left(\theta, \alpha; \frac{\ell+1}{2}, \frac{\ell^{\Delta}+1}{2}, \frac{\ell^{\cap}-1}{2}\right) \cdot \mathrm{c}\left(\frac{\ell+1}{2}; t-(j'+j''-j)\right)$$
$$\times \mathrm{c}\left(\frac{\ell^{\Delta}+1}{2}; j'+j''-2j\right) \cdot \mathrm{c}\left(\frac{\ell^{\cap}-1}{2}; j\right) \cdot \binom{j'+j''-2j-1}{j'-j}$$

pairs (A, B) *such that*

$$|\{1, t\} \cap A| = 0, \quad \textit{and} \quad \{1, t\} \cap B = \{t\} ;$$

(x)

$$\mathfrak{T}\left(\theta, \beta; \frac{\ell+1}{2}, \frac{\ell^{\triangle}-1}{2}, \frac{\ell^{\cap}+1}{2}\right) \cdot \mathrm{c}\left(\frac{\ell+1}{2}; t-(j'+j''-j)\right)$$
$$\times \mathrm{c}\left(\frac{\ell^{\triangle}-1}{2}; j'+j''-2j\right) \cdot \mathrm{c}\left(\frac{\ell^{\cap}+1}{2}; j\right) \cdot \binom{j'+j''-2j}{j'-j}$$

pairs (A, B) *such that*

$$\{1, t\} \cap A = \{1, t\} \cap B = \{t\}\ .$$

Proof. For each pair (A, B) in the family (7.21), pick an *ascending system of distinct representatives* $(e_1 < e_2 < \ldots < e_{\varrho(E_t-(A\cup B))+\varrho(A\triangle B)+\varrho(A\cap B)}) \subseteq E_t$ of the intervals composing the sets $(E_t - (A \cup B))$, $(A\triangle B)$ and $(A \cap B)$. By making the substitutions

$$e_i \mapsto \begin{cases} \theta\,, & \text{if } e_i \in E_t - (A \cup B), \\ \alpha\,, & \text{if } e_i \in A\triangle B, \\ \beta\,, & \text{if } e_i \in A \cap B, \end{cases}$$

$$1 \le i \le \varrho(E_t - (A \cup B)) + \varrho(A\triangle B) + \varrho(A \cap B)\,,$$

we obtain some ternary Smirnov words over the ordered alphabet (θ, α, β). Given any such Smirnov word, we find the number of pairs (A, B) in the corresponding subfamily of the family (7.21) by means of a product of the form (7.22).

(i) Since $\{1, t\} \cap A = \{1, t\} \cap B = \{1\}$, we have

$$\varrho(A\triangle B) = \frac{\ell^{\triangle}-1}{2}\,,$$

by Proposition 5.9(iii), and

$$\varrho(A \cap B) = \frac{\ell^{\cap}+1}{2}\,, \quad \varrho(A \cup B) = \frac{\ell+1}{2}\,,$$

by Proposition 5.9(i). Note that $\varrho(E_t - (A \cup B)) = \varrho(A \cup B)$, that is,

$$\varrho\,(E_t - (A \cup B)) = \frac{\ell+1}{2}\,.$$

(ii) Since $\{1, t\} \cap A = \{1\}$, and $\{1, t\} \cap B = \{1, t\}$, we have

$$\varrho(A\triangle B) = \frac{\ell^{\triangle}+1}{2}\,,$$

by Proposition 5.9(iv), and

$$\varrho(A \cap B) = \frac{\ell^{\cap}+1}{2}\,, \quad \varrho(A \cup B) = \frac{\ell+1}{2}\,,$$

by Proposition 5.9(i)(ii). Note that $\varrho(E_t - (A \cup B)) = \varrho(A \cup B) - 1$, that is,

$$\varrho\left(E_t - (A \cup B)\right) = \frac{\ell - 1}{2}.$$

(iii) Since $\{1, t\} \cap A = \{1\}$, and $|\{1, t\} \cap B| = 0$, we have

$$\varrho(A \triangle B) = \frac{\ell^{\triangle} + 1}{2},$$

by Proposition 5.9(i), and

$$\varrho(A \cap B) = \frac{\ell^{\cap} - 1}{2}, \quad \varrho(A \cup B) = \frac{\ell + 1}{2},$$

by Proposition 5.9(iii)(i). Note that

$$\varrho\left(E_t - (A \cup B)\right) = \frac{\ell + 1}{2}.$$

(iv) Since $\{1, t\} \cap A = \{1\}$, and $\{1, t\} \cap B = \{t\}$, we have

$$\varrho(A \triangle B) = \frac{\ell^{\triangle} + 1}{2},$$

by Proposition 5.9(ii), and

$$\varrho(A \cap B) = \frac{\ell^{\cap} - 1}{2}, \quad \varrho(A \cup B) = \frac{\ell + 1}{2},$$

by Proposition 5.9(iii)(ii). Note that

$$\varrho\left(E_t - (A \cup B)\right) = \frac{\ell - 1}{2}.$$

(v) Since $\{1, t\} \cap A = \{1, t\} \cap B = \{1, t\}$, we have

$$\varrho(A \triangle B) = \frac{\ell^{\triangle} - 1}{2},$$

by Proposition 5.9(iii), and

$$\varrho(A \cap B) = \frac{\ell^{\cap} + 1}{2}, \quad \varrho(A \cup B) = \frac{\ell + 1}{2},$$

by Proposition 5.9(ii). Note that

$$\varrho\left(E_t - (A \cup B)\right) = \frac{\ell - 1}{2}.$$

(vi) Since $\{1, t\} \cap A = \{1, t\}$, and $|\{1, t\} \cap B| = 0$, we have

$$\varrho(A \triangle B) = \frac{\ell^{\triangle} + 1}{2},$$

by Proposition 5.9(ii), and

$$\varrho(A \cap B) = \frac{\ell^{\cap} - 1}{2}, \quad \varrho(A \cup B) = \frac{\ell + 1}{2},$$

by Proposition 5.9(iii)(ii). Note that

$$\varrho\left(E_t - (A \cup B)\right) = \frac{\ell - 1}{2}.$$

(vii) Since $\{1, t\} \cap A = \{1, t\}$, and $\{1, t\} \cap B = \{t\}$, we have

$$\varrho(A \triangle B) = \frac{\ell^{\triangle} + 1}{2},$$

by Proposition 5.9(i), and

$$\varrho(A \cap B) = \frac{\ell^{\cap} + 1}{2}, \quad \varrho(A \cup B) = \frac{\ell + 1}{2},$$

by Proposition 5.9(iv)(ii). Note that

$$\varrho\left(E_t - (A \cup B)\right) = \frac{\ell - 1}{2}.$$

(viii) Since $|\{1, t\} \cap A| = |\{1, t\} \cap B| = 0$, we have

$$\varrho(A \triangle B) = \frac{\ell^{\triangle} - 1}{2},$$

and

$$\varrho(A \cap B) = \frac{\ell^{\cap} - 1}{2}, \quad \varrho(A \cup B) = \frac{\ell - 1}{2},$$

by Proposition 5.9(iii). Note that $\varrho(E_t - (A \cup B)) = \varrho(A \cup B) + 1$, that is,

$$\varrho\left(E_t - (A \cup B)\right) = \frac{\ell + 1}{2}.$$

(ix) Since $|\{1, t\} \cap A| = 0$, and $\{1, t\} \cap B = \{t\}$, we have

$$\varrho(A \triangle B) = \frac{\ell^{\triangle} + 1}{2},$$

by Proposition 5.9(iv), and

$$\varrho(A \cap B) = \frac{\ell^{\cap} - 1}{2}, \quad \varrho(A \cup B) = \frac{\ell + 1}{2},$$

by Proposition 5.9(iii)(iv). Note that $\varrho(E_t - (A \cup B)) = \varrho(A \cup B)$, that is,

$$\varrho\left(E_t - (A \cup B)\right) = \frac{\ell + 1}{2}.$$

(x) Since $\{1, t\} \cap A = \{1, t\} \cap B = \{t\}$, we have

$$\varrho(A \triangle B) = \frac{\ell^{\triangle} - 1}{2},$$

by Proposition 5.9(iii), and

$$\varrho(A \cap B) = \frac{\ell^{\cap} + 1}{2}, \quad \varrho(A \cup B) = \frac{\ell + 1}{2},$$

by Proposition 5.9(iv). Note that

$$\varrho\left(E_t - (A \cup B)\right) = \frac{\ell + 1}{2}. \qquad \square$$

Example 7.12. (We will only present explicit calculations for the case (ix) on page 200). Suppose $t := 5$.

(i) The family

$$\Big\{(A, B) \in \mathbf{2}^{[t]} \times \mathbf{2}^{[t]} : \ j' := |A| = 2\,, \ j'' := |B| = 2\,, \ j := |A \cap B| = 1\,, \ \ell^{\triangle} := \mathfrak{q}(A \triangle B) = 3\,, \ \ell^{\cap} := \mathfrak{q}(A \cap B) = 1\,, \ \ell := \mathfrak{q}(A \cup B) = 3\Big\}$$

consists of

$$\begin{aligned}
&\mathfrak{T}\left(\beta, \theta; \frac{\ell+1}{2}, \frac{\ell^{\triangle}-1}{2}, \frac{\ell^{\cap}+1}{2}\right) \cdot \mathrm{c}\left(\frac{\ell+1}{2}; t - (j' + j'' - j)\right) \\
&\quad \times \mathrm{c}\left(\frac{\ell^{\triangle}-1}{2}; j' + j'' - 2j\right) \cdot \mathrm{c}\left(\frac{\ell^{\cap}+1}{2}; j\right) \cdot \tbinom{j'+j''-2j}{j'-j} \\
&:= \mathfrak{T}\left(\beta, \theta; \frac{3+1}{2}, \frac{3-1}{2}, \frac{1+1}{2}\right) \cdot \mathrm{c}\left(\frac{3+1}{2}; 5 - (2 + 2 - 1)\right) \\
&\quad \times \mathrm{c}\left(\frac{3-1}{2}; 2 + 2 - 2 \cdot 1\right) \cdot \mathrm{c}\left(\frac{1+1}{2}; 1\right) \cdot \tbinom{2+2-2\cdot 1}{2-1} \\
&= \underbrace{\mathfrak{T}(\beta, \theta; 2, 1, 1)}_{\substack{\text{the same as } \mathfrak{T}(\theta, \beta; 2, 1, 1); \\ \text{certainly, } 1 \cdot 1! = 1; \\ \text{but see (A.8)}}} \cdot \mathrm{c}(2; 2) \cdot \mathrm{c}(1; 2) \cdot \mathrm{c}(1; 1) \cdot \tbinom{2}{1} \\
&= 1 \cdot 1 \cdot 1 \cdot 1 \cdot 2 = 2
\end{aligned}$$

pairs (A, B) of sets A and B such that

$$\{1, t\} \cap A = \{1, t\} \cap B = \{1\}.$$

One such pair is

$$\begin{aligned}
(\ A &:= \{1, \ \ , \ \ , 4, \ \ \}, \\
B &:= \{1, \ \ , 3, \ \ , \ \ \}\ ),
\end{aligned}$$

for which we have

$$A\triangle B = \{\ ,\ \ ,3,\ 4,\ \ \}\,,$$
$$A\cap B = \{1,\ \ ,\ \ ,\ \ ,\ \ \}\,,$$
$$A\cup B = \{1,\ \ ,3,\ 4,\ \ \}\,,$$

and

$$\begin{aligned}
\boldsymbol{x}(A\triangle B) &= (1,\quad 0, -1,\quad 0,\quad 1)\,,\\
\boldsymbol{Q}({}_{-(A\triangle B)}\mathrm{T}^{(+)},\ \boldsymbol{R}) &= \{R^0,\ R^4,\ R^7\}\,,\\
\boldsymbol{x}(A\cap B) &= (0,\quad 1,\quad 0,\quad 0,\quad 0)\,,\\
\boldsymbol{Q}({}_{-(A\cap B)}\mathrm{T}^{(+)},\ \boldsymbol{R}) &= \{R^1\}\,,\\
\boldsymbol{x}(A\cup B) &= (0,\quad 1, -1,\quad 0,\quad 1)\,,\\
\boldsymbol{Q}({}_{-(A\cup B)}\mathrm{T}^{(+)},\ \boldsymbol{R}) &= \{R^1,\ R^4,\ R^7\}\,;
\end{aligned}$$

see Figure 7.21.

(ii) The family

$$\begin{aligned}
\big\{(A,\ B)\in \mathbf{2}^{[t]}\times\mathbf{2}^{[t]}:\ \ & j' := |A| = 3\,,\ \ j'' := |B| = 3\,,\\
& j := |A\cap B| = 2\,,\ \ \ell^{\triangle} := \mathrm{q}(A\triangle B) = 1\,,\\
& \ell^{\cap} := \mathrm{q}(A\cap B) = 3\,,\ \ \ell := \mathrm{q}(A\cup B) = 3\big\}
\end{aligned}$$

consists of

$$\begin{aligned}
&\mathfrak{T}\left(\beta,\alpha;\frac{\ell-1}{2},\frac{\ell^{\triangle}+1}{2},\frac{\ell^{\cap}+1}{2}\right)\cdot \mathrm{c}\left(\frac{\ell-1}{2};t-(j'+j''-j)\right)\\
&\qquad\times \mathrm{c}\left(\frac{\ell^{\triangle}+1}{2};j'+j''-2j\right)\cdot \mathrm{c}\left(\frac{\ell^{\cap}+1}{2};j\right)\cdot\tbinom{j'+j''-2j-1}{j'-j}\\
&\quad := \mathfrak{T}\left(\beta,\alpha;\frac{3-1}{2},\frac{1+1}{2},\frac{3+1}{2}\right)\cdot \mathrm{c}\left(\frac{3-1}{2};5-(3+3-2)\right)\\
&\qquad\times \mathrm{c}\left(\frac{1+1}{2};3+3-2\cdot 2\right)\cdot \mathrm{c}\left(\frac{1+1}{2};2\right)\cdot\tbinom{3+3-2\cdot 2-1}{3-2}\\
&\quad = \underbrace{\mathfrak{T}(\beta,\alpha;1,1,2)}_{\substack{\text{certainly, } 1\cdot 1! = 1;\\ \text{but cf. (A.8)}}}\cdot \mathrm{c}(1;1)\cdot \mathrm{c}(1;2)\cdot \mathrm{c}(1;2)\cdot\tbinom{1}{1}\\
&\quad = 1\cdot 1\cdot 1\cdot 1\cdot 1 = 1
\end{aligned}$$

pair $(A,\ B)$ such that

$$\{1,\ t\}\cap A = \{1\}\,,\quad\text{and}\quad \{1,\ t\}\cap B = \{1,\ t\}\,.$$

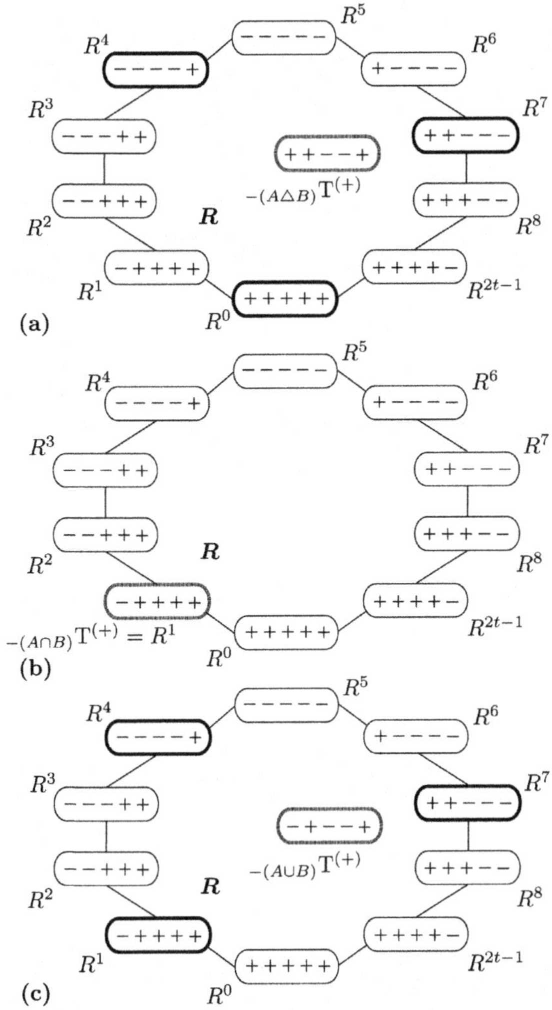

Figure 7.21 Three instances of the same distinguished symmetric cycle $\boldsymbol{R} := (R^0, R^1, \ldots, R^{2t-1}, R^0)$ in the hypercube graph $\boldsymbol{H}((t := 5), 2)$, defined by (7.1)(7.2).
Pick the subsets $A := \{1, 4\}$ and $B := \{1, 3\}$ of the set E_t.
Note that $\{1, t\} \cap A = \{1, t\} \cap B = \{1\}$.
(a) The vertex $_{-(A\triangle B)}\mathrm{T}^{(+)}$ of the graph $\boldsymbol{H}(t, 2)$, with the corresponding decomposition set $\boldsymbol{Q}(_{-(A\triangle B)}\mathrm{T}^{(+)}, \boldsymbol{R}) = \{R^0, R^4, R^7\}$.
(b) The vertex $_{-(A\cap B)}\mathrm{T}^{(+)} = R^1 \in \mathrm{V}(\boldsymbol{R})$.
(c) The vertex $_{-(A\cup B)}\mathrm{T}^{(+)}$, with its decomposition set $\boldsymbol{Q}(_{-(A\cup B)}\mathrm{T}^{(+)}, \boldsymbol{R}) = \{R^1, R^4, R^7\}$.

This pair is

$$\begin{aligned}(\ A &:= \{1, \quad , 3,\ 4, \quad \}\,, \\ B &:= \{1, \quad , 3, \quad , t\} \ )\,,\end{aligned}$$

for which we have

$$\begin{aligned} A \triangle B &= \{ \ , \quad , \quad , 4,\ t\}\,, \\ A \cap B &= \{1, \quad , 3, \quad , \ \}\,, \\ A \cup B &= \{1, \quad , 3,\ 4,\ t\}\,, \end{aligned}$$

and

$$\begin{aligned} \boldsymbol{x}(A \triangle B) &= (\ \ 0, \quad 0, \quad 0, -1, \quad 0)\,, \\ \boldsymbol{Q}({}_{-(A \triangle B)}\mathrm{T}^{(+)}, \boldsymbol{R}) &= \{R^8\}\,, \\ \boldsymbol{x}(A \cap B) &= (\ \ 0, \quad 1, -1, \quad 1, \quad 0)\,, \\ \boldsymbol{Q}({}_{-(A \cap B)}\mathrm{T}^{(+)}, \boldsymbol{R}) &= \{R^1, R^3, R^7\}\,, \\ \boldsymbol{x}(A \cup B) &= (-1, \quad 1, -1, \quad 0, \quad 0)\,, \\ \boldsymbol{Q}({}_{-(A \cup B)}\mathrm{T}^{(+)}, \boldsymbol{R}) &= \{R^1, R^5, R^7\}\,; \end{aligned}$$

see Figure 7.22.

(iii) The family

$$\begin{aligned}\big\{(A, B) \in \mathbf{2}^{[t]} \times \mathbf{2}^{[t]} :\ \ j' := |A| = 3\,,\ \ j'' := |B| = 2\,, \\ j := |A \cap B| = 1\,,\ \ \ell^{\triangle} := \mathfrak{q}(A \triangle B) = 1\,, \\ \ell^{\cap} := \mathfrak{q}(A \cap B) = 3\,,\ \ \ell := \mathfrak{q}(A \cup B) = 1\big\}\end{aligned}$$

consists of

$$\begin{aligned} &\mathfrak{T}\left(\alpha, \theta; \frac{\ell+1}{2}, \frac{\ell^{\triangle}+1}{2}, \frac{\ell^{\cap}-1}{2}\right) \cdot \mathrm{c}\left(\frac{\ell+1}{2}; t - (j' + j'' - j)\right) \\ &\quad \times \mathrm{c}\left(\frac{\ell^{\triangle}+1}{2}; j' + j'' - 2j\right) \cdot \mathrm{c}\left(\frac{\ell^{\cap}-1}{2}; j\right) \cdot \tbinom{j'+j''-2j-1}{j''-j} \\ &:= \mathfrak{T}\left(\alpha, \theta; \frac{1+1}{2}, \frac{1+1}{2}, \frac{3-1}{2}\right) \cdot \mathrm{c}\left(\frac{1+1}{2}; 5 - (3+2-1)\right) \\ &\quad \times \mathrm{c}\left(\frac{1+1}{2}; 3+2-2\cdot 1\right) \cdot \mathrm{c}\left(\frac{3-1}{2}; 1\right) \cdot \tbinom{3+2-2\cdot 1-1}{2-1} \\ &= \underbrace{\mathfrak{T}(\alpha, \theta; 1, 1, 1)}_{\substack{\text{certainly, } 1! = 1; \\ \text{but see (6.16)}}} \cdot \mathrm{c}(1;1) \cdot \mathrm{c}(1;3) \cdot \mathrm{c}(1;1) \cdot \tbinom{2}{1} \\ &= 1 \cdot 1 \cdot 1 \cdot 1 \cdot 2 = 2 \end{aligned}$$

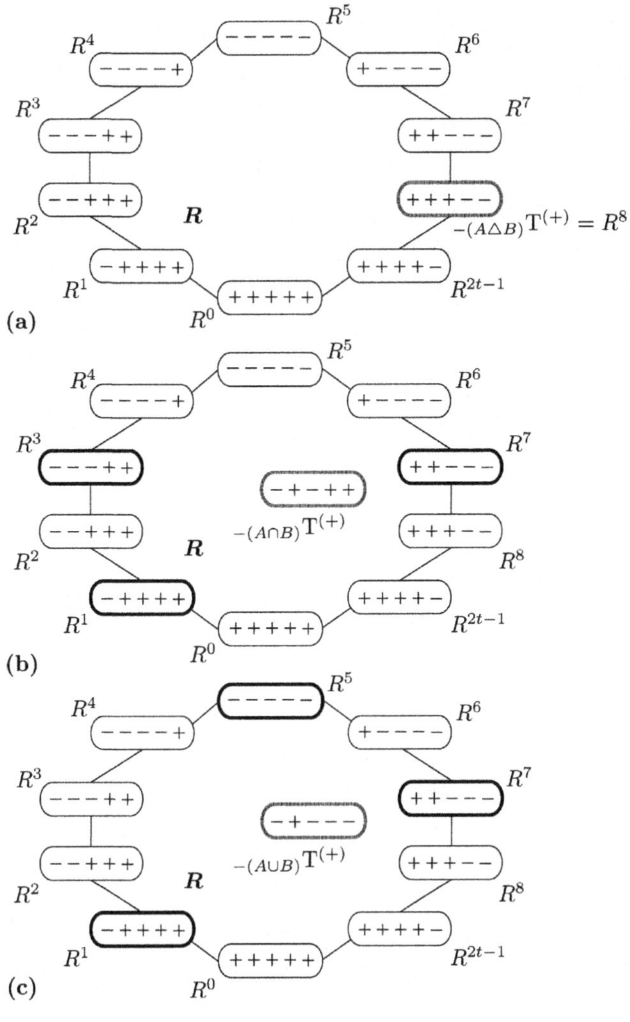

Figure 7.22 Three instances of the same distinguished symmetric cycle $\boldsymbol{R} := (R^0, R^1, \ldots, R^{2t-1}, R^0)$ in the hypercube graph $\boldsymbol{H}((t := 5), 2)$, defined by (7.1)(7.2).
Pick the subsets $A := \{1, 3, 4\}$ and $B := \{1, 3, t\}$ of the set E_t.
Note that $\{1, t\} \cap A = \{1\}$, and $\{1, t\} \cap B = \{1, t\}$.
(a) The vertex $_{-(A\triangle B)}\mathrm{T}^{(+)} = R^8 \in \mathrm{V}(\boldsymbol{R})$.
(b) The vertex $_{-(A\cap B)}\mathrm{T}^{(+)}$, with the corresponding decomposition set $\boldsymbol{Q}(_{-(A\cap B)}\mathrm{T}^{(+)}, \boldsymbol{R}) = \{R^1, R^3, R^7\}$.
(c) The vertex $_{-(A\cup B)}\mathrm{T}^{(+)}$, with its decomposition set $\boldsymbol{Q}(_{-(A\cup B)}\mathrm{T}^{(+)}, \boldsymbol{R}) = \{R^1, R^5, R^7\}$.

pairs $(A,\ B)$ such that

$$\{1,\ t\}\cap A=\{1\}\ ,\quad \text{and}\quad |\{1,\ t\}\cap B|=0\ .$$

One such pair is

$$\begin{aligned}(\ A&:=\{1,\quad,\ 3,\ 4,\ \}\ ,\\ B&:=\{\ ,\ 2,\quad,\ 4,\ \}\ )\ ,\end{aligned}$$

for which we have

$$\begin{aligned}A\triangle B&=\{1,\ 2,\ 3,\quad,\ \}\ ,\\ A\cap B&=\{\ ,\quad,\quad,\ 4,\ \}\ ,\\ A\cup B&=\{1,\ 2,\ 3,\ 4,\ \}\ ,\end{aligned}$$

and

$$\begin{aligned}\boldsymbol{x}(A\triangle B)&=(0,\quad 0,\quad 0,\quad 1,\quad 0)\ ,\\ \boldsymbol{Q}({}_{-(A\triangle B)}\mathrm{T}^{(+)},\ \boldsymbol{R})&=\{R^3\}\ ,\\ \boldsymbol{x}(A\cap B)&=(1,\quad 0,\quad 0,\,-1,\quad 1)\ ,\\ \boldsymbol{Q}({}_{-(A\cap B)}\mathrm{T}^{(+)},\ \boldsymbol{R})&=\{R^0,\ R^4,\ R^8\}\ ,\\ \boldsymbol{x}(A\cup B)&=(0,\quad 0,\quad 0,\quad 0,\quad 1)\ ,\\ \boldsymbol{Q}({}_{-(A\cup B)}\mathrm{T}^{(+)},\ \boldsymbol{R})&=\{R^4\}\ ;\end{aligned}$$

see Figure 7.23.

(iv) The family

$$\begin{aligned}\big\{(A,\ B)\in\mathbf{2}^{[t]}\times\mathbf{2}^{[t]}:\ \ j'&:=|A|=3\ ,\ \ j'':=|B|=3\ ,\\ j:=|A\cap B|&=2\ ,\ \ \ell^{\triangle}:=\mathfrak{q}(A\triangle B)=3\ ,\\ \ell^{\cap}&:=\mathfrak{q}(A\cap B)=5\ ,\ \ \ell:=\mathfrak{q}(A\cup B)=3\big\}\end{aligned}$$

consists of

$$\begin{aligned}&\mathfrak{T}\left(\alpha,\alpha;\frac{\ell-1}{2},\frac{\ell^{\triangle}+1}{2},\frac{\ell^{\cap}-1}{2}\right)\cdot\mathsf{c}\left(\frac{\ell-1}{2};t-(j'+j''-j)\right)\\ &\qquad\times\mathsf{c}\left(\frac{\ell^{\triangle}+1}{2};j'+j''-2j\right)\cdot\mathsf{c}\left(\frac{\ell^{\cap}-1}{2};j\right)\cdot\tbinom{j'+j''-2j-2}{j'-j-1}\\ &:=\mathfrak{T}\left(\alpha,\alpha;\frac{3-1}{2},\frac{3+1}{2},\frac{5-1}{2}\right)\cdot\mathsf{c}\left(\frac{3-1}{2};5-(3+3-2)\right)\\ &\qquad\times\mathsf{c}\left(\frac{3+1}{2};3+3-2\cdot 2\right)\cdot\mathsf{c}\left(\frac{5-1}{2};2\right)\cdot\tbinom{3+3-2\cdot2-2}{3-2-1}\\ &:=\underbrace{\mathfrak{T}(\alpha,\alpha;1,\ 2,\ 2)}_{\substack{\text{certainly, }1!=1;\\ \text{but see (6.15)}}}\cdot\mathsf{c}(1;1)\cdot\mathsf{c}(2;2)\cdot\mathsf{c}(2;2)\cdot\tbinom{0}{0}\\ &=1\cdot 1\cdot 1\cdot 1\cdot 1=1\end{aligned}$$

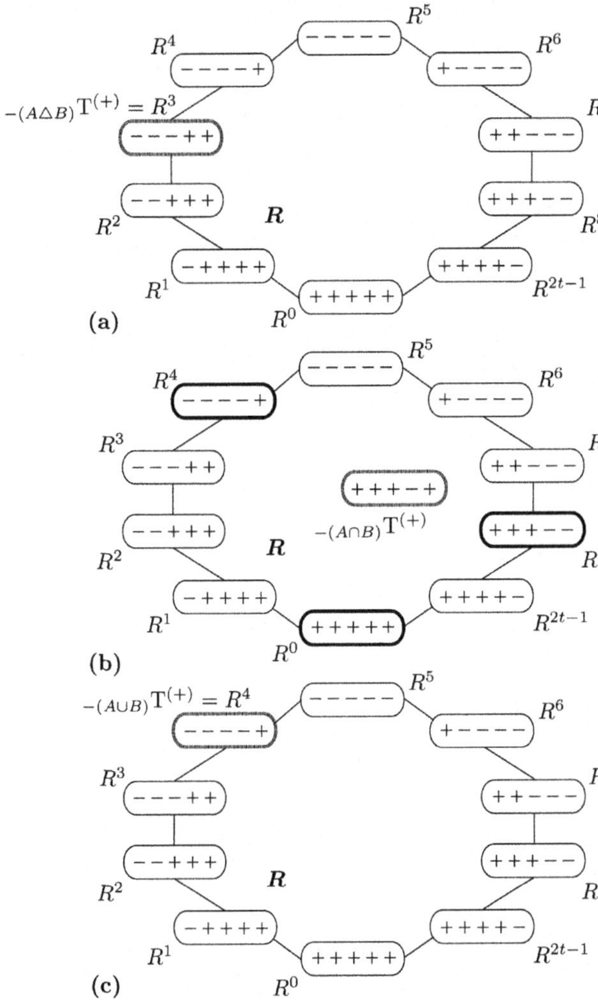

Figure 7.23 Three instances of the same distinguished symmetric cycle $\boldsymbol{R} := (R^0, R^1, \ldots, R^{2t-1}, R^0)$ in the hypercube graph $\boldsymbol{H}((t := 5), 2)$, defined by (7.1)(7.2).
Pick the subsets $A := \{1, 3, 4\}$ and $B := \{2, 4\}$ of the set E_t.
Note that $\{1, t\} \cap A = \{1\}$, and $|\{1, t\} \cap B| = 0$.
(a) The vertex $_{-(A\triangle B)}\mathrm{T}^{(+)}$ of the graph $\boldsymbol{H}(t, 2)$, with the corresponding decomposition set $\boldsymbol{Q}(_{-(A\triangle B)}\mathrm{T}^{(+)}, \boldsymbol{R}) = \{R^3\}$.
(b) The vertex $_{-(A\cap B)}\mathrm{T}^{(+)}$, with its decomposition set $\boldsymbol{Q}(_{-(A\cap B)}\mathrm{T}^{(+)}, \boldsymbol{R}) = \{R^0, R^4, R^8\}$.
(c) The vertex $_{-(A\cup B)}\mathrm{T}^{(+)} = R^4 \in \mathrm{V}(\boldsymbol{R})$.

pair $(A,\,B)$ such that

$$\{1,\,t\}\cap A=\{1\}\,,\quad\text{and}\quad\{1,\,t\}\cap B=\{t\}\,.$$

This pair is

$$\begin{aligned}(\;A &:= \{1,\ 2,\ \ ,\ 4,\ \ \}\,,\\ B &:= \{\ \ ,\ 2,\ \ ,\ 4,\ t\}\;)\,,\end{aligned}$$

for which we have

$$\begin{aligned}A\triangle B &= \{1,\ \ ,\ \ ,\ \ ,\ t\}\,,\\ A\cap B &= \{\ \ ,\ 2,\ \ ,\ 4,\ \ \}\,,\\ A\cup B &= \{1,\ 2,\ \ ,\ 4,\ t\}\,,\end{aligned}$$

and

$$\begin{aligned}\boldsymbol{x}(A\triangle B) &= (-1,\quad 1,\quad 0,\quad 0,\,-1)\,,\\ \boldsymbol{Q}({}_{-(A\triangle B)}\mathrm{T}^{(+)},\,\boldsymbol{R}) &= \{R^1,\,R^5,\,R^{2t-1}\}\,,\\ \boldsymbol{x}(A\cap B) &= (\ \ 1,\,-1,\quad 1,\,-1,\quad 1)\,,\\ \boldsymbol{Q}({}_{-(A\cap B)}\mathrm{T}^{(+)},\,\boldsymbol{R}) &= \{R^0,\,R^2,\,R^4,\,R^6,\,R^8\}\,,\\ \boldsymbol{x}(A\cup B) &= (-1,\quad 0,\quad 1,\,-1,\quad 0)\,,\\ \boldsymbol{Q}({}_{-(A\cup B)}\mathrm{T}^{(+)},\,\boldsymbol{R}) &= \{R^2,\,R^5,\,R^8\}\,;\end{aligned}$$

see Figure 7.24.

(v) In the family

$$\begin{aligned}\big\{(A,\,B)\in\mathbf{2}^{[t]}\times\mathbf{2}^{[t]}:\ \ j' &:= |A|=3\,,\ \ j'':=|B|=3\,,\\ j &:= |A\cap B|=2\,,\ \ \ell^{\triangle}:=\mathfrak{q}(A\triangle B)=3\,,\\ \ell^{\cap} &:= \mathfrak{q}(A\cap B)=3\,,\ \ \ell:=\mathfrak{q}(A\cup B)=3\big\}\end{aligned}$$

there are

$$\begin{aligned}&\mathfrak{T}\left(\beta,\,\beta;\frac{\ell-1}{2},\,\frac{\ell^{\triangle}-1}{2},\,\frac{\ell^{\cap}+1}{2}\right)\cdot\mathrm{c}\left(\frac{\ell-1}{2};t-(j'+j''-j)\right)\\ &\qquad\times\mathrm{c}\left(\frac{\ell^{\triangle}-1}{2};j'+j''-2j\right)\cdot\mathrm{c}\left(\frac{\ell^{\cap}+1}{2};j\right)\cdot\tbinom{j'+j''-2j}{j'-j}\\ &:=\mathfrak{T}\left(\beta,\,\beta;\frac{3-1}{2},\,\frac{3-1}{2},\,\frac{3+1}{2}\right)\cdot\mathrm{c}\left(\frac{3-1}{2};5-(3+3-2)\right)\\ &\qquad\times\mathrm{c}\left(\frac{3-1}{2};3+3-2\cdot 2\right)\cdot\mathrm{c}\left(\frac{3+1}{2};2\right)\cdot\tbinom{3+3-2\cdot 2}{3-2}\\ &=\underbrace{\mathfrak{T}(\beta,\,\beta;1,\,1,\,2)}_{\substack{\text{certainly, }2!\,=\,2;\\ \text{but cf. (A.7)}}}\cdot\mathrm{c}\,(1;1)\cdot\mathrm{c}\,(1;2)\cdot\mathrm{c}\,(2;2)\cdot\tbinom{2}{1}\\ &=2\cdot 1\cdot 1\cdot 1\cdot 2=4\end{aligned}$$

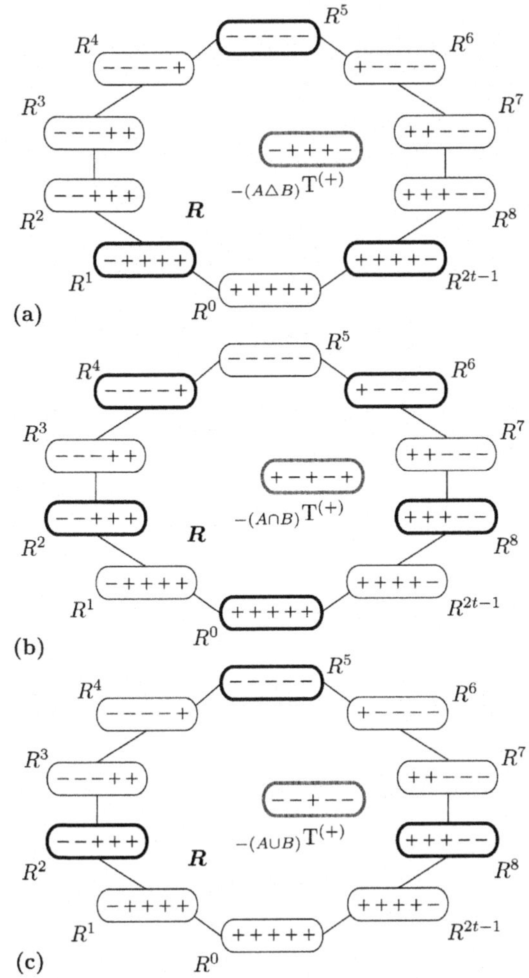

Figure 7.24 Three instances of the same distinguished symmetric cycle $\boldsymbol{R} := (R^0, R^1, \ldots, R^{2t-1}, R^0)$ in the hypercube graph $\boldsymbol{H}((t := 5), 2)$, defined by (7.1)(7.2).
Pick the subsets $A := \{1, 2, 4\}$ and $B := \{2, 4, t\}$ of the set E_t.
Note that $\{1, t\} \cap A = \{1\}$, and $\{1, t\} \cap B = \{t\}$.
(a) The vertex ${}_{-(A\triangle B)}\mathrm{T}^{(+)}$ of the graph $\boldsymbol{H}(t, 2)$, with the corresponding decomposition set $\boldsymbol{Q}({}_{-(A\triangle B)}\mathrm{T}^{(+)}, \boldsymbol{R}) = \{R^1, R^5, R^{2t-1}\}$.
(b) The vertex ${}_{-(A\cap B)}\mathrm{T}^{(+)}$, with its decomposition set $\boldsymbol{Q}({}_{-(A\cap B)}\mathrm{T}^{(+)}, \boldsymbol{R}) = \{R^0, R^2, R^4, R^6, R^8\}$.
(c) The vertex ${}_{-(A\cup B)}\mathrm{T}^{(+)}$, with its decomposition set $\boldsymbol{Q}({}_{-(A\cup B)}\mathrm{T}^{(+)}, \boldsymbol{R}) = \{R^2, R^5, R^8\}$.

pairs $(A,\ B)$ such that

$$\{1,\ t\}\cap A=\{1,\ t\}\cap B=\{1,\ t\}\ .$$

One such pair is

$$\begin{aligned}(\ A&:=\{1,\ \ ,\ 3,\ \ ,\ t\}\ ,\\ B&:=\{1,\ 2,\ \ ,\ \ ,\ t\}\ )\ ,\end{aligned}$$

for which we have

$$\begin{aligned}A\triangle B&=\{\ \ ,\ 2,\ 3,\ \ ,\ \ \}\ ,\\ A\cap B&=\{1,\ \ ,\ \ ,\ \ ,\ t\}\ ,\\ A\cup B&=\{1,\ 2,\ 3,\ \ ,\ t\}\ ,\end{aligned}$$

and

$$\begin{aligned}\boldsymbol{x}(A\triangle B)&=(\ \ 1,-1,\ \ \ 0,\ \ \ 1,\ \ \ 0)\ ,\\ \boldsymbol{Q}({}_{-(A\triangle B)}\mathrm{T}^{(+)},\ \boldsymbol{R})&=\{R^0,\ R^3,\ R^6\}\ ,\\ \boldsymbol{x}(A\cap B)&=(-1,\ \ \ 1,\ \ \ 0,\ \ \ 0,-1)\ ,\\ \boldsymbol{Q}({}_{-(A\cap B)}\mathrm{T}^{(+)},\ \boldsymbol{R})&=\{R^1,\ R^5,\ R^{2t-1}\}\ ,\\ \boldsymbol{x}(A\cup B)&=(-1,\ \ \ 0,\ \ \ 0,\ \ \ 1,-1)\ ,\\ \boldsymbol{Q}({}_{-(A\cup B)}\mathrm{T}^{(+)},\ \boldsymbol{R})&=\{R^3,\ R^5,\ R^{2t-1}\}\ ;\end{aligned}$$

see Figure 7.25.

(vi) In the family

$$\begin{aligned}\big\{(A,\ B)\in\mathbf{2}^{[t]}\times\mathbf{2}^{[t]}:\ \ j'&:=|A|=3\ ,\ \ j'':=|B|=2\ ,\\ j:=|A\cap B|=1\ ,\ \ \ell^{\triangle}&:=\mathfrak{q}(A\triangle B)=3\ ,\\ \ell^{\cap}:=\mathfrak{q}(A\cap B)=3\ ,\ \ \ell&:=\mathfrak{q}(A\cup B)=3\big\}\end{aligned}$$

there are

$$\begin{aligned}&\mathfrak{T}\left(\alpha,\alpha;\frac{\ell-1}{2},\frac{\ell^{\triangle}+1}{2},\frac{\ell^{\cap}-1}{2}\right)\cdot\mathsf{c}\left(\frac{\ell-1}{2};t-(j'+j''-j)\right)\\ &\quad\times\mathsf{c}\left(\frac{\ell^{\triangle}+1}{2};j'+j''-2j\right)\cdot\mathsf{c}\left(\frac{\ell^{\cap}-1}{2};j\right)\cdot\tbinom{j'+j''-2j-2}{j''-j}\\ &:=\mathfrak{T}\left(\alpha,\alpha;\frac{3-1}{2},\frac{3+1}{2},\frac{3-1}{2}\right)\cdot\mathsf{c}\left(\frac{3-1}{2};5-(3+2-1)\right)\\ &\quad\times\mathsf{c}\left(\frac{3+1}{2};3+2-2\cdot1\right)\cdot\mathsf{c}\left(\frac{3-1}{2};1\right)\cdot\tbinom{3+2-2\cdot1-2}{2-1}\\ &=\underbrace{\mathfrak{T}(\alpha,\alpha;1,\ 2,\ 1)}_{\substack{\text{certainly, }2!=2;\\ \text{but see (6.15)}}}\cdot\mathsf{c}(1;1)\cdot\mathsf{c}(2;3)\cdot\mathsf{c}(1;1)\cdot\tbinom{1}{1}\\ &=2\cdot1\cdot2\cdot1\cdot1=4\end{aligned}$$

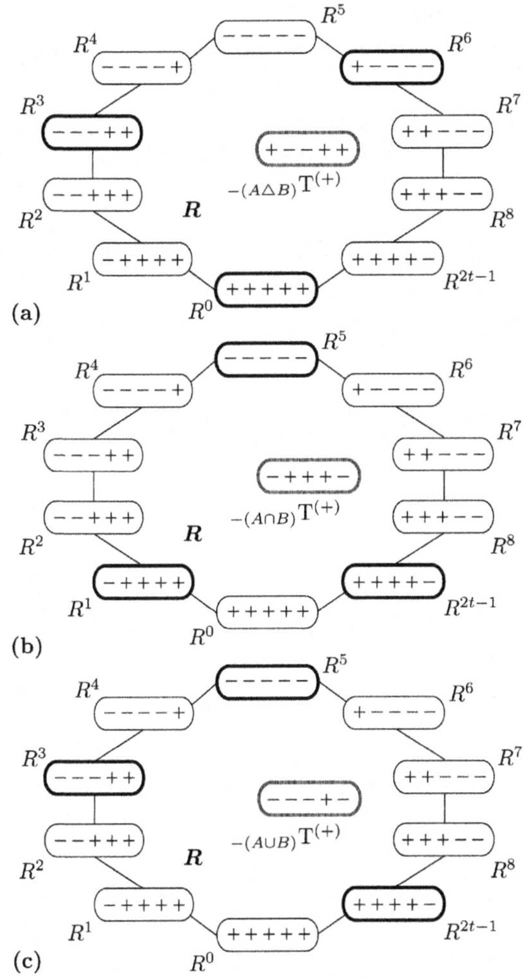

Figure 7.25 Three instances of the same distinguished symmetric cycle $\boldsymbol{R} := (R^0, R^1, \ldots, R^{2t-1}, R^0)$ in the hypercube graph $\boldsymbol{H}((t := 5), 2)$, defined by (7.1)(7.2).
Pick the subsets $A := \{1, 3, t\}$ and $B := \{1, 2, t\}$ of the set E_t.
Note that $\{1, t\} \cap A = \{1, t\} \cap B = \{1, t\}$.
(a) The vertex ${}_{-(A \triangle B)}\mathrm{T}^{(+)}$ of the graph $\boldsymbol{H}(t, 2)$, with the corresponding decomposition set $\boldsymbol{Q}({}_{-(A\triangle B)}\mathrm{T}^{(+)}, \boldsymbol{R}) = \{R^0, R^3, R^6\}$.
(b) The vertex ${}_{-(A\cap B)}\mathrm{T}^{(+)}$, with its decomposition set $\boldsymbol{Q}({}_{-(A\cap B)}\mathrm{T}^{(+)}, \boldsymbol{R}) = \{R^1, R^5, R^{2t-1}\}$.
(c) The vertex ${}_{-(A\cup B)}\mathrm{T}^{(+)}$, with its decomposition set $\boldsymbol{Q}({}_{-(A\cup B)}\mathrm{T}^{(+)}, \boldsymbol{R}) = \{R^3, R^5, R^{2t-1}\}$.

pairs $(A,\ B)$ such that

$$\{1,\ t\}\cap A=\{1,\ t\}\,,\quad\text{and}\quad |\{1,\ t\}\cap B|=0\,.$$

One such pair is

$$\begin{aligned}(\ A &:= \{1,\quad ,\ 3,\quad ,\ t\}\,,\\ B &:= \{\quad ,\ 2,\ 3,\quad ,\quad \}\ )\,,\end{aligned}$$

for which we have

$$\begin{aligned}A\triangle B &= \{1,\ 2,\quad ,\quad ,\ t\}\,,\\ A\cap B &= \{\quad ,\quad ,\ 3,\quad ,\quad \}\,,\\ A\cup B &= \{1,\ 2,\ 3,\quad ,\ t\}\,,\end{aligned}$$

and

$$\begin{aligned}\boldsymbol{x}(A\triangle B) &= (-1,\quad 0,\quad 1,\quad 0,\ -1)\,,\\ \boldsymbol{Q}({}_{-(A\triangle B)}\mathrm{T}^{(+)},\ \boldsymbol{R}) &= \{R^{2},\ R^{5},\ R^{2t-1}\}\,,\\ \boldsymbol{x}(A\cap B) &= (\ \ 1,\quad 0,\ -1,\quad 1,\quad 0)\,,\\ \boldsymbol{Q}({}_{-(A\cap B)}\mathrm{T}^{(+)},\ \boldsymbol{R}) &= \{R^{0},\ R^{3},\ R^{7}\}\,,\\ \boldsymbol{x}(A\cup B) &= (-1,\quad 0,\quad 0,\quad 1,\ -1)\,,\\ \boldsymbol{Q}({}_{-(A\cup B)}\mathrm{T}^{(+)},\ \boldsymbol{R}) &= \{R^{3},\ R^{5},\ R^{2t-1}\}\,;\end{aligned}$$

see Figure 7.26.

(vii) The family

$$\big\{(A,\ B)\in\mathbf{2}^{[t]}\times\mathbf{2}^{[t]}:\ j':=|A|=2\,,\ j'':=|B|=2\,,\\ j:=|A\cap B|=1\,,\ \ell^{\triangle}:=\mathfrak{q}(A\triangle B)=3\,,\\ \ell^{\cap}:=\mathfrak{q}(A\cap B)=1\,,\ \ell:=\mathfrak{q}(A\cup B)=3\big\}$$

consists of

$$\begin{aligned}&\mathfrak{T}\left(\alpha,\beta;\frac{\ell-1}{2},\frac{\ell^{\triangle}+1}{2},\frac{\ell^{\cap}+1}{2}\right)\cdot\mathrm{c}\left(\frac{\ell-1}{2};t-(j'+j''-j)\right)\\ &\quad\times\mathrm{c}\left(\frac{\ell^{\triangle}+1}{2};j'+j''-2j\right)\cdot\mathrm{c}\left(\frac{\ell^{\cap}+1}{2};j\right)\cdot\tbinom{j'+j''-2j-1}{j''-j}\\ &:=\mathfrak{T}\left(\alpha,\beta;\frac{3-1}{2},\frac{3+1}{2},\frac{1+1}{2}\right)\cdot\mathrm{c}\left(\frac{3-1}{2};5-(2+2-1)\right)\\ &\quad\times\mathrm{c}\left(\frac{3+1}{2};2+2-2\cdot1\right)\cdot\mathrm{c}\left(\frac{1+1}{2};j\right)\cdot\tbinom{2+2-2\cdot1-1}{2-1}\\ &=\underbrace{\mathfrak{T}(\alpha,\beta;1,2,1)}_{\substack{\text{certainly, }1!=1;\\ \text{but see (6.17)}}}\cdot\mathrm{c}(1;2)\cdot\mathrm{c}(2;2)\cdot\mathrm{c}(1;1)\cdot\tbinom{1}{1}\\ &=1\cdot1\cdot1\cdot1\cdot1=1\end{aligned}$$

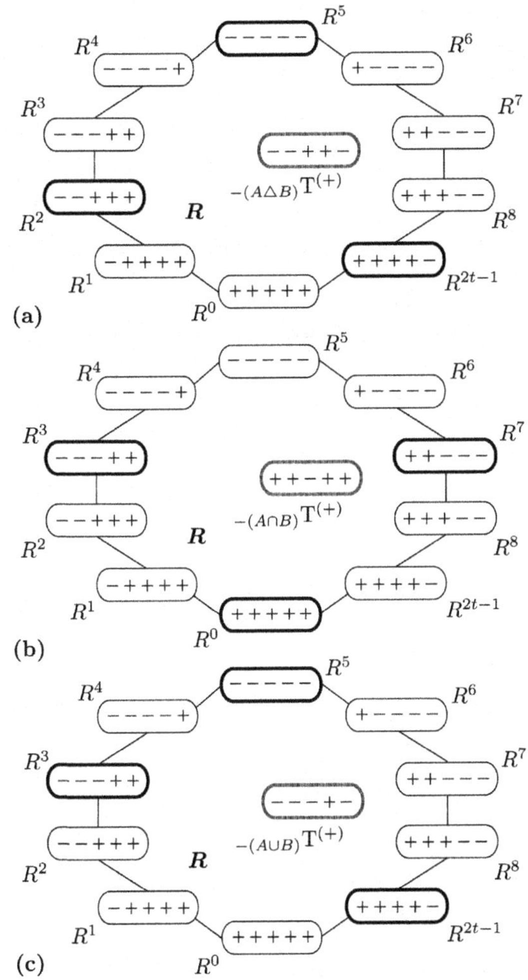

Figure 7.26 Three instances of the same distinguished symmetric cycle $\boldsymbol{R} := (R^0, R^1, \ldots, R^{2t-1}, R^0)$ in the hypercube graph $\boldsymbol{H}((t := 5), 2)$, defined by (7.1)(7.2).
Pick the subsets $A := \{1, 3, t\}$ and $B := \{2, 3\}$ of the set E_t.
Note that $\{1, t\} \cap A = \{1, t\}$, and $|\{1, t\} \cap B| = 0$.
(a) The vertex ${}_{-(A\triangle B)}\mathrm{T}^{(+)}$ of the graph $\boldsymbol{H}(t, 2)$, with the corresponding decomposition set $\boldsymbol{Q}({}_{-(A\triangle B)}\mathrm{T}^{(+)}, \boldsymbol{R}) = \{R^2, R^5, R^{2t-1}\}$.
(b) The vertex ${}_{-(A\cap B)}\mathrm{T}^{(+)}$, with its decomposition set $\boldsymbol{Q}({}_{-(A\cap B)}\mathrm{T}^{(+)}, \boldsymbol{R}) = \{R^0, R^3, R^7\}$.
(c) The vertex ${}_{-(A\cup B)}\mathrm{T}^{(+)}$, with its decomposition set $\boldsymbol{Q}({}_{-(A\cup B)}\mathrm{T}^{(+)}, \boldsymbol{R}) = \{R^3, R^5, R^{2t-1}\}$.

pair (A, B) such that

$$\{1, t\} \cap A = \{1, t\}\,, \quad \text{and} \quad \{1, t\} \cap B = \{t\}\,.$$

This pair is

$$\begin{aligned}(\ A &:= \{1, \ \ , \ \ , \ \ , t\}\,, \\ B &:= \{\ \ , \ \ , \ \ , 4, t\} \ )\,,\end{aligned}$$

for which we have

$$\begin{aligned} A \triangle B &= \{1, \ \ , \ \ , 4, \ \ \}\,, \\ A \cap B &= \{\ \ , \ \ , \ \ , \ \ , t\}\,, \\ A \cup B &= \{1, \ \ , \ \ , 4, t\}\,, \end{aligned}$$

and

$$\begin{aligned} \boldsymbol{x}(A \triangle B) &= (\ \ 0, \ \ 1, \ \ 0, -1, \ \ 1)\,, \\ \boldsymbol{Q}({}_{-(A \triangle B)}\mathrm{T}^{(+)}, \boldsymbol{R}) &= \{R^1, R^4, R^8\}\,, \\ \boldsymbol{x}(A \cap B) &= (\ \ 0, \ \ 0, \ \ 0, \ \ 0, -1)\,, \\ \boldsymbol{Q}({}_{-(A \cap B)}\mathrm{T}^{(+)}, \boldsymbol{R}) &= \{R^{2t-1}\}\,, \\ \boldsymbol{x}(A \cup B)) &= (-1, \ \ 1, \ \ 0, -1, \ \ 0)\,, \\ \boldsymbol{Q}({}_{-(A \cup B)}\mathrm{T}^{(+)}, \boldsymbol{R}) &= \{R^1, R^5, R^8\}\,; \end{aligned}$$

see Figure 7.27.

(viii) In the family

$$\begin{aligned}\big\{(A, B) \in \mathbf{2}^{[t]} \times \mathbf{2}^{[t]} :\ & j' := |A| = 2\,,\ j'' := |B| = 2\,, \\ & j := |A \cap B| = 1\,,\ \ell^{\triangle} := \mathfrak{q}(A \triangle B) = 3\,, \\ & \ell^{\cap} := \mathfrak{q}(A \cap B) = 3\,,\ \ell := \mathfrak{q}(A \cup B) = 3\big\}\end{aligned}$$

there are

$$\begin{aligned} & \mathfrak{T}\left(\theta, \theta; \frac{\ell+1}{2}, \frac{\ell^{\triangle}-1}{2}, \frac{\ell^{\cap}-1}{2}\right) \cdot \mathrm{c}\left(\frac{\ell+1}{2}; t - (j' + j'' - j)\right) \\ & \quad \times \mathrm{c}\left(\frac{\ell^{\triangle}-1}{2}; j' + j'' - 2j\right) \cdot \mathrm{c}\left(\frac{\ell^{\cap}-1}{2}; j\right) \cdot \tbinom{j'+j''-2j}{j'-j} \\ & := \mathfrak{T}\left(\theta, \theta; \frac{3+1}{2}, \frac{3-1}{2}, \frac{3-1}{2}\right) \cdot \mathrm{c}\left(\frac{3+1}{2}; 5 - (2+2-1)\right) \\ & \quad \times \mathrm{c}\left(\frac{3-1}{2}; 2+2-2\cdot 1\right) \cdot \mathrm{c}\left(\frac{3-1}{2}; 1\right) \cdot \tbinom{2+2-2\cdot 1}{2-1} \\ & = \underbrace{\mathfrak{T}(\theta, \theta; 2, 1, 1)}_{\substack{\text{certainly, } 2! = 2; \\ \text{but see (A.7)}}} \cdot \mathrm{c}(2; 2) \cdot \mathrm{c}(1; 2) \cdot \mathrm{c}(1; 1) \cdot \tbinom{2}{1} \\ & = 2 \cdot 1 \cdot 1 \cdot 1 \cdot 2 = 4 \end{aligned}$$

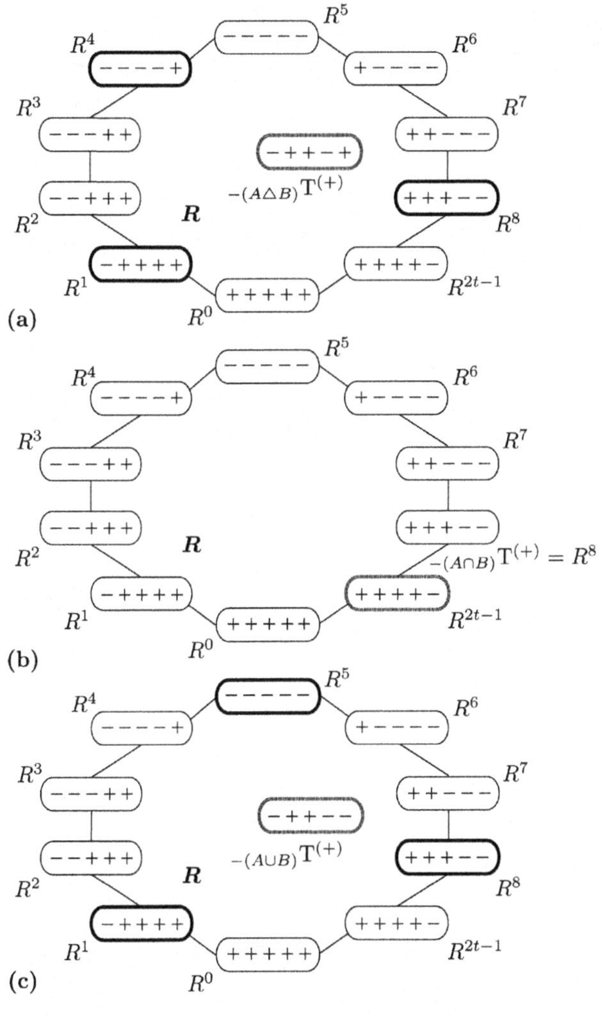

Figure 7.27 Three instances of the same distinguished symmetric cycle $\boldsymbol{R} := (R^0, R^1, \ldots, R^{2t-1}, R^0)$ in the hypercube graph $\boldsymbol{H}((t := 5), 2)$, defined by (7.1)(7.2).
Pick the subsets $A := \{1, t\}$ and $B := \{4, t\}$ of the set E_t.
Note that $\{1, t\} \cap A = \{1, t\}$, and $\{1, t\} \cap B = \{t\}$.
(a) The vertex $_{-(A\triangle B)}\mathrm{T}^{(+)}$ of the graph $\boldsymbol{H}(t, 2)$, with the corresponding decomposition set $\boldsymbol{Q}(_{-(A\triangle B)}\mathrm{T}^{(+)}, \boldsymbol{R}) = \{R^1, R^4, R^8\}$.
(b) The vertex $_{-(A\cap B)}\mathrm{T}^{(+)} = R^{2t-1} \in \mathrm{V}(\boldsymbol{R})$.
(c) The vertex $_{-(A\cup B)}\mathrm{T}^{(+)}$, with its decomposition set $\boldsymbol{Q}(_{-(A\cup B)}\mathrm{T}^{(+)}, \boldsymbol{R}) = \{R^1, R^5, R^8\}$.

pairs (A, B) such that

$$|\{1, t\} \cap A| = |\{1, t\} \cap B| = 0 .$$

One such pair is

$$\begin{aligned}(\ A &:= \{ \ \ , 2,\ 3, \ \ , \ \ \} , \\ B &:= \{ \ \ , 2, \ \ , 4, \ \ \} \ ) ,\end{aligned}$$

for which we have

$$\begin{aligned} A \triangle B &= \{ \ \ , \ \ , 3,\ 4, \ \ \} , \\ A \cap B &= \{ \ \ , 2, \ \ , \ \ , \ \ \} , \\ A \cup B &= \{ \ \ , 2,\ 3,\ 4, \ \ \} , \end{aligned}$$

and

$$\begin{aligned} \boldsymbol{x}(A \triangle B) &= (1, \quad 0, -1, \quad 0, \quad 1) , \\ \boldsymbol{Q}({}_{-(A \triangle B)}\mathrm{T}^{(+)}, \boldsymbol{R}) &= \{R^0, R^4, R^7\} , \\ \boldsymbol{x}(A \cap B) &= (1, -1, \quad 1, \quad 0, \quad 0) , \\ \boldsymbol{Q}({}_{-(A \cap B)}\mathrm{T}^{(+)}, \boldsymbol{R}) &= \{R^0, R^2, R^6\} , \\ \boldsymbol{x}(A \cup B) &= (1, -1, \quad 0, \quad 0, \quad 1) , \\ \boldsymbol{Q}({}_{-(A \cup B)}\mathrm{T}^{(+)}, \boldsymbol{R}) &= \{R^0, R^4, R^6\} ; \end{aligned}$$

see Figure 7.28.

(ix) In the family

$$\begin{aligned}\Big\{(A, B) \in \mathbf{2}^{[t]} \times \mathbf{2}^{[t]} :\ & j' := |A| = 2 , \ \ j'' := |B| = 2 , \\ & j := |A \cap B| = 1 , \ \ \ell^{\triangle} := \mathfrak{q}(A \triangle B) = 3 , \\ & \ell^{\cap} := \mathfrak{q}(A \cap B) = 3 , \ \ \ell := \mathfrak{q}(A \cup B) = 3\Big\}\end{aligned}$$

there are

$$\begin{aligned} & \mathfrak{T}\left(\theta, \alpha; \frac{\ell+1}{2}, \frac{\ell^{\triangle}+1}{2}, \frac{\ell^{\cap}-1}{2}\right) \cdot \mathrm{c}\left(\frac{\ell+1}{2}; t - (j' + j'' - j)\right) \\ & \quad \times \mathrm{c}\left(\frac{\ell^{\triangle}+1}{2}; j' + j'' - 2j\right) \cdot \mathrm{c}\left(\frac{\ell^{\cap}-1}{2}; j\right) \cdot \tbinom{j'+j''-2j-1}{j'-j} \\ & := \mathfrak{T}\left(\theta, \alpha; \frac{3+1}{2}, \frac{3+1}{2}, \frac{3-1}{2}\right) \cdot \mathrm{c}\left(\frac{3+1}{2}; 5 - (2+2-1)\right) \\ & \quad \times \mathrm{c}\left(\frac{3+1}{2}; 2+2-2\cdot 1\right) \cdot \mathrm{c}\left(\frac{3-1}{2}; 1\right) \cdot \tbinom{2+2-2\cdot 1-1}{2-1} \end{aligned}$$

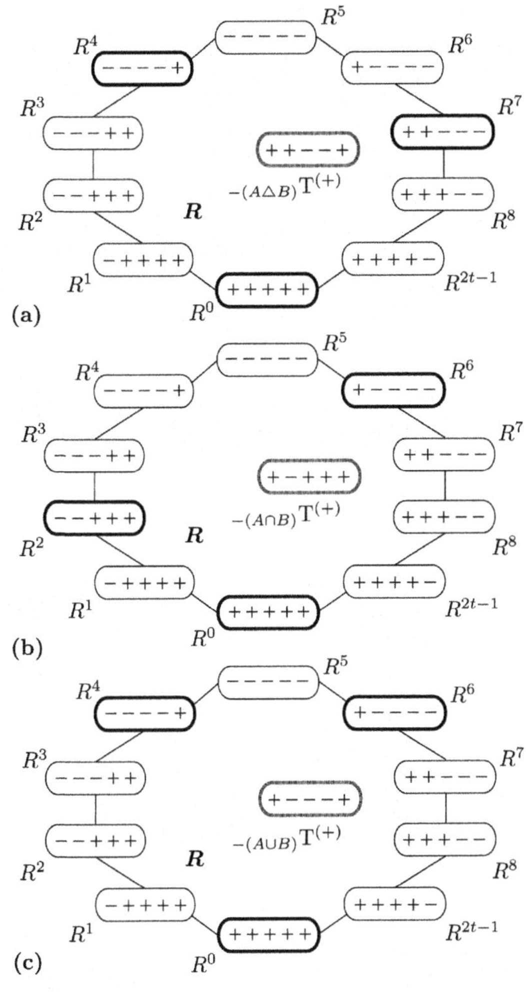

Figure 7.28 Three instances of the same distinguished symmetric cycle $\boldsymbol{R} := (R^0, R^1, \ldots, R^{2t-1}, R^0)$ in the hypercube graph $\boldsymbol{H}((t := 5), 2)$, defined by (7.1)(7.2).
Pick the subsets $A := \{2, 3\}$ and $B := \{2, 4\}$ of the set E_t.
Note that $|\{1, t\} \cap A| = |\{1, t\} \cap B| = 0$.
(a) The vertex $_{-(A\triangle B)}\mathrm{T}^{(+)}$ of the graph $\boldsymbol{H}(t, 2)$, with the corresponding decomposition set $\boldsymbol{Q}(_{-(A\triangle B)}\mathrm{T}^{(+)}, \boldsymbol{R}) = \{R^0, R^4, R^7\}$.
(b) The vertex $_{-(A\cap B)}\mathrm{T}^{(+)}$, with its decomposition set $\boldsymbol{Q}(_{-(A\cap B)}\mathrm{T}^{(+)}, \boldsymbol{R}) = \{R^0, R^2, R^6\}$.
(c) The vertex $_{-(A\cup B)}\mathrm{T}^{(+)}$, with its decomposition set $\boldsymbol{Q}(_{-(A\cup B)}\mathrm{T}^{(+)}, \boldsymbol{R}) = \{R^0, R^4, R^6\}$.

$$= \underbrace{\mathfrak{T}(\theta, \alpha; 2, 2, 1)}_{\substack{\text{the same as } \mathfrak{T}(\alpha, \theta; 2, 2, 1); \\ \text{we will apply (6.16)}}} \cdot \mathrm{c}(2; 2) \cdot \mathrm{c}(2; 2) \cdot \mathrm{c}(1; 1) \cdot \binom{1}{1}$$

$$= \left((2+2-1) \binom{2-1}{\frac{2+2-1-1}{2}} \binom{\frac{2+1+2-3}{2}}{2-1} \right) \cdot \binom{2-1}{2-1} \cdot \binom{2-1}{2-1} \cdot \binom{1-1}{1-1} \cdot \binom{1}{1}$$

$$= 3 \cdot 1 \cdot 1 \cdot 1 \cdot 1 = 3$$

pairs (A, B) such that

$$|\{1, t\} \cap A| = 0 \,, \quad \text{and} \quad \{1, t\} \cap B = \{t\} \,.$$

One such pair is

$$\begin{aligned} (\; A &:= \{ \;, 2, 3, \;, \; \} \,, \\ B &:= \{ \;, 2, \;, \;, t\} \;) \,, \end{aligned}$$

for which we have

$$\begin{aligned} A \triangle B &= \{ \;, \;, 3, \;, t\} \,, \\ A \cap B &= \{ \;, 2, \;, \;, \; \} \,, \\ A \cup B &= \{ \;, 2, 3, \;, t\} \,, \end{aligned}$$

and

$$\begin{aligned} \boldsymbol{x}(A \triangle B) &= (0, \quad 0, -1, \quad 1, -1) \,, \\ \boldsymbol{Q}({}_{-(A \triangle B)}\mathrm{T}^{(+)}, \boldsymbol{R}) &= \{R^3, R^7, R^{2t-1}\} \,, \\ \boldsymbol{x}(A \cap B) &= (1, -1, \quad 1, \quad 0, \quad 0) \,, \\ \boldsymbol{Q}({}_{-(A \cap B)}\mathrm{T}^{(+)}, \boldsymbol{R}) &= \{R^0, R^2, R^6\} \,, \\ \boldsymbol{x}(A \cup B) &= (0, -1, \quad 0, \quad 1, -1) \,, \\ \boldsymbol{Q}({}_{-(A \cup B)}\mathrm{T}^{(+)}, \boldsymbol{R}) &= \{R^3, R^6, R^{2t-1}\} \,; \end{aligned}$$

see Figure 7.29.

(x) The family

$$\begin{aligned} \big\{ (A, B) \in \mathbf{2}^{[t]} \times \mathbf{2}^{[t]} : \;\; & j' := |A| = 2 \,, \;\; j'' := |B| = 2 \,, \\ & j := |A \cap B| = 1 \,, \;\; \ell^{\triangle} := \mathfrak{q}(A \triangle B) = 3 \,, \\ & \ell^{\cap} := \mathfrak{q}(A \cap B) = 1 \,, \;\; \ell := \mathfrak{q}(A \cup B) = 3 \big\} \end{aligned}$$

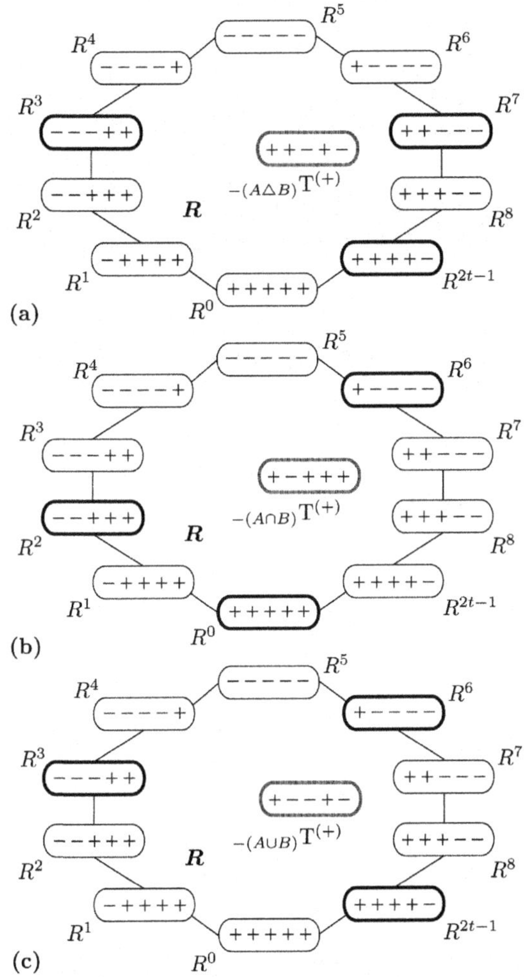

Figure 7.29 Three instances of the same distinguished symmetric cycle $\boldsymbol{R} := (R^0, R^1, \ldots, R^{2t-1}, R^0)$ in the hypercube graph $\boldsymbol{H}((t := 5), 2)$, defined by (7.1)(7.2).
Pick the subsets $A := \{2, 3\}$ and $B := \{2, t\}$ of the set E_t.
Note that $|\{1, t\} \cap A| = 0$, and $\{1, t\} \cap B = \{t\}$.
(a) The vertex ${}_{-(A\triangle B)}\mathrm{T}^{(+)}$ of the graph $\boldsymbol{H}(t, 2)$, with the corresponding decomposition set $\boldsymbol{Q}({}_{-(A\triangle B)}\mathrm{T}^{(+)}, \boldsymbol{R}) = \{R^3, R^7, R^{2t-1}\}$.
(b) The vertex ${}_{-(A\cap B)}\mathrm{T}^{(+)}$, with its decomposition set $\boldsymbol{Q}({}_{-(A\cap B)}\mathrm{T}^{(+)}, \boldsymbol{R}) = \{R^0, R^2, R^6\}$.
(c) The vertex ${}_{-(A\cup B)}\mathrm{T}^{(+)}$, with its decomposition set $\boldsymbol{Q}({}_{-(A\cup B)}\mathrm{T}^{(+)}, \boldsymbol{R}) = \{R^3, R^6, R^{2t-1}\}$.

consists of

$$\begin{aligned}
&\mathfrak{T}\left(\theta,\beta;\frac{\ell+1}{2},\frac{\ell^{\triangle}-1}{2},\frac{\ell^{\cap}+1}{2}\right)\cdot\mathrm{c}\left(\frac{\ell+1}{2};t-(j'+j''-j)\right)\\
&\qquad\times\mathrm{c}\left(\frac{\ell^{\triangle}-1}{2};j'+j''-2j\right)\cdot\mathrm{c}\left(\frac{\ell^{\cap}+1}{2};j\right)\cdot\tbinom{j'+j''-2j}{j'-j}\\
&:=\mathfrak{T}\left(\theta,\beta;\frac{3+1}{2},\frac{3-1}{2},\frac{1+1}{2}\right)\cdot\mathrm{c}\left(\frac{3+1}{2};5-(2+2-1)\right)\\
&\qquad\times\mathrm{c}\left(\frac{3-1}{2};2+2-2\cdot1\right)\cdot\mathrm{c}\left(\frac{1+1}{2};1\right)\cdot\tbinom{2+2-2\cdot1}{2-1}\\
&=\underbrace{\mathfrak{T}(\theta,\beta;2,1,1)}_{\substack{\text{certainly, } 1!=1;\\ \text{but see (A.8)}}}\cdot\mathrm{c}(2;2)\cdot\mathrm{c}(1;2)\cdot\mathrm{c}(1;1)\cdot\tbinom{2}{1}\\
&=1\cdot1\cdot1\cdot1\cdot2=2
\end{aligned}$$

pairs (A, B) such that

$$\{1,t\}\cap A=\{1,t\}\cap B=\{t\}\,.$$

One such pair is

$$\begin{aligned}
(\ A&:=\{\ ,\ \ ,3,\ \ ,t\}\,,\\
B&:=\{\ ,2,\ \ ,\ \ ,t\}\ )\,,
\end{aligned}$$

for which we have

$$\begin{aligned}
A\triangle B&=\{\ ,2,3,\ \ ,\ \}\,,\\
A\cap B&=\{\ ,\ \ ,\ \ ,\ \ ,t\}\,,\\
A\cup B&=\{\ ,2,3,\ \ ,t\}\,,
\end{aligned}$$

and

$$\begin{aligned}
\boldsymbol{x}(A\triangle B)&=(1,-1,\ \ 0,\ \ 1,\ \ 0)\,,\\
\boldsymbol{Q}({}_{-(A\triangle B)}\mathrm{T}^{(+)},\boldsymbol{R})&=\{R^0,R^3,R^6\}\,,\\
\boldsymbol{x}(A\cap B)&=(0,\ \ 0,\ \ 0,\ \ 0,-1)\,,\\
\boldsymbol{Q}({}_{-(A\cap B)}\mathrm{T}^{(+)},\boldsymbol{R})&=\{R^{2t-1}\}\,,\\
\boldsymbol{x}(A\cup B)&=(0,-1,\ \ 0,\ \ 1,-1)\,,\\
\boldsymbol{Q}({}_{-(A\cup B)}\mathrm{T}^{(+)},\boldsymbol{R})&=\{R^3,R^6,R^{2t-1}\}\,;
\end{aligned}$$

see Figure 7.30.

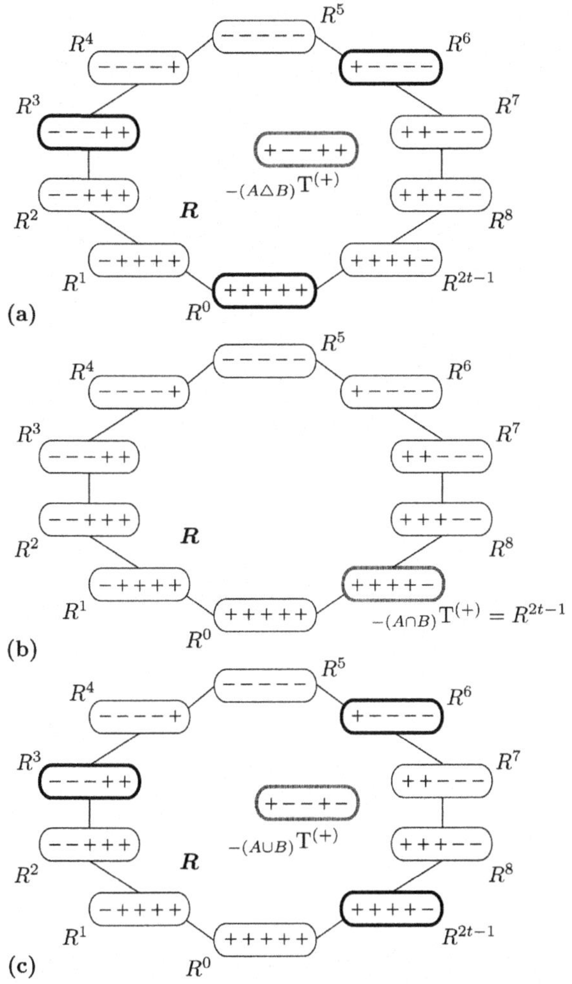

Figure 7.30 Three instances of the same distinguished symmetric cycle $\boldsymbol{R} := (R^0, R^1, \ldots, R^{2t-1}, R^0)$ in the hypercube graph $\boldsymbol{H}((t := 5), 2)$, defined by (7.1)(7.2).
Pick the subsets $A := \{3, t\}$ and $B := \{2, t\}$ of the set E_t.
Note that $\{1, t\} \cap A = \{1, t\} \cap B = \{t\}$.
(a) The vertex $_{-(A\triangle B)}\mathrm{T}^{(+)}$ of the graph $\boldsymbol{H}(t, 2)$, with the corresponding decomposition set $\boldsymbol{Q}(_{-(A\triangle B)}\mathrm{T}^{(+)}, \boldsymbol{R}) = \{R^0, R^3, R^6\}$.
(b) The vertex $_{-(A\cap B)}\mathrm{T}^{(+)} = R^{2t-1} \in \mathrm{V}(\boldsymbol{R})$.
(c) The vertex $_{-(A\cup B)}\mathrm{T}^{(+)}$, with its decomposition set $\boldsymbol{Q}(_{-(A\cup B)}\mathrm{T}^{(+)}, \boldsymbol{R}) = \{R^3, R^6, R^{2t-1}\}$.

Chapter 8

Vertices, Their Relabeled Opposites, and Distinguished Symmetric Cycles in Hypercube Graphs

In this chapter, we consider innocent-looking transformations of vertices of hypercube graphs that have interesting applications discussed later in Chapter 9.

Given a vertex $T \in \{1, -1\}^t$ of the hypercube graph $\boldsymbol{H}(t, 2)$, we define its '*relabeled opposite*' $\mathtt{ro}(T) \in \{1, -1\}^t$ by

$$\mathtt{ro}(T) := -T\,\overline{\mathbf{U}}(t)\,, \tag{8.1}$$

where $\overline{\mathbf{U}}(t)$ denotes[1] the square *backward identity matrix* (with the rows and columns indexed starting with 1) of order t whose $(i,\ j)$th entry is the Kronecker delta $\delta_{i+j,t+1}$.

For a vertex $\widetilde{T}$ of the discrete hypercube $\{0, 1\}^t$, the counterpart of the vertex $\mathtt{ro}(T)$ in the discrete hypercube $\{1, -1\}^t$ is the so called '*relabeled negation*' $\mathtt{rn}(\widetilde{T}) \in \{0, 1\}^t$ of the vertex $\widetilde{T}$, defined by

$$\mathtt{rn}(\widetilde{T}) := \mathrm{T}^{(+)} - \widetilde{T}\,\overline{\mathbf{U}}(t)\,. \tag{8.2}$$

[1]In Section 4.1 of the present monograph, and in Section 2.1 of Ref. [5], the similar notation $\mathbf{U}(t)$ was used to denote the backward identity matrix of order $(t+1)$ whose rows and columns were indexed starting with zero.

Symmetric Cycles
Andrey O. Matveev

ISBN 978-981-4968-81-2 (Hardcover), 978-1-003-43832-8 (eBook)
www.jennystanford.com

For example, suppose

$$\begin{aligned} T &:= (1,\ -1,\quad 1,\ -1,\ -1) \in \{1,\ -1\}^5\ , \\ \widetilde{T} &:= (0,\quad 1,\quad 0,\quad 1,\quad 1) \in \{0,\quad 1\}^5\ . \end{aligned}$$

Then we have

$$\mathrm{ro}(T) = (1,\quad 1,\ -1,\quad 1,\ -1)\ ,$$

and

$$\mathrm{rn}(\widetilde{T}) = (0,\quad 0,\quad 1,\quad 0,\quad 1)\ .$$

As earlier, we define a distinguished symmetric cycle $\boldsymbol{R} := (R^0, R^1, \ldots, R^{2t-1}, R^0)$ of the graph $\boldsymbol{H}(t, 2)$ by

$$\begin{aligned} R^0 &:= \mathrm{T}^{(+)}\ , \\ R^s &:= {}_{-[s]}R^0\ , \quad 1 \leq s \leq t-1\ , \end{aligned} \tag{8.3}$$

and

$$R^{t+k} := -R^k\ , \quad 0 \leq k \leq t-1\ . \tag{8.4}$$

For any vertex $T \in \{1, -1\}^t$ of the graph $\boldsymbol{H}(t, 2)$, there exists a *unique inclusion-minimal* subset

$$\boldsymbol{Q}(T, \boldsymbol{R}) \subset \mathrm{V}(\boldsymbol{R}) := \{R^0, R^1, \ldots, R^{2t-1}\} \tag{8.5}$$

of the vertex set $\mathrm{V}(\boldsymbol{R})$ of the cycle $\boldsymbol{R}$, such that

$$T = \sum_{Q \in \boldsymbol{Q}(T, \boldsymbol{R})} Q\ ; \tag{8.6}$$

see Subsection 2.1.4. This subset $\boldsymbol{Q}(T, \boldsymbol{R}) \subset \mathbb{R}^t$ is *linearly independent*, and it contains an *odd* number $\mathfrak{q}(T) := \mathfrak{q}(T, \boldsymbol{R}) := |\boldsymbol{Q}(T, \boldsymbol{R})|$ of vertices.

Now we proceed to compare the decomposition sets $\boldsymbol{Q}(T, \boldsymbol{R})$ and $\boldsymbol{Q}(\mathrm{ro}(T), \boldsymbol{R})$ for vertices T and $\mathrm{ro}(T)$ of the graph $\boldsymbol{H}(t, 2)$, with respect to the symmetric cycle $\boldsymbol{R}$ defined by (8.3)(8.4).

8.1 Vertices, Their Relabeled Opposites, and Decompositions

We consider vertices T of the discrete hypercube $\{1,-1\}^t$, their relabeled opposites $\mathtt{ro}(T)$ defined by (8.1), and we discuss basic properties of the decompositions $\boldsymbol{Q}(T,\boldsymbol{R})$ and $\boldsymbol{Q}(\mathtt{ro}(T),\boldsymbol{R})$ of vertices T and $\mathtt{ro}(T)$ with respect to the distinguished symmetric cycle $\boldsymbol{R}$ in the graph $\boldsymbol{H}(t,2)$, defined by (8.3)(8.4).
- Definitions (8.1) and (8.2) determine the maps

$$\{1,-1\}^t \to \{1,-1\}^t: \quad T \mapsto \mathtt{ro}(T) := \quad -T\,\overline{\mathbf{U}}(t)\,, \tag{8.7}$$

$$\{0,1\}^t \to \{0,1\}^t: \quad \widetilde{T} \mapsto \mathtt{rn}(\widetilde{T}) := \mathrm{T}^{(+)} - \widetilde{T}\,\overline{\mathbf{U}}(t)\,, \tag{8.8}$$

and since we deal with the standard one-to-one correspondences between the vertex sets of the discrete hypercubes $\{1,-1\}^t$ and $\{0,1\}^t$, established by means of the maps (2.20) and (2.19), we mention the mappings

$$\{1,-1\}^t \ni \mathtt{ro}(T) \overset{(2.20)}{\mapsto} \mathtt{rn}(\widetilde{T}) = \tfrac{1}{2}(\mathrm{T}^{(+)} + T\,\overline{\mathbf{U}}(t)) \in \{0,1\}^t\,,$$

and

$$\{0,1\}^t \ni \mathtt{rn}(\widetilde{T}) \overset{(2.19)}{\mapsto} \mathtt{ro}(T) = -\mathrm{T}^{(+)} + 2\widetilde{T}\,\overline{\mathbf{U}}(t) \in \{1,-1\}^t\,. \tag{8.9}$$

- Certainly the maps (8.7) and (8.8) are both *involutions*:

$$\{1,-1\}^t \ni T = \mathtt{ro}(\mathtt{ro}(T))\,, \quad \text{and} \quad \{0,1\}^t \ni \widetilde{T} = \mathtt{rn}(\mathtt{rn}(\widetilde{T}))\,.$$

- For a vertex $\widetilde{T}$ of the discrete hypercube $\{0,1\}^t$ we let $\mathtt{hwt}(\widetilde{T})$ denote, as earlier, its *Hamming weight*: $\mathtt{hwt}(\widetilde{T}) := |\operatorname{supp}(\widetilde{T})|$.

Note that we have

$$\{1,-1\}^t \ni T = \mathtt{ro}(T) \iff -T = T\,\overline{\mathbf{U}}(t)\,;$$

$$T = \mathtt{ro}(T) \implies |\mathfrak{n}(T)| = \tfrac{t}{2}\,.$$

We also have

$$\{0,1\}^t \ni \widetilde{T} = \mathtt{rn}(\widetilde{T}) \iff \mathrm{T}^{(+)} - \widetilde{T} = \widetilde{T}\,\overline{\mathbf{U}}(t)\,;$$

$$\widetilde{T} = \mathtt{rn}(\widetilde{T}) \implies \mathtt{hwt}(\widetilde{T}) = \tfrac{t}{2}\,.$$

Thus, if t is *odd*, then we always have $\mathtt{ro}(T) \neq T$, and $\mathtt{rn}(\widetilde{T}) \neq \widetilde{T}$.

- For vertices $\widetilde{T} \in \{0,1\}^t$ and $T := {}_{-\operatorname{supp}(\widetilde{T})}\mathrm{T}^{(+)} \in \{1,-1\}^t$, we have

$$\langle T, \mathtt{ro}(T)\rangle = -T\,\overline{\mathbf{U}}(t)T^\top = -\sum_{e\in[t]} T(e)T(t-e+1)$$

$$= \begin{cases} -1 - 2\sum_{e\in[(t-1)/2]} T(e)T(t-e+1)\,, & \text{if } t \text{ is odd,}\\ \quad - 2\sum_{e\in[t/2]} T(e)T(t-e+1)\,, & \text{if } t \text{ is even;} \end{cases}$$

$$\langle \widetilde{T}, \mathtt{rn}(\widetilde{T})\rangle := \langle \widetilde{T}, \mathrm{T}^{(+)} - \widetilde{T}\,\overline{\mathbf{U}}(t)\rangle$$

$$= \mathtt{hwt}(\widetilde{T}) - \sum_{e\in[t]} \widetilde{T}(e)\widetilde{T}(t-e+1)$$

$$= \mathtt{hwt}(\widetilde{T}) - \begin{cases} \widetilde{T}((t+1)/2) + 2\sum_{e\in[(t-1)/2]} \widetilde{T}(e)\widetilde{T}(t-e+1)\,, \\ \qquad\qquad \text{if } t \text{ is odd,}\\ 2\sum_{e\in[t/2]} \widetilde{T}(e)\widetilde{T}(t-e+1)\,, \\ \qquad\qquad \text{if } t \text{ is even.} \end{cases}$$

- Given two *tuples* $X, Y \in \{-1,0,1\}^t$, again we let $d(X,Y) := |\{e \in E_t\colon X(e) \neq Y(e)\}|$ denote the *Hamming distance* between them.

Since the equal distances $d(T,\mathtt{ro}(T)) = d(\widetilde{T},\mathtt{rn}(\widetilde{T}))$ can be calculated with the help of the formulas (see (2.19) and (8.9))

$$d(T,\mathtt{ro}(T)) = \tfrac{1}{2}\big(t - \langle T, \mathtt{ro}(T)\rangle\big)\,,$$
$$d(\widetilde{T},\mathtt{rn}(\widetilde{T})) = \tfrac{1}{2}\big(t - \langle \mathrm{T}^{(+)} - 2\widetilde{T}, -\mathrm{T}^{(+)} + 2\widetilde{T}\,\overline{\mathbf{U}}(t)\rangle\big)\,,$$

we see that

$$d(T,\mathtt{ro}(T)) = \tfrac{1}{2}\big(t + T\,\overline{\mathbf{U}}(t)T^\top\big) = \tfrac{1}{2}\Big(t + \sum_{e\in[t]} T(e)T(t-e+1)\Big)$$

$$= \frac{t}{2} + \begin{cases} \frac{1}{2} + \sum_{e\in[(t-1)/2]} T(e)T(t-e+1)\,, & \text{if } t \text{ is odd,}\\ \sum_{e\in[t/2]} T(e)T(t-e+1)\,, & \text{if } t \text{ is even,} \end{cases}$$

and

$$d(\widetilde{T},\mathtt{rn}(\widetilde{T})) = t - 2\cdot\mathtt{hwt}(\widetilde{T}) + 2\sum_{e\in[t]} \widetilde{T}(e)\widetilde{T}(t-e+1)$$

$$= t - 2\cdot\mathtt{hwt}(\widetilde{T})$$

$$+2\cdot\begin{cases} \widetilde{T}((t+1)/2) + 2\sum_{e\in[(t-1)/2]} \widetilde{T}(e)\widetilde{T}(t-e+1)\,, & \text{if } t \text{ is odd,}\\ 2\sum_{e\in[t/2]} \widetilde{T}(e)\widetilde{T}(t-e+1)\,, & \text{if } t \text{ is even.} \end{cases}$$

- Suppose that $4|t$ (i.e., t is divisible by 4). Note that

$$\langle T, \mathtt{ro}(T)\rangle = 0 \iff \sum_{e\in[t/2]} T(e)T(t-e+1) = 0 \, ;$$

$$\langle T, \mathtt{ro}(T)\rangle = 0 \iff \sum_{e\in[t/2]} \widetilde{T}(e)\widetilde{T}(t-e+1) = \frac{4\cdot \mathtt{hwt}(\widetilde{T}) - t}{8} \, .$$

- Considering the restriction of the map (8.7) to the vertex sequence $\vec{\mathrm{V}}(\boldsymbol{R}) := (R^0, R^1, \ldots, R^{2t-1})$ of the symmetric cycle $\boldsymbol{R}$ in the graph $\boldsymbol{H}(t, 2)$, defined by (8.3)(8.4), we have the mappings

$$R^i \overset{(8.7)}{\mapsto} \mathtt{ro}(R^i) = R^{(3t-i) \bmod 2t} = \begin{cases} R^{t-i}, & \text{if } 0 \le i \le t \, , \\ R^{3t-i}, & \text{if } t+1 \le i \le 2t-1 \, . \end{cases}$$

If t is *even*, then the following implication holds:

$$R^i \in \vec{\mathrm{V}}(\boldsymbol{R}),\ \mathtt{ro}(R^i) = R^i \implies i \in \{\tfrac{t}{2}, \tfrac{3t}{2}\} \, .$$

Remark 8.1. Let $\boldsymbol{R}$ be the distinguished symmetric cycle in the hypercube graph $\boldsymbol{H}(t, 2)$, defined by (8.3)(8.4). Given a vertex $T \in \{1, -1\}^t$ of $\boldsymbol{H}(t, 2)$, suppose that

$$(R^0, R^1, \ldots, R^{2t-1}) =: \vec{\mathrm{V}}(\boldsymbol{R}) \supset \boldsymbol{Q}(T, \boldsymbol{R}) = (R^{i_0}, R^{i_1}, \ldots, R^{i_{\mathfrak{q}(T)-1}}) \, ,$$

for some indices $i_0 < i_1 < \cdots < \mathfrak{q}(T) - 1$.

(i) We have

$$\boldsymbol{Q}(\mathtt{ro}(T), \boldsymbol{R}) = (R^{(3t-i_0) \bmod 2t}, \\ R^{(3t-i_1) \bmod 2t}, \ldots, R^{(3t-i_{\mathfrak{q}(T)-1}) \bmod 2t}) \, ,$$

or, in other words,

$$\boldsymbol{Q}(\mathtt{ro}(T), \boldsymbol{R}) = \{R^{(3t-i) \bmod 2t} : R^i \in \boldsymbol{Q}(T, \boldsymbol{R})\} \, .$$

(ii) If t is *even*, then

$$\mathtt{ro}(T) = T \iff \Big(Q \in \boldsymbol{Q}(T, \boldsymbol{R}) \implies \mathtt{ro}(Q) \in \boldsymbol{Q}(T, \boldsymbol{R})\Big) \, .$$

Note that the following implication holds:

$$\mathtt{ro}(T) = T \implies |\{R^{t/2}, R^{3t/2}\} \cap \boldsymbol{Q}(T, \boldsymbol{R})| = 1 \, .$$

– We will now give an explicit description of the decomposition sets $\boldsymbol{Q}(T, \boldsymbol{R})$ and $\boldsymbol{Q}(\mathrm{ro}(T), \boldsymbol{R})$ via $\boldsymbol{x}$-vectors.

Let $\overline{\mathbf{T}}(t)$ denote[2] the square *forward shift matrix* of order t whose (i, j)th entry is $\delta_{j-i,1}$.

Proposition 8.2 (Proposition 5.9, extended). *Let $\boldsymbol{R}$ be the distinguished symmetric cycle in the hypercube graph $\boldsymbol{H}(t, 2)$, defined by* (8.3)(8.4).

Let A be a nonempty subset of the ground set E_t, regarded as a disjoint union

$$A = [i_1, j_1] \dot{\cup} [i_2, j_2] \dot{\cup} \cdots \dot{\cup} [i_{\varrho-1}, j_{\varrho-1}] \dot{\cup} [i_\varrho, j_\varrho]$$

of inclusion-maximal intervals of the set E_t, such that

$$j_1 + 2 \leq i_2, \quad j_2 + 2 \leq i_3, \quad \ldots, \quad j_{\varrho-2} + 2 \leq i_{\varrho-1}, \quad j_{\varrho-1} + 2 \leq i_\varrho \,,$$

for some $\varrho := \varrho(A)$.

(i) (a) *If* $\{1, t\} \cap A = \{\underset{\substack{\uparrow \\ i_1}}{1}\}$, *then we have*

$$|\boldsymbol{Q}({}_{-A}\mathrm{T}^{(+)}, \boldsymbol{R})| = 2\varrho - 1 \,,$$
$$\boldsymbol{x}({}_{-A}\mathrm{T}^{(+)}, \boldsymbol{R}) = \sum_{1 \leq k \leq \varrho} \boldsymbol{\sigma}(j_k + 1) - \sum_{2 \leq \ell \leq \varrho} \boldsymbol{\sigma}(i_\ell) \,.$$

(b) *Since*

$$\{t - e + 1 \colon e \in E_t - A\} = [\underset{\substack{\uparrow \\ i_1}}{1}, t - j_\varrho] \dot{\cup} [t - i_\varrho + 2, t - j_{\varrho-1}]$$
$$\dot{\cup} \cdots \dot{\cup} [t - i_3 + 2, t - j_2] \dot{\cup} [t - i_2 + 2, t - j_1] \,,$$

and $\{1, t\} \cap \{t - e + 1 \colon e \in E_t - A\} = \{1\}$, *we see that*

$$|\boldsymbol{Q}(\mathrm{ro}({}_{-A}\mathrm{T}^{(+)}), \boldsymbol{R})| = 2\varrho - 1 \,,$$
$$\boldsymbol{x}(\mathrm{ro}({}_{-A}\mathrm{T}^{(+)}), \boldsymbol{R}) = \sum_{1 \leq k \leq \varrho} \boldsymbol{\sigma}(t - j_k + 1) - \sum_{2 \leq \ell \leq \varrho} \boldsymbol{\sigma}(t - i_\ell + 2) \,.$$

(c) *Note that*

$$\boldsymbol{x}(\mathrm{ro}({}_{-A}\mathrm{T}^{(+)}), \boldsymbol{R}) = \boldsymbol{x}({}_{-A}\mathrm{T}^{(+)}, \boldsymbol{R}) \cdot \overline{\mathbf{U}}(t) \cdot \overline{\mathbf{T}}(t) \,.$$

[2]In Section 4.1 of the present monograph, and in Section 2.1 of Ref. [5], the similar notation $\mathbf{T}(t)$ was used to denote the forward shift matrix of order $(t + 1)$ whose rows and columns were indexed starting with zero.

(ii) (a) *If* $\{1, t\} \cap A = \{\underset{i_1}{\underset{\uparrow}{1}}, \underset{j_\varrho}{\underset{\uparrow}{t}}\}$, *then*

$$|\boldsymbol{Q}({}_{-A}\mathrm{T}^{(+)}, \boldsymbol{R})| = 2\varrho - 1\,,$$

$$\boldsymbol{x}({}_{-A}\mathrm{T}^{(+)}, \boldsymbol{R}) = -\boldsymbol{\sigma}(1) + \sum_{1 \le k \le \varrho - 1} \boldsymbol{\sigma}(j_k + 1) - \sum_{2 \le \ell \le \varrho} \boldsymbol{\sigma}(i_\ell)\,.$$

(b) *Since*

$$\{t - e + 1 \colon e \in E_t - A\} = [t - i_\varrho + 2, t - j_{\varrho-1}] \,\dot\cup\, [t - i_{\varrho-1} + 2, t - j_{\varrho-2}] \,\dot\cup \cdots \dot\cup\, [t - i_3 + 2, t - j_2] \,\dot\cup\, [t - i_2 + 2, t - j_1]\,,$$

and $|\{1, t\} \cap \{t - e + 1 \colon e \in E_t - A\}| = 0$, *we have*

$$|\boldsymbol{Q}(\mathtt{ro}({}_{-A}\mathrm{T}^{(+)}), \boldsymbol{R})| = 2\varrho - 1\,,$$

$$\boldsymbol{x}(\mathtt{ro}({}_{-A}\mathrm{T}^{(+)}), \boldsymbol{R}) = \boldsymbol{\sigma}(1) + \sum_{1 \le k \le \varrho - 1} \boldsymbol{\sigma}(t - j_k + 1) - \sum_{2 \le \ell \le \varrho} \boldsymbol{\sigma}(t - i_\ell + 2)\,.$$

(c) *Note that*

$$\boldsymbol{x}(\mathtt{ro}({}_{-A}\mathrm{T}^{(+)}), \boldsymbol{R}) = \boldsymbol{\sigma}(1) + \boldsymbol{x}({}_{-A}\mathrm{T}^{(+)}, \boldsymbol{R}) \cdot \overline{\mathbf{U}}(t) \cdot \overline{\mathbf{T}}(t)\,.$$

(iii) (a) *If* $|\{1, t\} \cap A| = 0$, *then*

$$|\boldsymbol{Q}({}_{-A}\mathrm{T}^{(+)}, \boldsymbol{R})| = 2\varrho + 1\,,$$

$$\boldsymbol{x}({}_{-A}\mathrm{T}^{(+)}, \boldsymbol{R}) = \boldsymbol{\sigma}(1) + \sum_{1 \le k \le \varrho} \boldsymbol{\sigma}(j_k + 1) - \sum_{1 \le \ell \le \varrho} \boldsymbol{\sigma}(i_\ell)\,.$$

(b) *Since*

$$\{t - e + 1 \colon e \in E_t - A\} = [1, t - j_\varrho] \,\dot\cup\, [t - i_\varrho + 2, t - j_{\varrho-1}] \,\dot\cup \cdots \dot\cup\, [t - i_2 + 2, t - j_1] \,\dot\cup\, [t - i_1 + 2, t]\,,$$

and $\{1, t\} \cap \{t - e + 1 \colon e \in E_t - A\} = \{1, t\}$, *we have*

$$|\boldsymbol{Q}(\mathtt{ro}({}_{-A}\mathrm{T}^{(+)}), \boldsymbol{R})| = 2\varrho + 1\,,$$

$$\boldsymbol{x}(\mathtt{ro}({}_{-A}\mathrm{T}^{(+)}), \boldsymbol{R}) = -\boldsymbol{\sigma}(1) + \sum_{1 \le k \le \varrho} \boldsymbol{\sigma}(t - j_k + 1) - \sum_{1 \le \ell \le \varrho} \boldsymbol{\sigma}(t - i_\ell + 2)\,.$$

(c) *Note that*

$$\boldsymbol{x}(-({}_{-A}\mathrm{T}^{(+)}) \cdot \overline{\mathbf{U}}(t), \boldsymbol{R}) = -\boldsymbol{\sigma}(1) + \boldsymbol{x}({}_{-A}\mathrm{T}^{(+)}, \boldsymbol{R}) \cdot \overline{\mathbf{U}}(t) \cdot \overline{\mathbf{T}}(t)\,.$$

(iv) (a) *If* $\{1, t\} \cap A = \{\underset{j_\varrho}{\underset{\uparrow}{t}}\}$, *then*

$$|\boldsymbol{Q}({}_{-A}\mathrm{T}^{(+)}, \boldsymbol{R})| = 2\varrho - 1\,,$$
$$\boldsymbol{x}({}_{-A}\mathrm{T}^{(+)}, \boldsymbol{R}) = \sum_{1 \le k \le \varrho - 1} \boldsymbol{\sigma}(j_k + 1) - \sum_{1 \le \ell \le \varrho} \boldsymbol{\sigma}(i_\ell)\,.$$

(b) *Since*

$$\{t - e + 1 \colon e \in E_t - A\} = [t - i_\varrho + 2, t - j_{\varrho-1}] \dot\cup [t - i_{\varrho-1} + 2, t - j_{\varrho-2}] \dot\cup \cdots \dot\cup [t - i_2 + 2, t - j_1] \dot\cup [t - i_1 + 2, \underset{j_\varrho}{\underset{\uparrow}{t}}]\,,$$

and $\{1, t\} \cap \{t - e + 1 \colon e \in E_t - A\} = \{t\}$, *we see that*

$$|\boldsymbol{Q}(\mathtt{ro}({}_{-A}\mathrm{T}^{(+)}), \boldsymbol{R})| = 2\varrho - 1\,,$$
$$\boldsymbol{x}(\mathtt{ro}({}_{-A}\mathrm{T}^{(+)}), \boldsymbol{R}) = \sum_{1 \le k \le \varrho - 1} \boldsymbol{\sigma}(t - j_k + 1) - \sum_{1 \le \ell \le \varrho} \boldsymbol{\sigma}(t - i_\ell + 2)\,.$$

(c) *Note that*

$$\boldsymbol{x}(\mathtt{ro}({}_{-A}\mathrm{T}^{(+)}), \boldsymbol{R}) = \boldsymbol{x}({}_{-A}\mathrm{T}^{(+)}, \boldsymbol{R}) \cdot \overline{\mathbf{U}}(t) \cdot \overline{\mathbf{T}}(t)\,.$$

Corollary 8.3. *Let* $\boldsymbol{R}$ *be the distinguished symmetric cycle in the hypercube graph* $\boldsymbol{H}(t, 2)$, *defined by* (8.3)(8.4). *For any vertex* $T \in \{1, -1\}^t$ *of the graph* $\boldsymbol{H}(t, 2)$ *we have*

$$\mathfrak{q}(\mathtt{ro}(T)) := |\operatorname{supp}(\boldsymbol{x}(\mathtt{ro}(T), \boldsymbol{R}))| = |\operatorname{supp}(\boldsymbol{x}(T, \boldsymbol{R}))| =: \mathfrak{q}(T)\,.$$

(i) *If* $|\mathfrak{n}(T) \cap \{1, t\}| = 1$, *then*

$$\boldsymbol{x}(\mathtt{ro}(T)) = \boldsymbol{x}(T) \cdot \overline{\mathbf{U}}(t) \cdot \overline{\mathbf{T}}(t)\,.$$

(ii) *If* $|\mathfrak{n}(T) \cap \{1, t\}| = 2$, *then*

$$\boldsymbol{x}(\mathtt{ro}(T)) = \boldsymbol{\sigma}(1) + \boldsymbol{x}(T) \cdot \overline{\mathbf{U}}(t) \cdot \overline{\mathbf{T}}(t)\,.$$

(iii) *If* $|\mathfrak{n}(T) \cap \{1, t\}| = 0$, *then*

$$\boldsymbol{x}(\mathtt{ro}(T)) = -\boldsymbol{\sigma}(1) + \boldsymbol{x}(T) \cdot \overline{\mathbf{U}}(t) \cdot \overline{\mathbf{T}}(t)\,.$$

Example 8.4. Suppose $t := 5$. Let $\boldsymbol{R} := (R^0, R^1, \ldots, R^{2t-1}, R^0)$ be the distinguished symmetric cycle in the hypercube graph $\boldsymbol{H}(t, 2)$, defined by (8.3)(8.4).

(i)(a) For the vertex $T := (-1, -1, 1, -1, 1) = {}_{-\{1,2,4\}}\mathrm{T}^{(+)} \in \{1, -1\}^t$ of the graph $\boldsymbol{H}(t, 2)$ we have $\{1, t\} \cap \mathfrak{n}(T) := \{1, t\} \cap \{1, 2, 4\} = \{1\}$, and

$$\boldsymbol{x}(T) = (0,\quad 0,\quad 1, -1,\quad 1)\,, \quad \boldsymbol{Q}(T, \boldsymbol{R}) = \{R^2, R^4, R^8\}\,.$$

For the relabeled opposite $\mathtt{ro}(T) = (-1, 1, -1, 1, 1)$ of the vertex T, by Corollary 8.3(i) we have

$$\begin{aligned}
\boldsymbol{x}(\mathtt{ro}(T)) &= \boldsymbol{x}(T) \cdot \overline{\mathbf{U}}(t) \cdot \overline{\mathbf{T}}(t) \\
&:= \begin{pmatrix} 0 & 0 & 1 & -1 & 1 \end{pmatrix} \cdot \begin{pmatrix} 0&0&0&0&1\\ 0&0&0&1&0\\ 0&0&1&0&0\\ 0&1&0&0&0\\ 1&0&0&0&0 \end{pmatrix} \cdot \begin{pmatrix} 0&1&0&0&0\\ 0&0&1&0&0\\ 0&0&0&1&0\\ 0&0&0&0&1\\ 0&0&0&0&0 \end{pmatrix} \\
&= \begin{pmatrix} 1 & -1 & 1 & 0 & 0 \end{pmatrix} \cdot \begin{pmatrix} 0&1&0&0&0\\ 0&0&1&0&0\\ 0&0&0&1&0\\ 0&0&0&0&1\\ 0&0&0&0&0 \end{pmatrix} \\
&= (0,\quad 1, -1,\quad 1,\quad 0)\,, \quad \boldsymbol{Q}(\mathtt{ro}(T), \boldsymbol{R}) = \{R^1, R^3, R^7\}\,;
\end{aligned}$$

see Figure 8.1.

(b) Pick the vertex $T := (1, -1, -1, 1, -1) = {}_{-\{2,3,t\}}\mathrm{T}^{(+)} \in \{1, -1\}^t$ of the graph $\boldsymbol{H}(t, 2)$. We have $\{1, t\} \cap \mathfrak{n}(T) := \{1, t\} \cap \{2, 3, t\} = \{t\}$, and

$$\boldsymbol{x}(T) = (0, -1,\quad 0,\quad 1, -1)\,, \quad \boldsymbol{Q}(T, \boldsymbol{R}) = \{R^3, R^6, R^{2t-1}\}\,.$$

For the relabeled opposite $\mathtt{ro}(T) = (1, -1, 1, 1, -1)$ of the vertex T, by Corollary 8.3(i) we have

$$\begin{aligned}
\boldsymbol{x}(\mathtt{ro}(T)) &= \boldsymbol{x}(T) \cdot \overline{\mathbf{U}}(t) \cdot \overline{\mathbf{T}}(t) \\
&= (0, -1,\quad 1,\quad 0, -1)\,, \quad \boldsymbol{Q}(\mathtt{ro}(T), \boldsymbol{R}) = \{R^2, R^6, R^{2t-1}\}\,;
\end{aligned}$$

see Figure 8.2.

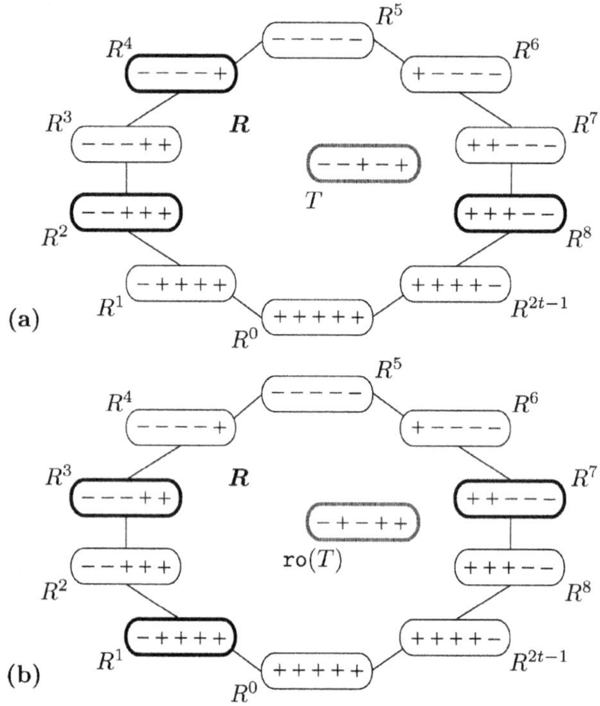

Figure 8.1 Two instances of the same distinguished symmetric cycle $\boldsymbol{R} := (R^0, R^1, \ldots, R^{2t-1}, R^0)$ in the hypercube graph $\boldsymbol{H}((t := 5), 2)$, defined by (8.3)(8.4), and some vertices T and $\mathtt{ro}(T)$.
(a) The vertex $T := {}_{-\{1,2,4\}}\mathrm{T}^{(+)}$, with the corresponding decomposition set $\boldsymbol{Q}(T, \boldsymbol{R}) = \{R^2, R^4, R^8\}$.
(b) The relabeled opposite $\mathtt{ro}(T) = {}_{-\{1,3\}}\mathrm{T}^{(+)}$, with its decomposition set $\boldsymbol{Q}(\mathtt{ro}(T), \boldsymbol{R}) = \{R^1, R^3, R^7\}$.

(ii) Pick the vertex $T := (-1, -1, 1, 1, -1) = {}_{-\{1,2,t\}}\mathrm{T}^{(+)} \in \{1, -1\}^t$ of the graph $\boldsymbol{H}(t, 2)$. We have $\{1, t\} \cap \mathbf{n}(T) := \{1, t\} \cap \{1, 2, t\} = \{1, t\}$, and

$$\boldsymbol{x}(T) = (-1, \quad 0, \quad 1, \quad 0, -1)\,, \quad \boldsymbol{Q}(T, \boldsymbol{R}) = \{R^2, R^5, R^{2t-1}\}\,.$$

For the relabeled opposite $\mathtt{ro}(T) = (1, -1, -1, 1, 1)$ of the vertex T, by Corollary 8.3(ii) we have

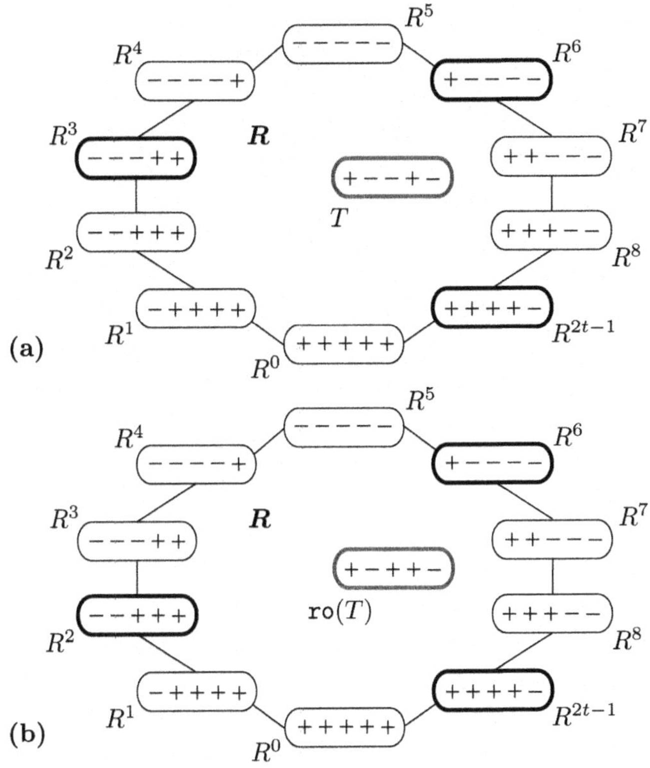

Figure 8.2 Two instances of the same distinguished symmetric cycle $\boldsymbol{R} := (R^0, R^1, \ldots, R^{2t-1}, R^0)$ in the hypercube graph $\boldsymbol{H}((t := 5), 2)$, defined by (8.3)(8.4), and some vertices T and $\mathtt{ro}(T)$.
(a) The vertex $T := {}_{-\{2,3,t\}}\mathrm{T}^{(+)}$, with the corresponding decomposition set $\boldsymbol{Q}(T, \boldsymbol{R}) = \{R^3, R^6, R^{2t-1}\}$.
(b) The relabeled opposite $\mathtt{ro}(T) = {}_{-\{2,t\}}\mathrm{T}^{(+)}$, with its decomposition set $\boldsymbol{Q}(\mathtt{ro}(T), \boldsymbol{R}) = \{R^2, R^6, R^{2t-1}\}$.

$$\begin{aligned}\boldsymbol{x}(\mathtt{ro}(T)) &= \boldsymbol{\sigma}(1) + \boldsymbol{x}(T) \cdot \overline{\mathbf{U}}(t) \cdot \overline{\mathbf{T}}(t) \\ &= (\ 1, -1,\ \ 0,\ \ 1,\ \ 0)\,, \quad \boldsymbol{Q}(\mathtt{ro}(T), \boldsymbol{R}) = \{R^0, R^3, R^6\}\,;\end{aligned}$$

see Figure 8.3.

(iii) Pick the vertex $T := (1, -1, -1, -1, 1) = {}_{-\{2,3,4\}}\mathrm{T}^{(+)} \in \{1, -1\}^t$ of the graph $\boldsymbol{H}(t, 2)$. We have $|\{1, t\} \cap \mathfrak{n}(T)| = 0$, and

$$\boldsymbol{x}(T) = (\ 1, -1,\ \ 0,\ \ 0,\ \ 1)\,, \quad \boldsymbol{Q}(T, \boldsymbol{R}) = \{R^0, R^4, R^6\}\,.$$

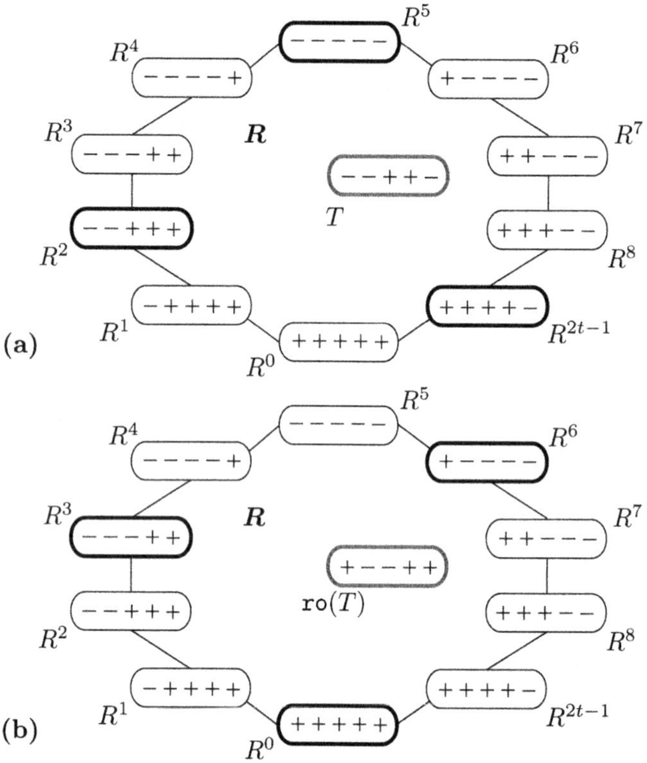

Figure 8.3 Two instances of the same distinguished symmetric cycle $\boldsymbol{R} := (R^0, R^1, \ldots, R^{2t-1}, R^0)$ in the hypercube graph $\boldsymbol{H}((t := 5), 2)$, defined by (8.3)(8.4), and some vertices T and $\mathtt{ro}(T)$.
(a) The vertex $T := {}_{-\{1,2,t\}}\mathrm{T}^{(+)}$, with the corresponding decomposition set $\boldsymbol{Q}(T, \boldsymbol{R}) = \{R^2, R^5, R^{2t-1}\}$.
(b) The relabeled opposite $\mathtt{ro}(T) = {}_{-\{2,3\}}\mathrm{T}^{(+)}$, with its decomposition set $\boldsymbol{Q}(\mathtt{ro}(T), \boldsymbol{R}) = \{R^0, R^3, R^6\}$.

For the relabeled opposite $\mathtt{ro}(T) = (-1, 1, 1, 1, -1)$ of the vertex T, by Corollary 8.3(iii) we have

$$\begin{aligned} \boldsymbol{x}(\mathtt{ro}(T)) &= -\boldsymbol{\sigma}(1) + \boldsymbol{x}(T) \cdot \overline{\mathbf{U}}(t) \cdot \overline{\mathbf{T}}(t) \\ &= (-1,\ \ 1,\ \ 0,\ \ 0, -1)\,, \quad \boldsymbol{Q}(\mathtt{ro}(T), \boldsymbol{R}) = \{R^1, R^5, R^{2t-1}\}\,; \end{aligned}$$

see Figure 8.4.

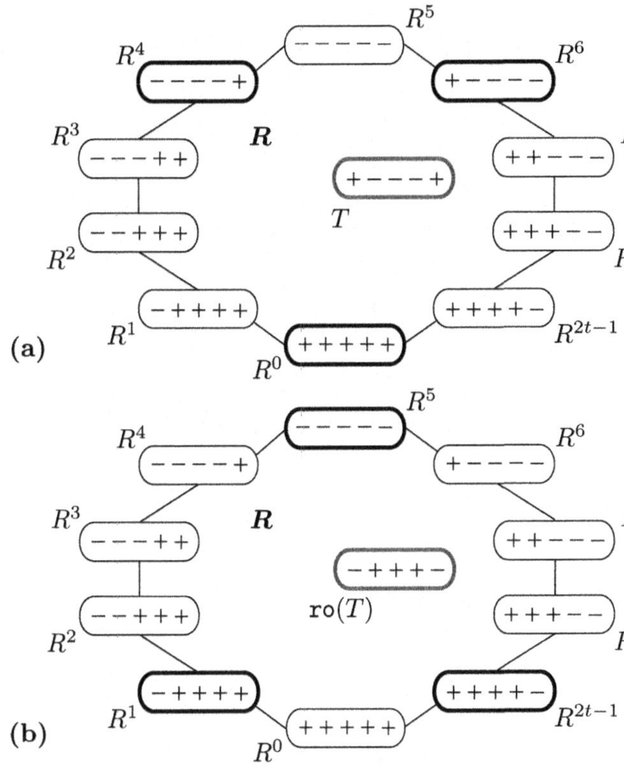

Figure 8.4 Two instances of the same distinguished symmetric cycle $\boldsymbol{R} := (R^0, R^1, \ldots, R^{2t-1}, R^0)$ in the hypercube graph $\boldsymbol{H}((t := 5), 2)$, defined by (8.3)(8.4), and some vertices T and $\mathtt{ro}(T)$.
(a) The vertex $T := {}_{-\{2,3,4\}}\mathrm{T}^{(+)}$, with the corresponding decomposition set $\boldsymbol{Q}(T, \boldsymbol{R}) = \{R^0, R^4, R^6\}$.
(b) The relabeled opposite $\mathtt{ro}(T) = {}_{-\{1,t\}}\mathrm{T}^{(+)}$, with its decomposition set $\boldsymbol{Q}(\mathtt{ro}(T), \boldsymbol{R}) = \{R^1, R^5, R^{2t-1}\}$.

Chapter 9

Set Families, Blocking Sets, Blockers, and Distinguished Symmetric Cycles in Hypercube Graphs

Each of the 2^t vertices $T := (T(1), \ldots, T(t)) \in \{1, -1\}^t$ of the hypercube graph $\boldsymbol{H}(t, 2)$ can be regarded as the '*characteristic covector*' of the *negative part* $\mathfrak{n}(T) := \{e \in E_t \colon T(e) = -1\}$ of the vertex T. Conversely, given an arbitrary subset A of the ground set E_t, we define the *characteristic covector* of A to be the *reorientation* ${}_{-A}\mathrm{T}^{(+)}$ of the *positive vertex* $\mathrm{T}^{(+)} := (1, \ldots, 1)$ on the subset A; recall that by notational convention we have $\mathfrak{n}({}_{-A}\mathrm{T}^{(+)}) := A$.

Let $\boldsymbol{R} := (R^0, R^1, \ldots, R^{2t-1}, R^0)$ denote, as earlier, the distinguished *symmetric cycle* in the graph $\boldsymbol{H}(t, 2)$, where

$$\begin{aligned} R^0 &:= \mathrm{T}^{(+)} , \\ R^s &:= {}_{-[s]}R^0 , \quad 1 \le s \le t-1 , \end{aligned} \tag{9.1}$$

and

$$R^{t+k} := -R^k , \quad 0 \le k \le t-1 . \tag{9.2}$$

Recall that for any vertex $T \in \{1, -1\}^t$ of the graph $\boldsymbol{H}(t, 2)$, there exists a *unique inclusion-minimal* subset

$$\boldsymbol{Q}(T, \boldsymbol{R}) \subset \vec{\mathrm{V}}(\boldsymbol{R}) := (R^0, R^1, \ldots, R^{2t-1}) \tag{9.3}$$

Symmetric Cycles
Andrey O. Matveev

ISBN 978-981-4968-81-2 (Hardcover), 978-1-003-43832-8 (eBook)
www.jennystanford.com

of the vertex sequence $\vec{\mathrm{V}}(\boldsymbol{R})$ of the cycle $\boldsymbol{R}$, such that

$$T = \sum_{Q \in \boldsymbol{Q}(T,\boldsymbol{R})} Q \ ; \tag{9.4}$$

see Subsection 2.1.4. This subset $\boldsymbol{Q}(T, \boldsymbol{R}) \subset \mathbb{R}^t$ is *linearly independent*, and it contains an *odd* number $\mathrm{q}(T) := \mathrm{q}(T, \boldsymbol{R}) := |\boldsymbol{Q}(T, \boldsymbol{R})|$ of vertices.

In fact, the linear algebraic decomposition (9.4) is merely a way to describe a particular mechanism of *majority voting*.

Again let $\boldsymbol{\sigma}(e)$ denote the eth unit vector of the standard basis of the space $\mathbb{R}^t$, $e \in E_t$. The bijections

$$\{1, -1\}^t \to \{0, 1\}^t: \qquad T \mapsto \frac{1}{2}(\mathrm{T}^{(+)} - T)\ , \tag{9.5}$$

and

$$\{0, 1\}^t \to \{1, -1\}^t: \qquad \widetilde{T} \mapsto \mathrm{T}^{(+)} - 2\widetilde{T}\ , \tag{9.6}$$

between the vertex set $\{1, -1\}^t$ of the hypercube graph $\boldsymbol{H}(t, 2)$ and the vertex set $\{0, 1\}^t$ of the hypercube graph $\widetilde{\boldsymbol{H}}(t, 2)$, mentioned in Section 2.2, allow us to associate with the symmetric cycle $\boldsymbol{R}$ in the graph $\boldsymbol{H}(t, 2)$ a distinguished symmetric cycle $\widetilde{\boldsymbol{R}} := (\widetilde{R}^0, \widetilde{R}^1, \ldots, \widetilde{R}^{2t-1}, \widetilde{R}^0)$ in the graph $\widetilde{\boldsymbol{H}}(t, 2)$, where

$$\widetilde{R}^0 := (0, \ldots, 0)\ ,$$
$$\widetilde{R}^s := \sum\nolimits_{e \in [s]} \boldsymbol{\sigma}(e)\ , \quad 1 \le s \le t-1\ ,$$

and

$$\widetilde{R}^{t+k} := \mathrm{T}^{(+)} - \widetilde{R}^k\ , \quad 0 \le k \le t-1\ .$$

Given a vertex $\widetilde{T}$ of the hypercube graph $\widetilde{\boldsymbol{H}}(t, 2)$, let us define a subset $\widetilde{\boldsymbol{Q}}(\widetilde{T}, \widetilde{\boldsymbol{R}}) \subset \vec{\mathrm{V}}(\widetilde{\boldsymbol{R}}) := (\widetilde{R}^0, \widetilde{R}^1, \ldots, \widetilde{R}^{2t-1})$ of the vertex sequence $\vec{\mathrm{V}}(\widetilde{\boldsymbol{R}})$ of the cycle $\widetilde{\boldsymbol{R}}$ indirectly, via the mapping

$$\widetilde{T} \overset{(9.6)}{\mapsto} T\ ,$$

and via the bijection

$$\boldsymbol{Q}(T, \boldsymbol{R}) \xrightarrow{(9.5)} \widetilde{\boldsymbol{Q}}(\widetilde{T}, \widetilde{\boldsymbol{R}})\ .$$

As noted in Remark 2.23, by involving the quantity $\mathfrak{q}(\widetilde{T}) := \mathfrak{q}(\widetilde{T}, \widetilde{\boldsymbol{R}}) := |\widetilde{\boldsymbol{Q}}(\widetilde{T}, \widetilde{\boldsymbol{R}})|$, with $\mathfrak{q}(\widetilde{T}) = \mathfrak{q}(T)$, we can write down the decomposition

$$\widetilde{T} = -\frac{1}{2}(\mathfrak{q}(\widetilde{T}) - 1) \cdot \mathrm{T}^{(+)} + \sum_{\substack{\widetilde{Q} \in \widetilde{\boldsymbol{Q}}(\widetilde{T}, \widetilde{\boldsymbol{R}}): \\ \widetilde{Q} \neq (0,\ldots,0) =: \widetilde{R}^0}} \widetilde{Q} \tag{9.7}$$

of the vertex $\widetilde{T}$ that describes yet another mechanism of *majority voting*,[1] but this decomposition has no essential meaning from the linear algebraic viewpoint, since the set $\widetilde{\boldsymbol{Q}}(\widetilde{T}, \widetilde{\boldsymbol{R}})$ can contain the origin $(0, \ldots, 0) =: \widetilde{R}^0$ of the space $\mathbb{R}^t$, which should be omitted in calculations.

Recall that a family $\mathcal{A} := \{A_1, \ldots, A_\alpha\} \subset \mathbf{2}^{[t]}$ of subsets[2] of the ground set E_t is called a *clutter*[3] if *no set* A_i from $\mathcal{A}$ contains another set A_j.

Given a family $\mathcal{F} \subseteq \mathbf{2}^{[t]}$, we let $\mathbf{min}\,\mathcal{F}$ denote the clutter composed of the *inclusion-minimal* sets in $\mathcal{F}$.

We say that a family of subsets $\mathcal{F} \subseteq \mathbf{2}^{[t]}$ is an *increasing family*[4] if the following implications hold:

$$A \in \mathcal{F}, \quad \mathbf{2}^{[t]} \ni B \supset A \implies B \in \mathcal{F}\,.$$

---

[1] Recall that maps

$$f\colon \{0,1\}^t \to \mathbb{R}\,,$$
$$-\frac{1}{2}(\mathfrak{q}(\widetilde{T}) - 1) \cdot \mathrm{T}^{(+)} + \sum_{\substack{\widetilde{Q} \in \widetilde{\boldsymbol{Q}}(\widetilde{T}, \widetilde{\boldsymbol{R}}): \\ \widetilde{Q} \neq (0,\ldots,0) =: \widetilde{R}^0}} \widetilde{Q} = \widetilde{T} \mapsto f(\widetilde{T})\,,$$

and

$$g\colon \{1,-1\}^t \to \mathbb{R}\,, \qquad \sum_{Q \in \boldsymbol{Q}(T, \boldsymbol{R})} Q = T \mapsto g(T)\,,$$

are *pseudo-Boolean functions*. Maps $\mathfrak{f}\colon \mathbf{2}^{[t]} \to \mathbb{R}$ are *set functions*. Maps $f\colon \{0,1\}^t \to \{0,1\}$, and $g\colon \{1,-1\}^t \to \{1,-1\}$ are *Boolean functions*.

[2] We denote by $\hat{0}$ the *empty subset* of the ground set E_t, and we let $\emptyset$ denote the *empty family* containing no sets.

Given a family $\mathcal{F} \subseteq \mathbf{2}^{[t]}$, such that $\emptyset \neq \mathcal{F} \not\ni \hat{0}$, the set $E_t := [t]$ is the *ground set* of $\mathcal{F}$, while the set $\mathrm{V}(\mathcal{F}) := \bigcup_{F \in \mathcal{F}} F \subseteq E_t$ is the *vertex set* of $\mathcal{F}$.

The families $\emptyset$ and $\{\hat{0}\}$ are the two *trivial clutters* on the ground set E_t. The other clutters on E_t are *nontrivial*.

[3] Or *Sperner family, antichain, simple hypergraph*.

[4] Or *up-set, upward-closed family of sets, filter of sets*.

If $C \subseteq E_t$, then the family $\{C\}^{\triangledown} := \{D \subseteq E_t\colon D \supseteq C\}$ is called the *principal* increasing family generated by the *one-member* clutter $\{C\}$. Conversely, an increasing family $\mathcal{F} \subseteq \mathbf{2}^{[t]}$ is said to be *principal* if $\#\,\mathbf{min}\,\mathcal{F} = 1$.

Given an arbitrary nonempty family $\mathcal{C} \subseteq \mathbf{2}^{[t]}$, we denote by $\mathcal{C}^{\triangledown}$ the *increasing family* on E_t, generated by $\mathcal{C}$:

$$\mathcal{C}^{\triangledown} := \bigcup_{C \in \mathcal{C}} \{C\}^{\triangledown} = \bigcup_{C \in \mathbf{min}\,\mathcal{C}} \{C\}^{\triangledown} .$$

'Decreasing' constructs are defined in the obvious similar way.[5]

The duality philosophy behind clutters and increasing families is that any clutter is the *blocker*[6] of a unique clutter, and any increasing family is the family of *blocking sets* of a unique clutter.

We often meet in the literature the *free distributive lattice* of antichains in the Boolean lattice of subsets of a finite nonempty set, ordered by containment of the corresponding generated *order ideals*, but an intrinsically related construct, the *free distributive lattice* of those antichains ordered by containment of the corresponding generated *order filters* has greater discrete mathematical expressiveness, because the latter lattice can be interpreted as the *lattice* of *blockers*, for which the *blocker map* is its anti-automorphism.[7]

---

[5]For a family $\mathcal{F} \subseteq \mathbf{2}^{[t]}$, we use the notation $\mathbf{max}\,\mathcal{F}$ to denote the clutter composed of the *inclusion-maximal* sets in $\mathcal{F}$.

A family $\mathcal{F} \subseteq \mathbf{2}^{[t]}$ is said to be a *decreasing family* (or *down-set, downward-closed family of sets, ideal of sets*) if the following implications hold:

$$B \in \mathcal{F},\ A \subset B \quad \Longrightarrow \quad A \in \mathcal{F} .$$

If $\emptyset \neq \mathcal{F} \neq \{\hat{0}\}$, then this decreasing family is the *abstract simplicial complex* on its *vertex set* $\bigcup_{M \in \mathbf{max}\,\mathcal{F}} M$, with the *facet* family $\mathbf{max}\,\mathcal{F}$.

If $D \subseteq E_t$, then the family $\{D\}^{\vartriangle} := \{C\colon C \subseteq D\}$ is called the *principal* decreasing family generated by the *one-member* clutter $\{D\}$. Conversely, a decreasing family $\mathcal{F} \subseteq \mathbf{2}^{[t]}$ is said to be *principal* if $\#\,\mathbf{max}\,\mathcal{F} = 1$.

Given an arbitrary nonempty family $\mathcal{D} \subseteq \mathbf{2}^{[t]}$, we denote by $\mathcal{D}^{\vartriangle}$ the *decreasing family* on E_t, generated by $\mathcal{D}$:

$$\mathcal{D}^{\vartriangle} := \bigcup_{D \in \mathcal{D}} \{D\}^{\vartriangle} = \bigcup_{D \in \mathbf{max}\,\mathcal{D}} \{D\}^{\vartriangle} .$$

[6]I enjoyed working with Ray [Fulkerson] and I coined the terms "clutter" and "blocker". Jack Edmonds (page 201 of Ref. [35])

[7]For this chapter, we chose the language of *power sets, clutters*, and *increasing* and *decreasing families*. A parallel exposition could be presented in poset-theoretic terms of *Boolean lattices, antichains*, and *order filters* and *ideals*.

Recall that a subset $B \subseteq E_t$ is called a *blocking set*[8] of a subset family $\mathcal{F} \subset \mathbf{2}^{[t]}$, where $\emptyset \neq \mathcal{F} \not\ni \hat{0}$, if we have

$$|B \cap F| > 0 ,$$

for each set $F \in \mathcal{F}$. The *blocker*[9] $\mathfrak{B}(\mathcal{F})$ of the family $\mathcal{F}$ is the family of all *inclusion-minimal blocking sets* of $\mathcal{F}$; note that we have $\mathfrak{B}(\mathcal{F}) = \mathfrak{B}(\mathbf{min}\,\mathcal{F})$. The notation $\mathfrak{B}(\mathcal{F})^{\triangledown}$ just means the increasing family of all blocking sets of the family $\mathcal{F}$.

For a nonempty family of subsets $\mathcal{F} \subseteq \mathbf{2}^{[t]}$, we define a family[10] of complements $\mathcal{F}^{\complement}$ by $\mathcal{F}^{\complement} := \{F^{\complement} \colon F \in \mathcal{F}\}$, where $F^{\complement} := E_t - F$.

Given a nontrivial clutter $\mathcal{A} \subset \mathbf{2}^{[t]}$, one associates with $\mathcal{A}$ the four extensively studied partitions of the power set of the ground set E_t:

$$\begin{aligned} \mathbf{2}^{[t]} &= \mathcal{A}^{\triangledown} \,\dot{\cup}\, (\mathfrak{B}(\mathcal{A})^{\complement})^{\vartriangle} , \\ \mathbf{2}^{[t]} &= \mathcal{A}^{\vartriangle} \,\dot{\cup}\, \mathfrak{B}(\mathcal{A}^{\complement})^{\triangledown} , \\ \mathbf{2}^{[t]} &= \mathfrak{B}(\mathcal{A})^{\triangledown} \,\dot{\cup}\, (\mathcal{A}^{\complement})^{\vartriangle} , \end{aligned} \tag{9.8}$$

and

$$\mathbf{2}^{[t]} = \mathfrak{B}(\mathcal{A})^{\vartriangle} \,\dot{\cup}\, \mathfrak{B}(\mathfrak{B}(\mathcal{A})^{\complement})^{\triangledown} .$$

In Section 9.1 of this chapter we make a few remarks on blocking sets and blockers.

In Section 9.2 we briefly review the classical set covering problems of combinatorial optimization on blocking sets and blockers.

Our interest in considering the *relabeled opposites* $\mathtt{ro}(T)$ and the *relabeled negations* $\mathtt{rn}(\widetilde{T})$ of vertices in hypercube graphs (see Chapter 8) lies in their application to combined *blocking/voting models* of increasing families of sets and models of clutters. In Section 9.3 we associate with set families their characteristic (co)vectors: First, we arrange the subsets of the ground set E_t in linear order, and we then turn to *characteristic vectors* $\boldsymbol{\gamma}(\mathcal{F}) \in \{0, 1\}^{2^t}$ of subset families $\mathcal{F} \subseteq \mathbf{2}^{[t]}$.

[8]Or *transversal, hitting set, vertex cover* (or *node cover*), *system of representatives.*

[9]Or *blocking hypergraph* (or *transversal hypergraph*), *blocking clutter, dual clutter, Alexander dual clutter.*

[10]Given a nonempty family of subsets $\mathcal{F} \subseteq \mathbf{2}^{[t]}$, we define a family of complements $\mathcal{F}^{\perp}$ by $\mathcal{F}^{\perp} := \{F^{\perp} \colon F \in \mathcal{F}\}$, where $F^{\perp} := \mathrm{V}(\mathcal{F}) - F$.

As noted in Section 9.4 of this chapter, if $\mathcal{A} \subset \mathbf{2}^{[t]}$ is a nontrivial clutter on its ground set E_t, then relation (9.8) reformulated in the form (cf. (8.2))

$$\boldsymbol{\gamma}(\mathfrak{B}(\mathcal{A})^{\triangledown}) = \mathrm{T}^{(+)}_{2^t} - \boldsymbol{\gamma}(\mathcal{A}^{\triangledown}) \cdot \overline{\mathbf{U}}(2^t)\,,$$

where $\mathrm{T}^{(+)}_{2^t} := (1, \ldots, 1)$ is the 2^t-dimensional row vector of all 1's, and $\overline{\mathbf{U}}(2^t)$ is the backward identity matrix of order 2^t, provides us with the characteristic vector

$$\boldsymbol{\gamma}(\mathfrak{B}(\mathcal{A})^{\triangledown}) = \mathrm{rn}(\boldsymbol{\gamma}(\mathcal{A}^{\triangledown}))$$

of the increasing family of blocking sets $\mathfrak{B}(\mathcal{A})^{\triangledown}$ of the clutter $\mathcal{A}$.

In Section 9.5 we mention a blocking/voting connection of the characteristic vectors $\boldsymbol{\gamma}(\mathcal{A}^{\triangledown})$ and $\boldsymbol{\gamma}(\mathfrak{B}(\mathcal{A})^{\triangledown})$ with the decompositions of the corresponding characteristic covectors of the increasing families $\mathcal{A}^{\triangledown}$ and $\mathfrak{B}(\mathcal{A})^{\triangledown}$ with respect to a distinguished symmetric cycle in the hypercube graph $\widetilde{\boldsymbol{H}}(2^t, 2)$.

9.1 Blocking Sets and Blockers

Blocking sets and the blockers of set families (families are often regarded as the *hyperedge* families of *hypergraphs*) are discussed, e.g., in the monographs [5, 18, 36, 37, 38, 39, 40, 41, 42, 43, 44, 45, 46, 47, 48, 49, 50, 51, 52, 53, 54], and in the works [55, 56, 57, 58, 59, 60, 61, 62, 63, 64, 65, 66, 67, 68, 69, 70, 71, 72, 73, 74, 75, 76, 77, 78, 79, 80, 81, 82, 83, 84, 85, 86, 87, 88, 89, 90, 91, 92, 93, 94, 95, 96].

- Let

$$\mho([t]) := \{\mathcal{A} \subset \mathbf{2}^{[t]} \colon\ \mathcal{A} = \mathbf{min}\,\mathcal{A} = \mathbf{max}\,\mathcal{A}\}$$

denote the family of all clutters on the ground set E_t. The map

$$\mho([t]) \to \mho([t])\,, \quad \mathcal{A} \mapsto \mathfrak{B}(\mathcal{A})\,, \tag{9.9}$$

is called the *blocker map* on clutters; see Ref. [74].

- If the (*abstract simplicial*) *complex* $\Delta := (\mathfrak{B}(\mathcal{A})^{\mathsf{c}})^{\vartriangle}$ appearing in (9.8), as well as the *complex* $\Delta^{\vee} := \{F^{\mathsf{c}} \colon F \in \mathcal{A}^{\triangledown}\}$, both have the same vertex set E_t, then the complex $\Delta^{\vee}$ is called the *Alexander dual* of the complex Δ; see, e.g., Refs.[54, 97, 98] and Ref. [99] on *combinatorial Alexander duality*.

- Given a clutter $\mathcal{A}$, the quantity

$$\tau(\mathcal{A}) := \min\{|B|: B \in \mathfrak{B}(\mathcal{A})\}$$

is called the *transversal number*[11] of $\mathcal{A}$.

- Recall a classical result in combinatorial optimization: For any clutter $\mathcal{A}$ we have

$$\mathfrak{B}(\mathfrak{B}(\mathcal{A})) = \mathcal{A}\,;$$

see Refs. [100, 101, 102, 103].

- For a nontrivial clutter $\mathcal{A} \subset \mathbf{2}^{[t]}$ on the ground set E_t, we have

$$\#\mathcal{A}^{\triangledown} + \#\mathfrak{B}(\mathcal{A})^{\triangledown} = 2^t\,. \tag{9.10}$$

More precisely, for any s, such that $0 \leq s \leq t$, we have

$$\#(\mathfrak{B}(\mathcal{A})^{\triangledown} \cap \tbinom{E_t}{s}) + \#(\mathcal{A}^{\triangledown} \cap \tbinom{E_t}{t-s}) = \tbinom{t}{s}\,, \tag{9.11}$$

where

$$\tbinom{E_t}{s} := \{F \subseteq E_t\colon |F| = s\}$$

is the *complete s-uniform clutter* (i.e., the family of all s-subsets) on the vertex set E_t.

The increasing families $\mathcal{A}^{\triangledown}$ and $\mathfrak{B}(\mathcal{A})^{\triangledown}$ are *comparable by containment*: Either we have

$$\mathfrak{B}(\mathcal{A})^{\triangledown} \subseteq \mathcal{A}^{\triangledown}\,, \quad \text{or} \quad \mathfrak{B}(\mathcal{A})^{\triangledown} \supseteq \mathcal{A}^{\triangledown}\,.$$

The following implications hold:

$$\mathfrak{B}(\mathcal{A})^{\triangledown} \subsetneqq \mathcal{A}^{\triangledown} \iff \#\mathcal{A}^{\triangledown} > 2^{t-1}\,;$$
$$\mathfrak{B}(\mathcal{A})^{\triangledown} \supsetneqq \mathcal{A}^{\triangledown} \iff \#\mathcal{A}^{\triangledown} < 2^{t-1}\,.$$

Note also that the following implications hold:

$$\#\mathcal{A}^{\triangledown} > 2^{t-1} \implies \min\{|A|: A \in \mathcal{A}\} \leq \min\{|B|: B \in \mathfrak{B}(\mathcal{A})\}\,;$$
$$\#\mathcal{A}^{\triangledown} < 2^{t-1} \implies \min\{|A|: A \in \mathcal{A}\} \geq \min\{|B|: B \in \mathfrak{B}(\mathcal{A})\}\,.$$

- A clutter $\mathcal{A}$ is called *self-dual* (see Section 5.7 of Ref. [5], and Chapter 9 of Ref. [48]) or *identically self-blocking* (see Refs. [104, 105]) if

$$\mathfrak{B}(\mathcal{A}) = \mathcal{A}\,;$$

[11] Or *covering number, vertex cover number, blocking number.*

see also Section 2.1 in the early Ref. [36]. In other words, the self-dual clutters $\mathcal{A} \subset \mathbf{2}^{[t]}$ on the ground set E_t are the *fixed points* of the *blocker map* (9.9); for each of them we also have

$$\mathfrak{B}(\mathcal{A})^{\triangledown} = \mathcal{A}^{\triangledown} .$$

As noted in Corollary 5.28(i) of Ref. [5], one criterion for a *clutter* $\mathcal{A} \subset \mathbf{2}^{[t]}$ on the ground set E_t to be *self-dual* is as follows:

$$\mathfrak{B}(\mathcal{A}) = \mathcal{A} \iff \#\mathcal{A}^{\triangledown} = 2^{t-1} .$$

Appendix B of the present monograph is devoted to a more detailed treatment of self-dual clutters.

- Let X be a subset of the ground set E_t. Given a nontrivial clutter $\mathcal{A}$ on E_t, its *deletion* $\mathcal{A} \setminus X$ is defined to be the clutter

$$\mathcal{A} \setminus X := \{A \in \mathcal{A} \colon |A \cap X| = 0\} .$$

The *contraction* $\mathcal{A} / X$ is defined to be the clutter

$$\mathcal{A} / X := \mathbf{min}\{A - X : A \in \mathcal{A}\} .$$

A classical result in combinatorial optimization is as follows:

$$\mathfrak{B}(\mathcal{A}) \setminus X = \mathfrak{B}(\mathcal{A} / X) , \quad \text{and} \quad \mathfrak{B}(\mathcal{A}) / X = \mathfrak{B}(\mathcal{A} \setminus X) ;$$

see Ref. [106].

We also have

$$(\mathfrak{B}(\mathcal{A}) \setminus X)^{\triangledown} = \mathfrak{B}(\mathcal{A} / X)^{\triangledown} \subseteq \mathfrak{B}(\mathcal{A})^{\triangledown} \subseteq (\mathfrak{B}(\mathcal{A}) / X)^{\triangledown} = \mathfrak{B}(\mathcal{A} \setminus X)^{\triangledown} ;$$

cf. Eq. (5.4) in Ref. [5]. Further,

$$\begin{aligned} \#(\mathcal{A} \setminus X)^{\triangledown} + \#(\mathfrak{B}(\mathcal{A}) / X)^{\triangledown} &= 2^t , \\ \#(\mathcal{A} / X)^{\triangledown} + \#(\mathfrak{B}(\mathcal{A}) \setminus X)^{\triangledown} &= 2^t ; \end{aligned}$$

see Corollary 5.28(ii) of Ref. [5]. More precisely, for any s, where $0 \leq s \leq t$, we have

$$\begin{aligned} \#\big((\mathfrak{B}(\mathcal{A}) / X)^{\triangledown} \cap \tbinom{E_t}{s}\big) + \#\big((\mathcal{A} \setminus X)^{\triangledown} \cap \tbinom{E_t}{t-s}\big) &= \tbinom{t}{s} , \\ \#\big((\mathfrak{B}(\mathcal{A}) \setminus X)^{\triangledown} \cap \tbinom{E_t}{s}\big) + \#\big((\mathcal{A} / X)^{\triangledown} \cap \tbinom{E_t}{t-s}\big) &= \tbinom{t}{s} . \end{aligned}$$

- Let p be a rational number such that $0 \leq p < 1$. Given a nontrivial clutter $\mathcal{A} := \{A_1, \ldots, A_\alpha\} \subset \mathbf{2}^{[t]}$ on the ground set E_t, a subset $B \subseteq E_t$ is called a *p-committee*[12] of the clutter $\mathcal{A}$, if we have

$$|B \cap A_i| > p \cdot |B| ,$$

[12] By convention, a $\frac{1}{2}$-committee of a clutter $\mathcal{A}$ is called its *committee*.

for each $i \in [\alpha]$. The 0-*committees* of the clutter $\mathcal{A}$ are its *blocking sets*.

- For a nontrivial clutter $\mathcal{A} := \{A_1, \ldots, A_\alpha\} \subset \mathbf{2}^{[t]}$ on the ground set E_t, we have

$$\#\big(\mathfrak{B}(\mathcal{A})^{\triangledown} \cap \tbinom{E_t}{k}\big) = \tbinom{t}{k} + \sum_{S\subseteq[\alpha]:\ |S|>0} (-1)^{|S|} \cdot \tbinom{t-|\bigcup_{s\in S} A_s|}{k}\,, \quad 1 \le k \le t\,.$$

Several ways to count the blocking k-sets of clutters are mentioned in Ref. [5].

9.2 Increasing Families of Blocking Sets, and Blockers: Set Covering Problems

In this section we recall the set covering problem(s); see, e.g., Section 2.4 of Ref. [38], and Chapter 1 of Ref. [39].

Let $\boldsymbol{\chi}(A) := (\chi_1(A), \ldots, \chi_t(A)) \in \{0, 1\}^t$ denote the familiar row *characteristic vector* of a subset A of the ground set E_t, defined for each element $j \in E_t$ by

$$\chi_j(A) := \begin{cases} 1\,, & \text{if } j \in A, \\ 0\,, & \text{if } j \notin A. \end{cases}$$

If $\mathcal{A} := \{A_1, \ldots, A_\alpha\} \subset \mathbf{2}^{[t]}$ is a nontrivial clutter on E_t, then

$$\boldsymbol{A} := \boldsymbol{A}(\mathcal{A}) := \begin{pmatrix} \boldsymbol{\chi}(A_1) \\ \vdots \\ \boldsymbol{\chi}(A_\alpha) \end{pmatrix}$$

is its *incidence matrix*.

Consider[13] the *set covering* $\{0, 1\}$-*collection*

$$\widetilde{\boldsymbol{S}} := \widetilde{\boldsymbol{S}}^{\mathrm{C}}(\boldsymbol{A}) := \big\{\tilde{\mathbf{z}} \in \{0, 1\}^t\colon \boldsymbol{A}\tilde{\mathbf{z}}^\top \geq \mathbb{1}\big\}\,, \tag{9.12}$$

[13] We will denote by $\mathbb{1}$ and $\mathbb{2}$ the α-dimensional column vectors $\begin{pmatrix} 1 \\ \vdots \\ 1 \end{pmatrix}$ and $\begin{pmatrix} 2 \\ \vdots \\ 2 \end{pmatrix}$, respectively.

which is the collection of *characteristic vectors* of the *blocking sets* of the clutter $\mathcal{A}$, that is,

$$\widetilde{\boldsymbol{\mathcal{S}}} = \{\chi(B)\colon B \in \mathfrak{B}(\mathcal{A})^{\triangledown}\}\,, \quad \text{and} \quad \mathfrak{B}(\mathcal{A})^{\triangledown} = \{\operatorname{supp}(\tilde{\mathbf{z}})\colon \tilde{\mathbf{z}} \in \widetilde{\boldsymbol{\mathcal{S}}}\}\,.$$

The latter expression just rephrases the convention according to which the *supports* of the vectors in the collection $\widetilde{\boldsymbol{\mathcal{S}}} \subset \{0, 1\}^t$ are the *blocking sets* of the clutter $\mathcal{A}$.

Let us redefine the collection

$$\widetilde{\boldsymbol{\mathcal{S}}} := \left\{\tilde{\mathbf{z}} \in \{0, 1\}^t\colon \begin{pmatrix} \chi(A_1) \\ \vdots \\ \chi(A_\alpha) \end{pmatrix} \tilde{\mathbf{z}}^\top \geq \mathbb{1}\right\}$$

as

$$\widetilde{\boldsymbol{\mathcal{S}}} := \left\{\frac{1}{2}(\mathrm{T}^{(+)} - \mathbf{z}) \in \{0, 1\}^t\colon \begin{pmatrix} \frac{1}{2}(\mathrm{T}^{(+)} - T^1) \\ \vdots \\ \frac{1}{2}(\mathrm{T}^{(+)} - T^\alpha) \end{pmatrix} \cdot \frac{1}{2}(\mathrm{T}^{(+)} - \mathbf{z})^\top \geq \mathbb{1}\right\}, \tag{9.13}$$

where the vertices T^i of the discrete hypercube $\{1, -1\}^t$ and the vector of unknowns $\mathbf{z} \in \{1, -1\}^t$ are given by

$$T^i := {}_{-A_i}\mathrm{T}^{(+)} = \mathrm{T}^{(+)} - 2\chi(A_i)\,, \quad i \in [\alpha]\,,$$

and

$$\mathbf{z} := \mathrm{T}^{(+)} - 2\tilde{\mathbf{z}}\,.$$

Let us now associate with the collection $\widetilde{\boldsymbol{\mathcal{S}}} \subset \{0, 1\}^t$, described in Eq. (9.13), a *set covering* $\{1, -1\}$*-collection* $\boldsymbol{\mathcal{S}} := \boldsymbol{\mathcal{S}}^{\mathrm{C}}(\boldsymbol{A}) \subset \{1, -1\}^t$, defined by

$$\boldsymbol{\mathcal{S}} := \left\{\mathbf{z} \in \{1, -1\}^t\colon \boldsymbol{A}\mathbf{z}^\top \leq \begin{pmatrix} |\mathfrak{n}(T^1)| \\ \vdots \\ |\mathfrak{n}(T^\alpha)| \end{pmatrix} - 2 \cdot \mathbb{1}\right\},$$

that is, the collection

$$\boldsymbol{\mathcal{S}} := \left\{\mathbf{z} \in \{1, -1\}^t\colon \boldsymbol{A}\mathbf{z}^\top \leq \begin{pmatrix} |A_1| \\ \vdots \\ |A_\alpha| \end{pmatrix} - 2\right\}. \tag{9.14}$$

We have defined the twin collections $\widetilde{\boldsymbol{\mathcal{S}}} \subset \{0, 1\}^t$ and $\boldsymbol{\mathcal{S}} \subset \{1, -1\}^t$, given in (9.12) and (9.14), respectively, that are equipped with the *bijections* $\widetilde{\boldsymbol{\mathcal{S}}} \to \boldsymbol{\mathcal{S}}$: $\widetilde{T} \mapsto \mathrm{T}^{(+)} - 2\widetilde{T}$, and $\boldsymbol{\mathcal{S}} \to \widetilde{\boldsymbol{\mathcal{S}}}$: $T \mapsto \frac{1}{2}(\mathrm{T}^{(+)} - T)$; see Example 9.1.

Example 9.1. Consider the clutter $\mathcal{A} := \{A_1, A_{\alpha:=2}\} := \{\{1, 2\}, \{2, 3\}\}$, on the ground set $E_{t:=3} := \{1, 2, 3\}$, with its incidence matrix

$$\boldsymbol{A} := \boldsymbol{A}(\mathcal{A}) = \begin{pmatrix} 1 & 1 & 0 \\ 0 & 1 & 1 \end{pmatrix} .$$

The *set covering* $\{0, 1\}$*-collection*

$$\begin{aligned} \widetilde{\boldsymbol{\mathcal{S}}} &:= \{\tilde{\mathbf{z}} \in \{0, 1\}^t : \boldsymbol{A}\tilde{\mathbf{z}}^\top \geq \mathbb{1}\} \\ &= \left\{\tilde{\mathbf{z}} \in \{0, 1\}^3 : \begin{pmatrix} 1 & 1 & 0 \\ 0 & 1 & 1 \end{pmatrix} \tilde{\mathbf{z}}^\top \geq \begin{pmatrix} 1 \\ 1 \end{pmatrix}\right\} \end{aligned}$$

is the collection

$$\begin{aligned} \widetilde{\boldsymbol{\mathcal{S}}} &= \{(0\ 1\ 0), (1\ 1\ 0), (1\ 0\ 1), (0\ 1\ 1), (1\ 1\ 1)\} \\ &= \{\boldsymbol{\chi}(\{2\}), \boldsymbol{\chi}(\{1, 2\}), \boldsymbol{\chi}(\{1, 3\}), \boldsymbol{\chi}(\{2, 3\}), \boldsymbol{\chi}(\{1, 2, 3\})\} . \end{aligned}$$

The *set covering* $\{1, -1\}$*-collection*

$$\begin{aligned} \boldsymbol{\mathcal{S}} &:= \left\{\mathbf{z} \in \{1, -1\}^t : \boldsymbol{A}\mathbf{z}^\top \leq \begin{pmatrix} |A_1| \\ \vdots \\ |A_\alpha| \end{pmatrix} - 2\right\} \\ &= \left\{\mathbf{z} \in \{1, -1\}^3 : \begin{pmatrix} 1 & 1 & 0 \\ 0 & 1 & 1 \end{pmatrix} \mathbf{z}^\top \leq \begin{pmatrix} 2 \\ 2 \end{pmatrix} - 2 = \begin{pmatrix} 0 \\ 0 \end{pmatrix}\right\} \end{aligned}$$

is the collection

$$\begin{aligned} \boldsymbol{\mathcal{S}} &= \{(1\ {-1}\ 1), ({-1}\ {-1}\ 1), ({-1}\ 1\ {-1}), (1\ {-1}\ {-1}), ({-1}\ {-1}\ {-1})\} \\ &= \{{}_{-\{2\}}\mathrm{T}^{(+)}, {}_{-\{1,2\}}\mathrm{T}^{(+)}, {}_{-\{1,3\}}\mathrm{T}^{(+)}, {}_{-\{2,3\}}\mathrm{T}^{(+)}, {}_{-\{1,2,3\}}\mathrm{T}^{(+)}\} . \end{aligned}$$

- Let $\boldsymbol{w} \in \mathbb{R}^t$ be a row vector of *nonnegative* weights. The *set covering* problems are

$$\min\{\boldsymbol{w}\tilde{\mathbf{z}}^\top : \tilde{\mathbf{z}} \in \widetilde{\boldsymbol{\mathcal{S}}}\} = \min\left\{\boldsymbol{w} \cdot \frac{1}{2}(\mathrm{T}^{(+)} - \mathbf{z})^\top : \mathbf{z} \in \boldsymbol{\mathcal{S}}\right\} .$$

- Suppose $\boldsymbol{w} := \mathrm{T}^{(+)}$, and consider the (*unweighted*) set covering problems

$$\begin{aligned} \tau(\mathcal{A}) :&= \min\{\mathtt{hwt}(\tilde{\mathbf{z}}) : \tilde{\mathbf{z}} \in \widetilde{\boldsymbol{\mathcal{S}}}\} = \min\{\mathrm{T}^{(+)}\tilde{\mathbf{z}}^\top : \tilde{\mathbf{z}} \in \widetilde{\boldsymbol{\mathcal{S}}}\} \\ &= \min\left\{\mathrm{T}^{(+)} \cdot \frac{1}{2}(\mathrm{T}^{(+)} - \mathbf{z})^\top : \mathbf{z} \in \boldsymbol{\mathcal{S}}\right\} \\ &= \min\left\{\frac{1}{2}(t - \underbrace{\mathrm{T}^{(+)}\mathbf{z}^\top}_{t-2|\mathfrak{n}(\mathbf{z})|}) : \mathbf{z} \in \boldsymbol{\mathcal{S}}\right\} \\ &= \min\{|\mathfrak{n}(\mathbf{z})| : \mathbf{z} \in \boldsymbol{\mathcal{S}}\} =: \tau(\mathcal{A}) , \end{aligned}$$

that is, the problem

$$\underbrace{\min\{\mathrm{T}^{(+)}\tilde{\mathbf{z}}^\top : \tilde{\mathbf{z}} \in \widetilde{\boldsymbol{\mathcal{S}}}\}}_{\tau(\mathcal{A}):=\min\{\mathtt{hwt}(\tilde{\mathbf{z}}):\,\tilde{\mathbf{z}}\in\widetilde{\boldsymbol{\mathcal{S}}}\}} := \min\{\mathrm{T}^{(+)}\tilde{\mathbf{z}}^\top : \tilde{\mathbf{z}} \in \{0,1\}^t,\ \boldsymbol{A}\tilde{\mathbf{z}} \geq \mathbb{1}\}\,, \quad (9.15)$$

and the problem

$$\underbrace{\frac{1}{2}\min\left\{t - \mathrm{T}^{(+)}\mathbf{z}^\top : \mathbf{z} \in \boldsymbol{\mathcal{S}}\right\}}_{\tau(\mathcal{A}):=\min\left\{|\,\mathfrak{n}(\mathbf{z})|:\,\mathbf{z}\in\boldsymbol{\mathcal{S}}\right\}}$$

$$:= \frac{1}{2}\min\left\{t - \mathrm{T}^{(+)}\mathbf{z}^\top : \mathbf{z} \in \{1,-1\}^t,\ \boldsymbol{A}\mathbf{z}^\top \leq \begin{pmatrix}|A_1|\\ \vdots \\ |A_\alpha|\end{pmatrix} - \mathbb{2}\right\}$$

$$= \frac{1}{2}\cdot\left(t - \max\left\{\mathrm{T}^{(+)}\mathbf{z}^\top : \mathbf{z} \in \{1,-1\}^t,\ \boldsymbol{A}\mathbf{z}^\top \leq \begin{pmatrix}|A_1|\\ \vdots \\ |A_\alpha|\end{pmatrix} - \mathbb{2}\right\}\right)$$

$$=: \underbrace{\frac{1}{2}\cdot\left(t - \max\left\{\mathrm{T}^{(+)}\mathbf{z}^\top : \mathbf{z} \in \boldsymbol{\mathcal{S}}\right\}\right)}_{\tau(\mathcal{A}):=\min\left\{|\,\mathfrak{n}(\mathbf{z})|:\,\mathbf{z}\in\boldsymbol{\mathcal{S}}\right\}}\,. \quad (9.16)$$

For vectors $\tilde{\mathbf{z}} \in \widetilde{\boldsymbol{\mathcal{S}}}$ and $\mathbf{z} \in \boldsymbol{\mathcal{S}}$, where $\tilde{\mathbf{z}} := \frac{1}{2}(\mathrm{T}^{(+)} - \mathbf{z})$, we have the inclusions

$$\tilde{\mathbf{z}} \in \arg\min\{\mathrm{T}^{(+)}\tilde{\mathbf{z}}^\top : \tilde{\mathbf{z}} \in \widetilde{\boldsymbol{\mathcal{S}}}\}\,,$$
$$\mathbf{z} \in \arg\max\{\mathrm{T}^{(+)}\mathbf{z}^\top : \mathbf{z} \in \boldsymbol{\mathcal{S}}\}\,,$$

that is, $\tilde{\mathbf{z}}$ and $\mathbf{z}$ provide the solution to the problems (9.15) and (9.16), respectively, if and only if the member

$$B := \mathrm{supp}(\tilde{\mathbf{z}}) = \mathfrak{n}(\mathbf{z}) \in \mathfrak{B}(\mathcal{A})$$

of the blocker of the clutter $\mathcal{A}$ has the *minimum* cardinality

$$|B| = \tau(\mathcal{A})\,.$$

– We conclude this section by noting that the rows of incidence matrices $\boldsymbol{A}$, as well as the vectors in the set covering collections $\widetilde{\boldsymbol{\mathcal{S}}} \subset \{0,1\}^t$ and $\boldsymbol{\mathcal{S}} \subset \{1,-1\}^t$, admit their decompositions with respect to symmetric cycles in the corresponding hypercube graphs $\widetilde{\boldsymbol{H}}(t,2)$ and $\boldsymbol{H}(t,2)$.

9.3 (Sub)set Families: Characteristic Vectors and Characteristic Covectors

We begin this section by noting that the *generation* of fundamental combinatorial objects is extensively treated in Ref. [107].

- Consider the *complete s-uniform clutter* $\binom{E_t}{s}$, for some s, where $0 \leq s \leq t$. We denote this family of all s-subsets $L^s_j \subseteq E_t$, ordered *lexicographically*, by $\overrightarrow{\binom{E_t}{s}} =: (L^s_1, \ldots, L^s_{\binom{t}{s}})$.

For an *s-uniform clutter* $\mathcal{G} := \{G_1, \ldots, G_k\} \subseteq \binom{E_t}{s}$, we define its row *characteristic vector* $\boldsymbol{\gamma}^{(s)}(\mathcal{G}) := (\gamma^{(s)}_1(\mathcal{G}), \ldots, \gamma^{(s)}_{\binom{t}{s}}(\mathcal{G})) \in \{0,1\}^{\binom{t}{s}}$ in the familiar way: for each j, where $1 \leq j \leq \binom{t}{s}$, we set

$$\gamma^{(s)}_j(\mathcal{G}) := \begin{cases} 1\,, & \text{if } \overrightarrow{\binom{E_t}{s}} \ni L^s_j \in \mathcal{G}, \\ 0\,, & \text{if } \overrightarrow{\binom{E_t}{s}} \ni L^s_j \notin \mathcal{G}; \end{cases}$$

see (9.18)–(9.24) in Example 9.2.

Now, given an arbitrary family $\mathcal{F} \subseteq \mathbf{2}^{[t]}$, we set

$$\boldsymbol{\gamma}^{(s)}(\mathcal{F}) := \boldsymbol{\gamma}^{(s)}(\mathcal{F} \cap \binom{E_t}{s})\,, \quad 0 \leq s \leq t\,,$$

and in a natural way[14] we define the *characteristic vector* $\boldsymbol{\gamma}(\mathcal{F}) := (\gamma_1(\mathcal{F}), \ldots, \gamma_{2^t}(\mathcal{F})) \in \{0,1\}^{2^t}$ of the family $\mathcal{F}$ to be the *concatenation*

$$\boldsymbol{\gamma}(\mathcal{F}) := \boldsymbol{\gamma}^{(0)}(\mathcal{F}) \centerdot \boldsymbol{\gamma}^{(1)}(\mathcal{F}) \centerdot \cdots \centerdot \boldsymbol{\gamma}^{(t-1)}(\mathcal{F}) \centerdot \boldsymbol{\gamma}^{(t)}(\mathcal{F})\,;$$

see (9.25)–(9.40).

- The characteristic vector $\boldsymbol{\gamma}(\mathbf{2}^{[t]}) = \mathrm{T}^{(+)}_{2^t} := (1, \ldots, 1)$, whose components are all 1's, describes the linearly ordered *power set* $\mathbf{2}^{[t]}$ of the ground set E_t; see (9.30).

- The Hamming weights $\mathtt{hwt}(\boldsymbol{\gamma}^{(s)}(\mathcal{F}))$ of the vectors $\boldsymbol{\gamma}^{(s)}(\mathcal{F})$, $0 \leq s \leq t$, are the components $f_s(\mathcal{F}; t)$ of the *long f-vectors* $\boldsymbol{f}(\mathcal{F}; t)$ associated with families $\mathcal{F} \subseteq \mathbf{2}^{[t]}$; see Section B.1.

---

[14] Indeed, this is the most popular ordering of all subsets of a nonempty finite set, the *lexicographic ordering subordinated to cardinality*, that is, considering smaller subsets first, and in case of equal cardinality, the lexicographic ordering; see page 77 of Ref. [108].

- If $\mathcal{F}' \subseteq \mathbf{2}^{[t]}$ and $\mathcal{F}'' \subseteq \mathbf{2}^{[t]}$ are families of subsets of the ground set E_t, then we will use the *componentwise product* of their characteristic vectors
$\boldsymbol{\gamma}(\mathcal{F}') * \boldsymbol{\gamma}(\mathcal{F}'') := \big(\gamma_1(\mathcal{F}') \cdot \gamma_1(\mathcal{F}''), \ldots, \gamma_{2^t}(\mathcal{F}') \cdot \gamma_{2^t}(\mathcal{F}'')\big) \in \{0, 1\}^{2^t}$
to describe[15] the *intersection* of these families:

$$\boldsymbol{\gamma}(\mathcal{F}' \cap \mathcal{F}'') = \boldsymbol{\gamma}(\mathcal{F}') * \boldsymbol{\gamma}(\mathcal{F}'') .$$

- Let $\Gamma(k)$ denote the subset $A \subseteq E_t$, for which the characteristic vector of the corresponding one-member clutter $\{A\}$ on E_t by convention is the kth standard unit vector $\boldsymbol{\sigma}(k)$ of the space $\mathbb{R}^{2^t}$; we thus use the map

$$\Gamma : [2^t] \to \mathbf{2}^{[t]} , \quad k \mapsto A\colon \ \boldsymbol{\gamma}(\{A\}) = \boldsymbol{\sigma}(k) \in \{0, 1\}^{2^t} ;$$

see (9.41)–(9.43). Conversely, we denote by $\Gamma^{-1}(A)$, where $A \subseteq E_t$, the position number k such that the vector $\boldsymbol{\sigma}(k)$ is the characteristic vector of the one-member clutter $\{A\}$ on E_t:

$$\Gamma^{-1} : \mathbf{2}^{[t]} \to [2^t] , \quad A \mapsto k\colon \ \boldsymbol{\sigma}(k) = \boldsymbol{\gamma}(\{A\}) \in \{0, 1\}^{2^t} ;$$

see (9.41)–(9.43).

By construction, we have the implications

$$\ell', \ell'' \in [2^t] , \ \ell' < \ell'' \implies |\Gamma(\ell')| \leq |\Gamma(\ell'')| ; \tag{9.17}$$

$$A, B \in \mathbf{2}^{[t]} , \ |A| < |B| \implies \Gamma^{-1}(A) < \Gamma^{-1}(B) ,$$

and, in particular,

$$A, B \in \mathbf{2}^{[t]} , \ A \subsetneqq B \implies \Gamma^{-1}(A) < \Gamma^{-1}(B) .$$

Note also that for any index $\ell \in [2^t]$, the disjoint union

$$\Gamma(\ell) \ \dot{\cup} \ \Gamma(2^t - \ell + 1) = E_t$$

is a *partition* of the ground set.

Example 9.2. Suppose $t := 3$, and $E_t = \{1, 2, 3\}$. We have

$$\boldsymbol{\gamma}^{(0)}(\tbinom{E_t}{0}) := \boldsymbol{\gamma}^{(0)}(\{\hat{0}\}) = (1) \in \{0, 1\}^{\binom{t}{0}} , \tag{9.18}$$

$$\boldsymbol{\gamma}^{(1)}(\tbinom{E_t}{1}) := \boldsymbol{\gamma}^{(1)}(\{\{1\}, \{2\}, \{3\}\}) = (1, 1, 1) \in \{0, 1\}^{\binom{t}{1}} , \tag{9.19}$$

$$\boldsymbol{\gamma}^{(2)}(\tbinom{E_t}{2}) := \boldsymbol{\gamma}^{(2)}(\{\{1, 2\}, \{1, 3\}, \{2, 3\}\}) = (1, 1, 1) \in \{0, 1\}^{\binom{t}{2}} , \tag{9.20}$$

$$\boldsymbol{\gamma}^{(t)}(\tbinom{E_t}{t}) := \boldsymbol{\gamma}^{(t)}(\{\{1, 2, 3\}\}) = (1) \in \{0, 1\}^{\binom{t}{t}} , \tag{9.21}$$

$$\boldsymbol{\gamma}^{(1)}(\{\{2\}\}) = (0, 1, 0) \in \{0, 1\}^{\binom{t}{1}} , \tag{9.22}$$

$$\boldsymbol{\gamma}^{(2)}(\{\{1, 2\}, \{2, 3\}\}) = (1, 0, 1) \in \{0, 1\}^{\binom{t}{2}} , \tag{9.23}$$

$$\boldsymbol{\gamma}^{(2)}(\{\{1, 3\}\}) = (0, 1, 0) \in \{0, 1\}^{\binom{t}{2}} , \tag{9.24}$$

---

[15] The notation $\prod^*$ will be used to denote the *componentwise product* of several characteristic vectors.

and

$$\gamma(\emptyset) = (0, 0, 0, 0, 0, 0, 0, 0) \in \{0, 1\}^{2^t}, \quad (9.25)$$
$$\gamma(\tbinom{E_t}{0}) = (1, 0, 0, 0, 0, 0, 0, 0), \quad (9.26)$$
$$\gamma(\tbinom{E_t}{1}) = (0, 1, 1, 1, 0, 0, 0, 0), \quad (9.27)$$
$$\gamma(\tbinom{E_t}{2}) = (0, 0, 0, 0, 1, 1, 1, 0), \quad (9.28)$$
$$\gamma(\tbinom{E_t}{t}) = (0, 0, 0, 0, 0, 0, 0, 1), \quad (9.29)$$
$$\gamma(\mathbf{2}^{[t]}) = \mathrm{T}^{(+)}_{2^t} := (1, 1, 1, 1, 1, 1, 1, 1). \quad (9.30)$$

If $\mathcal{A} := \{A_1, A_2\}$ and $\mathcal{B} := \{B_1, B_2\}$ are clutters on E_t, where $A_1 := \{1, 2\}$, $A_2 := \{2, 3\}$, $B_1 := \{1, 3\}$, $B_2 := \{2\}$, and $\mathcal{B} = \mathfrak{B}(\mathcal{A})$, then we have

$$\gamma(\mathcal{A}) := \gamma(\{A_1, A_2\}) \quad (9.31)$$
$$:= \gamma(\{\{1, 2\}, \{2, 3\}\}) = (0, 0, 0, 0, 1, 0, 1, 0) \in \{0, 1\}^{2^t}, \quad (9.32)$$
$$\gamma(\mathcal{B}) := \gamma(\{B_1, B_2\}) \quad (9.33)$$
$$:= \gamma(\{\{1, 3\}, \{2\}\}) = (0, 0, 1, 0, 0, 1, 0, 0), \quad (9.34)$$
$$\gamma(\mathcal{A}^{\triangledown}) := \gamma(\{\{1, 2\}, \{2, 3\}\}^{\triangledown}) = (0, 0, 0, 0, 1, 0, 1, 1), \quad (9.35)$$
$$\gamma(\mathcal{B}^{\triangledown}) := \gamma(\{\{1, 3\}, \{2\}\}^{\triangledown}) = (0, 0, 1, 0, 1, 1, 1, 1), \quad (9.36)$$
$$\gamma(\{A_1\}^{\triangledown}) := \gamma(\{\{1, 2\}\}^{\triangledown}) = (0, 0, 0, 0, 1, 0, 0, 1), \quad (9.37)$$
$$\gamma(\{A_2\}^{\triangledown}) := \gamma(\{\{2, 3\}\}^{\triangledown}) = (0, 0, 0, 0, 0, 0, 1, 1), \quad (9.38)$$
$$\gamma(\{B_1\}^{\triangledown}) := \gamma(\{\{1, 3\}\}^{\triangledown}) = (0, 0, 0, 0, 0, 1, 0, 1), \quad (9.39)$$
$$\gamma(\{B_2\}^{\triangledown}) := \gamma(\{\{2\}\}^{\triangledown}) = (0, 0, 1, 0, 1, 0, 1, 1). \quad (9.40)$$

We have

$$\Gamma(3) = \{2\}, \qquad \gamma(\{\{2\}\}) = \sigma(3) \in \{0, 1\}^{2^t}, \qquad \Gamma^{-1}(\{2\}) = 3, \quad (9.41)$$
$$\Gamma(6) = \{1, 3\}, \qquad \gamma(\{\{1, 3\}\}) = \sigma(6) \in \{0, 1\}^{2^t}, \qquad \Gamma^{-1}(\{1, 3\}) = 6, \quad (9.42)$$
$$\Gamma(2^t) = E_t, \qquad \gamma(\{E_t\}) = \sigma(2^t) \in \{0, 1\}^{2^t}, \qquad \Gamma^{-1}(E_t) = 2^t. \quad (9.43)$$

- Given a family $\mathcal{F} \subseteq \mathbf{2}^{[t]}$ of subsets of the ground set E_t, we call the vertex

$$T_{\mathcal{F}} := {}_{-\operatorname{supp}(\gamma(\mathcal{F}))}\mathrm{T}^{(+)}_{2^t} = \mathrm{T}^{(+)}_{2^t} - 2\gamma(\mathcal{F}) \quad (9.44)$$

of the discrete hypercube $\{1, -1\}^{2^t}$ the *characteristic covector* of the family $\mathcal{F}$; see Example 9.3.

If we let $\mathrm{T}^{(+)}_{\binom{t}{s}} := (1, \ldots, 1) \in \mathbb{R}^{\binom{t}{s}}$ denote the $\binom{t}{s}$-dimensional row vector of all 1's, where $0 \leq s \leq t$, then

$$T^{(s)}_{\mathcal{F}} := {}_{-\operatorname{supp}(\boldsymbol{\gamma}^{(s)}(\mathcal{F}))}\mathrm{T}^{(+)}_{\binom{t}{s}} = \mathrm{T}^{(+)}_{\binom{t}{s}} - 2\boldsymbol{\gamma}^{(s)}(\mathcal{F})\,, \quad 0 \leq s \leq t\,.$$

Example 9.3. Suppose $t := 3$ and $E_t = \{1, 2, 3\}$. If $\mathcal{A}$ and $\mathcal{B} = \mathfrak{B}(\mathcal{A})$ are clutters on the ground set E_t, mentioned in Example 9.2 on page 235, then we have

$$\begin{aligned}
\boldsymbol{\gamma}(\mathcal{A}) &= (0,\ 0,\ 0,\ 0,\ 1,\ 0,\ 1,\ 0) \in \{0, 1\}^{2^t}\,,\\
T_{\mathcal{A}} &:= (1,\ 1,\ 1,\ 1, -1,\ 1, -1,\ 1) \in \{1, -1\}^{2^t}\,;\\
\boldsymbol{\gamma}(\mathcal{A}^{\triangledown}) &= (0,\ 0,\ 0,\ 0,\ 1,\ 0,\ 1,\ 1)\,,\\
T_{\mathcal{A}^{\triangledown}} &:= (1,\ 1,\ 1,\ 1, -1,\ 1, -1, -1)\,;\\
\boldsymbol{\gamma}(\mathcal{B}) &= (0,\ 0,\ 1,\ 0,\ 0,\ 1,\ 0,\ 0)\,,\\
T_{\mathcal{B}} &:= (1,\ 1, -1,\ 1,\ 1, -1,\ 1,\ 1)\,;\\
\boldsymbol{\gamma}(\mathcal{B}^{\triangledown}) &= (0,\ 0,\ 1,\ 0,\ 1,\ 1,\ 1,\ 1)\,,\\
T_{\mathcal{B}^{\triangledown}} &:= (1,\ 1, -1,\ 1, -1, -1, -1, -1)\,.
\end{aligned}$$

9.4 Increasing Families of Blocking Sets, and Blockers: Characteristic Vectors and Characteristic Covectors

In this section, we begin with somewhat sophisticated restatements of the simple basic observations (9.8), (9.10) and (9.11) on set families in terms of their characteristic vectors.

- For a nontrivial clutter $\mathcal{A} \subset \mathbf{2}^{[t]}$ on the ground set E_t, we have

$$\begin{aligned}
&\boldsymbol{\gamma}(\mathfrak{B}(\mathcal{A})^{\triangledown}) * \boldsymbol{\gamma}(\mathcal{A}^{\triangledown})\\
&\quad= \mathtt{rn}(\boldsymbol{\gamma}(\mathcal{A}^{\triangledown})) * \boldsymbol{\gamma}(\mathcal{A}^{\triangledown}) := \big(\mathrm{T}^{(+)}_{2^t} - \boldsymbol{\gamma}(\mathcal{A}^{\triangledown}) \cdot \overline{\mathbf{U}}(2^t)\big) * \boldsymbol{\gamma}(\mathcal{A}^{\triangledown})\\
&\quad= \boldsymbol{\gamma}(\mathfrak{B}(\mathcal{A})^{\triangledown}) * \mathtt{rn}(\boldsymbol{\gamma}(\mathfrak{B}(\mathcal{A})^{\triangledown}))\\
&\quad:= \boldsymbol{\gamma}(\mathfrak{B}(\mathcal{A})^{\triangledown}) * \big(\mathrm{T}^{(+)}_{2^t} - \boldsymbol{\gamma}(\mathfrak{B}(\mathcal{A})^{\triangledown}) \cdot \overline{\mathbf{U}}(2^t)\big)\\
&\quad= \begin{cases} \boldsymbol{\gamma}(\mathcal{A}^{\triangledown})\,, & \text{if } \#\mathcal{A}^{\triangledown} < 2^{t-1},\\ \boldsymbol{\gamma}(\mathfrak{B}(\mathcal{A})^{\triangledown})\,, & \text{if } \#\mathcal{A}^{\triangledown} > 2^{t-1},\\ \boldsymbol{\gamma}(\mathcal{A}^{\triangledown}) = \boldsymbol{\gamma}(\mathfrak{B}(\mathcal{A})^{\triangledown})\,, & \text{if } \#\mathcal{A}^{\triangledown} = 2^{t-1}. \end{cases}
\end{aligned}$$

Remark 9.4. Let $\mathcal{A} \subset \mathbf{2}^{[t]}$ be a nontrivial clutter on the ground set E_t. We have

$$\gamma_1(\mathfrak{B}(\mathcal{A})^{\triangledown}) = 0\,, \quad \text{and} \quad \gamma_{2^t}(\mathfrak{B}(\mathcal{A})^{\triangledown}) = 1\,;$$

$$\boldsymbol{\gamma}(\mathfrak{B}(\mathcal{A})^{\triangledown}) = \mathtt{rn}(\boldsymbol{\gamma}(\mathcal{A}^{\triangledown})) := \mathrm{T}^{(+)}_{2^t} - \boldsymbol{\gamma}(\mathcal{A}^{\triangledown}) \cdot \overline{\mathbf{U}}(2^t)\,; \tag{9.45}$$

$$T_{\mathfrak{B}(\mathcal{A})^{\triangledown}} = \mathtt{ro}(T_{\mathcal{A}^{\triangledown}}) := -T_{\mathcal{A}^{\triangledown}} \cdot \overline{\mathbf{U}}(2^t)\,; \tag{9.46}$$

$$\underbrace{\mathtt{hwt}(\boldsymbol{\gamma}(\mathfrak{B}(\mathcal{A})^{\triangledown}))}_{\#\mathfrak{B}(\mathcal{A})^{\triangledown}} + \underbrace{\mathtt{hwt}(\boldsymbol{\gamma}(\mathcal{A}^{\triangledown}))}_{\#\mathcal{A}^{\triangledown}} = 2^t\,;$$

$$\underbrace{|\mathfrak{n}(T_{\mathfrak{B}(\mathcal{A})^{\triangledown}})|}_{\#\mathfrak{B}(\mathcal{A})^{\triangledown}} + \underbrace{|\mathfrak{n}(T_{\mathcal{A}^{\triangledown}})|}_{\#\mathcal{A}^{\triangledown}} = 2^t\,;$$

$$\boldsymbol{\gamma}^{(s)}(\mathfrak{B}(\mathcal{A})^{\triangledown}) = \mathtt{rn}(\boldsymbol{\gamma}^{(t-s)}(\mathcal{A}^{\triangledown})) \tag{9.47}$$

$$:= \mathrm{T}^{(+)}_{\binom{t}{s}} - \boldsymbol{\gamma}^{(t-s)}(\mathcal{A}^{\triangledown}) \cdot \overline{\mathbf{U}}(\tbinom{t}{s})\,, \quad 0 \le s \le t\,; \tag{9.48}$$

$$T^{(s)}_{\mathfrak{B}(\mathcal{A})^{\triangledown}} = \mathtt{ro}(T^{(t-s)}_{\mathcal{A}^{\triangledown}}) := -T^{(t-s)}_{\mathcal{A}^{\triangledown}} \cdot \overline{\mathbf{U}}(\tbinom{t}{s})\,, \quad 0 \le s \le t\,; \tag{9.49}$$

$$\underbrace{\mathtt{hwt}(\boldsymbol{\gamma}^{(s)}(\mathfrak{B}(\mathcal{A})^{\triangledown}))}_{\#(\mathfrak{B}(\mathcal{A})^{\triangledown}\cap\binom{E_t}{s})} + \underbrace{\mathtt{hwt}(\boldsymbol{\gamma}^{(t-s)}(\mathcal{A}^{\triangledown}))}_{\#(\mathcal{A}^{\triangledown}\cap\binom{E_t}{t-s})} = \tbinom{t}{s}\,, \quad 0 \le s \le t\,;$$

$$\underbrace{|\mathfrak{n}(T^{(s)}_{\mathfrak{B}(\mathcal{A})^{\triangledown}})|}_{\#(\mathfrak{B}(\mathcal{A})^{\triangledown}\cap\binom{E_t}{s})} + \underbrace{|\mathfrak{n}(T^{(t-s)}_{\mathcal{A}^{\triangledown}})|}_{\#(\mathcal{A}^{\triangledown}\cap\binom{E_t}{t-s})} = \tbinom{t}{s}\,, \quad 0 \le s \le t\,.$$

In addition to (9.45) and (9.46), relations (9.47) and (9.49) imply that

$$\boldsymbol{\gamma}(\mathfrak{B}(\mathcal{A})^{\triangledown}) = \underbrace{\mathtt{rn}(\boldsymbol{\gamma}^{(t)}(\mathcal{A}^{\triangledown}))}_{(0)} \centerdot \mathtt{rn}(\boldsymbol{\gamma}^{(t-1)}(\mathcal{A}^{\triangledown}))$$

$$\centerdot \cdots \centerdot \mathtt{rn}(\boldsymbol{\gamma}^{(1)}(\mathcal{A}^{\triangledown})) \centerdot \underbrace{\mathtt{rn}(\boldsymbol{\gamma}^{(0)}(\mathcal{A}^{\triangledown}))}_{(1)}\,,$$

$$T^{\triangledown}_{\mathfrak{B}(\mathcal{A})} = \underbrace{\mathtt{ro}(T^{(t)}_{\mathcal{A}^{\triangledown}})}_{(1)} \centerdot \mathtt{ro}(T^{(t-1)}_{\mathcal{A}^{\triangledown}}) \centerdot \cdots \centerdot \mathtt{ro}(T^{(1)}_{\mathcal{A}^{\triangledown}}) \centerdot \underbrace{\mathtt{ro}(T^{(0)}_{\mathcal{A}^{\triangledown}})}_{(-1)}\,.$$

- In view of (9.17), if

$$\ell^{\star} := \min \operatorname{supp}(\boldsymbol{\gamma}(\mathfrak{B}(\mathcal{A})^{\triangledown})) = \min \mathfrak{n}(T_{\mathfrak{B}(\mathcal{A})^{\triangledown}})\,,$$

then the member $\Gamma(\ell^{\star})$ of the blocker $\mathfrak{B}(\mathcal{A})$ of a nontrivial clutter $\mathcal{A} \subset \mathbf{2}^{[t]}$ is a blocking set of *minimum* cardinality for $\mathcal{A}$, that is, the vectors $\boldsymbol{\chi}(\Gamma(\ell^{\star}))$ and ${}_{-\Gamma(\ell^{\star})}\mathrm{T}^{(+)}$ provide the solution $|\Gamma(\ell^{\star})| = \tau(\mathcal{A})$ to the set covering problems (9.15) and (9.16), respectively:

$$\boldsymbol{\chi}(\Gamma(\ell^{\star})) \in \arg\min\{\mathrm{T}^{(+)}\tilde{\mathbf{z}}^{\top} : \tilde{\mathbf{z}} \in \widetilde{\boldsymbol{\mathcal{S}}}\} ,$$
$${}_{-\Gamma(\ell^{\star})}\mathrm{T}^{(+)} \in \arg\max\{\mathrm{T}^{(+)}\mathbf{z}^{\top} : \mathbf{z} \in \boldsymbol{\mathcal{S}}\} .$$

9.4.1 A Clutter $\{\{a\}\}$

Let $\{\{a\}\}$ be a (nontrivial) clutter on the ground set E_t, whose only member is a *one-element* subset $\{a\} \subset E_t$.

9.4.1.1 The principal increasing family of blocking sets $\mathfrak{B}(\{\{a\}\})^{\triangledown} = \{\{a\}\}^{\triangledown}$

- The *increasing family* of *blocking sets* $\mathfrak{B}(\{\{a\}\})^{\triangledown}$ of the *self-dual* clutter $\{\{a\}\}$ coincides with the principal increasing family $\{\{a\}\}^{\triangledown}$.

- We will use the notation $\widetilde{\mathfrak{a}}(a) := \widetilde{\mathfrak{a}}(a; 2^t)$ and $\mathfrak{a}(a) := \mathfrak{a}(a; 2^t)$ to denote the characteristic vector and the characteristic covector, respectively, that are associated with the principal increasing family $\{\{a\}\}^{\triangledown} = \mathfrak{B}(\{\{a\}\})^{\triangledown}$:

$$\widetilde{\mathfrak{a}}(a) := \boldsymbol{\gamma}(\{\{a\}\}^{\triangledown}) = \boldsymbol{\gamma}(\mathfrak{B}(\{\{a\}\})^{\triangledown}) \in \{0, 1\}^{2^t} ,$$
$$\mathfrak{a}(a) := T_{\{\{a\}\}^{\triangledown}} = T_{\mathfrak{B}(\{\{a\}\})^{\triangledown}} \in \{1, -1\}^{2^t} .$$

We have

$$\widetilde{\mathfrak{a}}(a) = \underbrace{(0)}_{\widetilde{\mathfrak{a}}^{(0)}(a)} \centerdot \underbrace{\boldsymbol{\chi}(\{a\})}_{\widetilde{\mathfrak{a}}^{(1)}(a)} \centerdot \underbrace{\boldsymbol{\gamma}^{(2)}(\{\{a\}\}^{\triangledown})}_{\widetilde{\mathfrak{a}}^{(2)}(a)} \centerdot \cdots \centerdot \underbrace{\boldsymbol{\gamma}^{(t-1)}(\{\{a\}\}^{\triangledown})}_{\widetilde{\mathfrak{a}}^{(t-1)}(a)} \centerdot \underbrace{(1)}_{\widetilde{\mathfrak{a}}^{(t)}(a)} ,$$

see (9.51), (9.56) and (9.61) in Example 9.8;

$$\mathfrak{a}(a) = \underbrace{(1)}_{\mathfrak{a}^{(0)}(a)} \centerdot \underbrace{{}_{-\{a\}}\mathrm{T}^{(+)}}_{\mathfrak{a}^{(1)}(a)} \centerdot \underbrace{T^{(2)}_{\{\{a\}\}^{\triangledown}}}_{\mathfrak{a}^{(2)}(a)} \centerdot \cdots \centerdot \underbrace{T^{(t-1)}_{\{\{a\}\}^{\triangledown}}}_{\mathfrak{a}^{(t-1)}(a)} \centerdot \underbrace{(-1)}_{\mathfrak{a}^{(t)}(a)} ,$$

see (9.52), (9.57) and (9.62).

Remark 9.5. (see Remark 9.4, and cf. Remark 9.9). Note that

$$\widetilde{\mathfrak{a}}(a) = \mathtt{rn}(\widetilde{\mathfrak{a}}(a)) , \quad \text{and} \quad \mathfrak{a}(a) = \mathtt{ro}(\mathfrak{a}(a)) ; \tag{9.50}$$

$$\mathtt{hwt}(\widetilde{\mathfrak{a}}(a)) = |\mathfrak{n}(\mathfrak{a}(a))| = \#\{\{a\}\}^{\triangledown} = \#\mathfrak{B}(\{\{a\}\})^{\triangledown} = 2^{t-1}\ ;$$

$$\widetilde{\mathfrak{a}}^{(s)}(a) = \mathtt{rn}(\widetilde{\mathfrak{a}}^{(t-s)}(a))\ ,$$
$$\text{and} \quad \mathfrak{a}^{(s)}(a) = \mathtt{ro}(\mathfrak{a}^{(t-s)}(a))\ , \quad 0 \leq s \leq t\ ;$$

$$\mathtt{hwt}(\widetilde{\mathfrak{a}}^{(s)}(a)) = |\mathfrak{n}(\mathfrak{a}^{(s)}(a))| = \tbinom{t-1}{s-1}\ , \quad 0 \leq s \leq t\ .$$

9.4.1.2 The blocker $\mathfrak{B}(\{\{a\}\}) = \{\{a\}\}$

- The *blocker* $\mathfrak{B}(\{\{a\}\})$ coincides with the self-dual clutter $\{\{a\}\}$.

- We associate with the clutter $\mathfrak{B}(\{\{a\}\}) = \{\{a\}\}$ its characteristic vector $\boldsymbol{\gamma}(\{\{a\}\}) = \boldsymbol{\gamma}(\mathfrak{B}(\{\{a\}\})) \in \{0,1\}^{2^t}$ and its characteristic covector $T_{\{\{a\}\}} = T_{\mathfrak{B}(\{\{a\}\})} \in \{1,-1\}^{2^t}$, where

$$\boldsymbol{\gamma}(\{\{a\}\}) = \boldsymbol{\gamma}(\mathfrak{B}(\{\{a\}\})) = (0) \ \bullet\ \boldsymbol{\chi}(\{a\}) \ \bullet\ (0,\ldots,0) \ \bullet\ \cdots \ \bullet\ (0)\ ;$$

see (9.55), (9.60) and (9.65) in Example 9.8.

9.4.1.3 More on the principal increasing family $\mathfrak{B}(\{\{a\}\})^{\triangledown} = \{\{a\}\}^{\triangledown}$

In view of (9.50), we can make the following observation:

Remark 9.6. (cf. Remark 9.10). For any element $a \in E_t$, we have

$$\begin{aligned}
&\min\{i \in E_{2^t} : T_{\{\{a\}\}^{\triangledown}}(i) = -1\}\\
&\quad := \min\{i \in E_{2^t} : \gamma_i(\{\{a\}\}^{\triangledown}) = 1\} := \Gamma^{-1}(\{a\}) = 1 + a\ ,\\
&\max\{i \in E_{2^t} : T_{\{\{a\}\}^{\triangledown}}(i) = 1\}\\
&\quad := \max\{i \in E_{2^t} : \gamma_i(\{\{a\}\}^{\triangledown}) = 0\} = 2^t - \min\{a\} = 2^t - a\ ;\\
&\underbrace{\min\{j \in E_{2^t} : T_{\mathfrak{B}(\{\{a\}\})^{\triangledown}}(j) = -1\}}_{\min\{j \in E_{2^t} :\, T_{\mathfrak{B}(\{\{a\}\})}(j) = -1\}}\\
&\quad := \underbrace{\min\{j \in E_{2^t} : \gamma_j(\mathfrak{B}(\{\{a\}\})^{\triangledown}) = 1\}}_{\min\{j \in E_{2^t} :\, \gamma_j(\mathfrak{B}(\{\{a\}\})) = 1\}} = 1 + \min\{a\} = 1 + a\ ,\\
&\max\{j \in E_{2^t} : T_{\mathfrak{B}(\{\{a\}\})^{\triangledown}}(j) = 1\}\\
&\quad := \max\{j \in E_{2^t} : \gamma_j(\mathfrak{B}(\{\{a\}\})^{\triangledown}) = 0\}\\
&\quad = 1 + 2^t - \Gamma^{-1}(\{a\}) = 2^t - a\ .
\end{aligned}$$

- We have

$$\{\{a\}\}^{\triangledown} \,\dot{\cup}\, \{E_t - \{a\}\}^{\vartriangle} = \mathbf{2}^{[t]} \,.$$

Let us denote by $\widetilde{\mathfrak{c}}(a) := \widetilde{\mathfrak{c}}(a; 2^t)$ and $\mathfrak{c}(a) := \mathfrak{c}(a; 2^t)$ the characteristic vector and the characteristic covector, respectively, of the principal decreasing family $\{E_t - \{a\}\}^{\vartriangle}$:

$$\widetilde{\mathfrak{c}}(a) := \boldsymbol{\gamma}(\{E_t - \{a\}\}^{\vartriangle}) \in \{0, 1\}^{2^t} \,,$$

see (9.53), (9.58) and (9.63);

$$\mathfrak{c}(a) := T_{\{E_t - \{a\}\}^{\vartriangle}} \in \{1, -1\}^{2^t} \,,$$

see (9.54), (9.59) and (9.64). We have

$$\begin{aligned} \widetilde{\mathfrak{c}}(a) &= \mathrm{T}^{(+)}_{2^t} - \widetilde{\mathfrak{a}}(a) = \widetilde{\mathfrak{a}}(a) \cdot \overline{\mathbf{U}}(2^t) \,, \\ \mathfrak{c}(a) &= -\mathfrak{a}(a) = \mathfrak{a}(a) \cdot \overline{\mathbf{U}}(2^t) \,. \end{aligned}$$

- For any two-element subset $\{i, j\} \subset E_t$ of the ground set, we have

$$\#(\{\{i\}\}^{\triangledown} \cap \{\{j\}\}^{\triangledown}) = \#\{\{i, j\}\}^{\triangledown} = 2^{t-2} \,,$$

and

$$\#(\underbrace{(\mathbf{2}^{[t]} - \{\{i\}\}^{\triangledown})}_{\{E_t - \{i\}\}^{\vartriangle}} \cap \underbrace{(\mathbf{2}^{[t]} - \{\{j\}\}^{\triangledown})}_{\{E_t - \{j\}\}^{\vartriangle}}) = \#\{E_t - \{i, j\}\}^{\vartriangle} = 2^{t-2} \,.$$

Thus, if i and j are elements of the ground set E_t, and $i \neq j$, then we have

$$\begin{aligned} d\big(\widetilde{\mathfrak{a}}(i), \widetilde{\mathfrak{a}}(j)\big) = d\big(\mathfrak{a}(i), \mathfrak{a}(j)\big) = d\big(\widetilde{\mathfrak{c}}(i), \widetilde{\mathfrak{c}}(j)\big) &= d\big(\mathfrak{c}(i), \mathfrak{c}(j)\big) \\ &= 2^{t-1} \,. \end{aligned}$$

Remark 9.7. For any two elements i and j of the ground set E_t we have

$$\big\langle \mathfrak{a}(i), \mathfrak{a}(j) \big\rangle = \big\langle \mathfrak{c}(i), \mathfrak{c}(j) \big\rangle = \delta_{i,j} \cdot 2^t \,.$$

In other words, the sequences of t row vectors

$$\left(\frac{1}{\sqrt{2^t}} \cdot \mathfrak{a}(1), \; \frac{1}{\sqrt{2^t}} \cdot \mathfrak{a}(2), \; \ldots, \; \frac{1}{\sqrt{2^t}} \cdot \mathfrak{a}(t) \right) \subset \mathbb{R}^{2^t}$$

and

$$\left(\frac{1}{\sqrt{2^t}} \cdot \mathfrak{c}(t), \; \frac{1}{\sqrt{2^t}} \cdot \mathfrak{c}(t-1), \; \ldots, \; \frac{1}{\sqrt{2^t}} \cdot \mathfrak{c}(1) \right) \subset \mathbb{R}^{2^t}$$

are both *orthonormal*.

Example 9.8. Suppose $t := 3$, and $E_t = \{1, 2, 3\}$. We have

$$\{0, 1\}^{2^t} \ni \widetilde{\mathfrak{a}}(1) := \boldsymbol{\gamma}(\{\{1\}\}^{\triangledown}) = \boldsymbol{\gamma}(\mathfrak{B}(\{\{1\}\})^{\triangledown}) = (\ 0,\ 1,\ 0,\ 0,\ 1,\ 1,\ 0,\ 1)\,, \tag{9.51}$$

$$\{1, -1\}^{2^t} \ni \mathfrak{a}(1) := T_{\{\{1\}\}^{\triangledown}} = T_{\mathfrak{B}(\{\{1\}\})^{\triangledown}} = (\ 1, -1,\ 1,\ 1, -1, -1,\ 1, -1)\,, \tag{9.52}$$

$$\widetilde{\mathfrak{c}}(1) := \boldsymbol{\gamma}(\{E_t - \{1\}\}^{\vartriangle}) = (\ 1,\ 0,\ 1,\ 1,\ 0,\ 0,\ 1,\ 0)\,, \tag{9.53}$$

$$\mathfrak{c}(1) := T_{\{E_t - \{1\}\}^{\vartriangle}} = (-1,\ 1, -1, -1,\ 1,\ 1, -1,\ 1)\,, \tag{9.54}$$

$$\boldsymbol{\gamma}(\{\{1\}\}) = \boldsymbol{\gamma}(\mathfrak{B}(\{\{1\}\})) = (\ 0,\ 1,\ 0,\ 0,\ 0,\ 0,\ 0,\ 0)\,, \tag{9.55}$$

$$\widetilde{\mathfrak{a}}(2) := \boldsymbol{\gamma}(\{\{2\}\}^{\triangledown}) = \boldsymbol{\gamma}(\mathfrak{B}(\{\{2\}\})^{\triangledown}) = (\ 0,\ 0,\ 1,\ 0,\ 1,\ 0,\ 1,\ 1)\,, \tag{9.56}$$

$$\mathfrak{a}(2) := T_{\{\{2\}\}^{\triangledown}} = T_{\mathfrak{B}(\{\{2\}\})^{\triangledown}} = (\ 1,\ 1, -1,\ 1, -1,\ 1, -1, -1)\,, \tag{9.57}$$

$$\widetilde{\mathfrak{c}}(2) := \boldsymbol{\gamma}(\{E_t - \{2\}\}^{\vartriangle}) = (\ 1,\ 1,\ 0,\ 1,\ 0,\ 1,\ 0,\ 0)\,, \tag{9.58}$$

$$\mathfrak{c}(2) := T_{\{E_t - \{2\}\}^{\vartriangle}} = (-1, -1,\ 1, -1,\ 1, -1,\ 1,\ 1)\,, \tag{9.59}$$

$$\boldsymbol{\gamma}(\{\{2\}\}) = \boldsymbol{\gamma}(\mathfrak{B}(\{\{2\}\})) = (\ 0,\ 0,\ 1,\ 0,\ 0,\ 0,\ 0,\ 0)\,, \tag{9.60}$$

$$\widetilde{\mathfrak{a}}(t) := \boldsymbol{\gamma}(\{\{t\}\}^{\triangledown}) = \boldsymbol{\gamma}(\mathfrak{B}(\{\{t\}\})^{\triangledown}) = (\ 0,\ 0,\ 0,\ 1,\ 0,\ 1,\ 1,\ 1)\,, \tag{9.61}$$

$$\mathfrak{a}(t) := T_{\{\{t\}\}^{\triangledown}} = T_{\mathfrak{B}(\{\{t\}\})^{\triangledown}} = (\ \ 1, \ \ 1, \ \ 1, -1, \ \ 1, -1, -1, -1)\,, \tag{9.62}$$

$$\widetilde{\mathfrak{c}}(t) := \boldsymbol{\gamma}(\{E_t - \{t\}\}^{\vartriangle}) = (\ \ 1, \ \ 1, \ \ 1, \ \ 0, \ \ 1, \ \ 0, \ \ 0, \ \ 0)\,, \tag{9.63}$$

$$\mathfrak{c}(t) := T_{\{E_t - \{t\}\}^{\vartriangle}} = (-1, -1, -1, \ \ 1, -1, \ \ 1, \ \ 1, \ \ 1)\,, \tag{9.64}$$

$$\boldsymbol{\gamma}(\{\{t\}\}) = \boldsymbol{\gamma}(\mathfrak{B}(\{\{t\}\})) = (\ \ 0, \ \ 0, \ \ 0, \ \ 1, \ \ 0, \ \ 0, \ \ 0, \ \ 0)\,, \tag{9.65}$$

$$\boldsymbol{\gamma}(\{\{1, 2\}\}) = (\ \ 0, \ \ 0, \ \ 0, \ \ 0, \ \ 1, \ \ 0, \ \ 0, \ \ 0)\,, \tag{9.66}$$

$$\boldsymbol{\gamma}(\{\{1, 2\}\}^{\triangledown}) = (\ \ 0, \ \ 0, \ \ 0, \ \ 0, \ \ 1, \ \ 0, \ \ 0, \ \ 1)\,, \tag{9.67}$$

$$\boldsymbol{\gamma}(\mathfrak{B}(\{\{1, 2\}\})) = (\ \ 0, \ \ 1, \ \ 1, \ \ 0, \ \ 0, \ \ 0, \ \ 0, \ \ 0)\,, \tag{9.68}$$

$$\boldsymbol{\gamma}(\mathfrak{B}(\{\{1, 2\}\})^{\triangledown}) = (\ \ 0, \ \ 1, \ \ 1, \ \ 0, \ \ 1, \ \ 1, \ \ 1, \ \ 1)\,, \tag{9.69}$$

$$\boldsymbol{\gamma}(\{\{1, t\}\}) = (\ \ 0, \ \ 0, \ \ 0, \ \ 0, \ \ 0, \ \ 1, \ \ 0, \ \ 0)\,, \tag{9.70}$$

$$\boldsymbol{\gamma}(\{\{1, t\}\}^{\triangledown}) = (\ \ 0, \ \ 0, \ \ 0, \ \ 0, \ \ 0, \ \ 1, \ \ 0, \ \ 1)\,, \tag{9.71}$$

$$\boldsymbol{\gamma}(\mathfrak{B}(\{\{1, t\}\})) = (\ \ 0, \ \ 1, \ \ 0, \ \ 1, \ \ 0, \ \ 0, \ \ 0, \ \ 0)\,, \tag{9.72}$$

$$\boldsymbol{\gamma}(\mathfrak{B}(\{\{1, t\}\})^{\triangledown}) = (\ \ 0, \ \ 1, \ \ 0, \ \ 1, \ \ 1, \ \ 1, \ \ 1, \ \ 1)\,, \tag{9.73}$$

$$\gamma(\{\{2,t\}\}) = (\ 0,\ \ 0,\ \ 0,\ \ 0,\ \ 0,\ \ 0,\ \ 1,\ \ 0)\,, \tag{9.74}$$

$$\gamma(\{\{2,t\}\}^{\triangledown}) = (\ 0,\ \ 0,\ \ 0,\ \ 0,\ \ 0,\ \ 0,\ \ 1,\ \ 1)\,, \tag{9.75}$$

$$\gamma(\mathfrak{B}(\{\{2,t\}\})) = (\ 0,\ \ 0,\ \ 1,\ \ 1,\ \ 0,\ \ 0,\ \ 0,\ \ 0)\,, \tag{9.76}$$

$$\gamma(\mathfrak{B}(\{\{2,t\}\})^{\triangledown}) = (\ 0,\ \ 0,\ \ 1,\ \ 1,\ \ 1,\ \ 1,\ \ 1,\ \ 1)\,. \tag{9.77}$$

- Given a nontrivial clutter $\mathcal{A} := \{A_1, \ldots, A_\alpha\} \subset \mathbf{2}^{[t]}$ on the ground set E_t, such that $\mathcal{A} \neq \{E_t\}$, we have

$$\begin{aligned}
\gamma(\mathcal{A}) :&= \sum_{i\in[\alpha]} \sigma(\Gamma^{-1}(A_i)) = \gamma(\mathcal{A}^{\triangledown}) * \gamma(\mathcal{A}^{\vartriangle}) \\
&= \sum_{i\in[\alpha]} \big(\gamma(\{A_i\}^{\triangledown}) * \gamma(\{A_i\}^{\vartriangle})\big) \\
&= \sum_{i\in[\alpha]} \Big(\prod_{a^i\in A_i}^{*} \tilde{\mathfrak{a}}(a^i) \ * \prod_{c^i\in E_t-A_i}^{*} \tilde{\mathfrak{c}}(c^i)\Big) \\
&= \sum_{i\in[\alpha]} \Big(\prod_{a^i\in A_i}^{*} \tilde{\mathfrak{a}}(a^i) \ * \Big(\Big(\prod_{c^i\in E_t-A_i}^{*} \tilde{\mathfrak{a}}(c^i)\Big)\cdot \overline{\mathbf{U}}(2^t)\Big)\Big) \\
&= \sum_{i\in[\alpha]} \Big(\Big(\Big(\prod_{a^i\in A_i}^{*} \tilde{\mathfrak{c}}(a^i)\Big)\cdot \overline{\mathbf{U}}(2^t)\Big) \ * \prod_{c^i\in E_t-A_i}^{*} \tilde{\mathfrak{c}}(c^i)\Big)\,.
\end{aligned}$$

9.4.2 A Clutter $\{A\}$

Let $\{A\}$ be a (nontrivial) clutter on the ground set E_t, whose only member is a *nonempty subset* $A \subseteq E_t$.

9.4.2.1 The increasing family of blocking sets $\mathfrak{B}(\{A\})^{\triangledown} = \{\{a\}\colon a \in A\}^{\triangledown}$

- The *family* of *blocking sets* $\mathfrak{B}(\{A\})^{\triangledown}$ of the clutter $\{A\}$ is the increasing family $\{\{a\}\colon a \in A\}^{\triangledown}$.

We have

$$\{A\}^{\triangledown} = \bigcap_{a\in A} \{\{a\}\}^{\triangledown}\,, \quad \text{and} \quad \mathfrak{B}(\{A\})^{\triangledown} = \bigcup_{a\in A} \{\{a\}\}^{\triangledown}\,.$$

Let us associate with the increasing families $\{A\}^{\triangledown}$ and $\mathfrak{B}(\{A\})^{\triangledown}$ their characteristic vectors $\boldsymbol{\gamma}(\{A\}^{\triangledown}) \in \{0,1\}^{2^t}$ and $\boldsymbol{\gamma}(\mathfrak{B}(\{A\})^{\triangledown}) \in \{0,1\}^{2^t}$, and their characteristic covectors $T_{\{A\}^{\triangledown}} \in \{1,-1\}^{2^t}$ and $T_{\mathfrak{B}(\{A\})^{\triangledown}} \in \{1,-1\}^{2^t}$, where

$$\begin{aligned}\boldsymbol{\gamma}(\{A\}^{\triangledown}) &= \boldsymbol{\gamma}\Big(\bigcap_{a\in A}\{\{a\}\}^{\triangledown}\Big)\\ &= \underbrace{(0)}_{\boldsymbol{\gamma}^{(0)}(\{A\}^{\triangledown})} \,\boldsymbol{.}\, \cdots \,\boldsymbol{.}\, \underbrace{(0,\ldots,0)}_{\boldsymbol{\gamma}^{(|A|-1)}(\{A\}^{\triangledown})} \,\boldsymbol{.}\, \underbrace{\boldsymbol{\gamma}^{(|A|)}(\{A\})}_{\boldsymbol{\gamma}^{(|A|)}(\{A\}^{\triangledown})}\\ &\boldsymbol{.}\, \underbrace{\boldsymbol{\gamma}^{(|A|+1)}\Big(\bigcap_{a\in A}\{\{a\}\}^{\triangledown}\Big)}_{\boldsymbol{\gamma}^{(|A|+1)}(\{A\}^{\triangledown})} \,\boldsymbol{.}\, \cdots \,\boldsymbol{.}\, \underbrace{\boldsymbol{\gamma}^{(t-1)}\Big(\bigcap_{a\in A}\{\{a\}\}^{\triangledown}\Big)}_{\boldsymbol{\gamma}^{(t-1)}(\{A\}^{\triangledown})} \,\boldsymbol{.}\, \underbrace{(1)}_{\boldsymbol{\gamma}^{(t)}(\{A\}^{\triangledown})}\,,\end{aligned}$$

see (9.67), (9.71) and (9.75) in Example 9.8, and

$$\begin{aligned}\boldsymbol{\gamma}(\mathfrak{B}(\{A\})^{\triangledown}) &= \boldsymbol{\gamma}\Big(\bigcup_{a\in A}\{\{a\}\}^{\triangledown}\Big)\\ &= \underbrace{(0)}_{\boldsymbol{\gamma}^{(0)}(\mathfrak{B}(\{A\})^{\triangledown})} \,\boldsymbol{.}\, \underbrace{\chi(A)}_{\boldsymbol{\gamma}^{(1)}(\mathfrak{B}(\{A\})^{\triangledown})}\\ &\boldsymbol{.}\, \underbrace{\boldsymbol{\gamma}^{(2)}\big(\bigcup_{a\in A}\{\{a\}\}^{\triangledown}\big)}_{\boldsymbol{\gamma}^{(2)}(\mathfrak{B}(\{A\})^{\triangledown})} \,\boldsymbol{.}\, \cdots \,\boldsymbol{.}\, \underbrace{\boldsymbol{\gamma}^{(t-1)}\big(\bigcup_{a\in A}\{\{a\}\}^{\triangledown}\big)}_{\boldsymbol{\gamma}^{(t-1)}(\mathfrak{B}(\{A\})^{\triangledown})} \,\boldsymbol{.}\, \underbrace{(1)}_{\boldsymbol{\gamma}^{(t)}(\mathfrak{B}(\{A\})^{\triangledown})}\,,\end{aligned}$$

see (9.69), (9.73) and (9.77).

Remark 9.9. (see Remark 9.4, and cf. Remark 9.5). Note that

$$\begin{aligned}\boldsymbol{\gamma}(\mathfrak{B}(\{A\})^{\triangledown}) &= \mathtt{rn}(\boldsymbol{\gamma}(\{A\}^{\triangledown})) := \mathrm{T}^{(+)}_{2^t} - \boldsymbol{\gamma}(\{A\}^{\triangledown})\cdot\overline{\mathbf{U}}(2^t)\,;\\ T_{\mathfrak{B}(\{A\})^{\triangledown}} &= \mathtt{ro}(T_{\{A\}^{\triangledown}}) := -T_{\{A\}^{\triangledown}}\cdot\overline{\mathbf{U}}(2^t)\,;\\ \underbrace{\mathtt{hwt}(\boldsymbol{\gamma}(\mathfrak{B}(\{A\})^{\triangledown}))}_{\#\mathfrak{B}(\{A\})^{\triangledown}} &= \underbrace{|\,\mathfrak{n}(T_{\mathfrak{B}(\{A\})^{\triangledown}})|}_{\#\mathfrak{B}(\{A\})^{\triangledown}} = 2^t - 2^{t-|A|}\,;\\ \underbrace{\mathtt{hwt}(\boldsymbol{\gamma}(\{A\}^{\triangledown}))}_{\#\{A\}^{\triangledown}} &= \underbrace{|\,\mathfrak{n}(T_{\{A\}^{\triangledown}})|}_{\#\{A\}^{\triangledown}} = 2^{t-|A|}\,;\end{aligned}$$

$$\begin{aligned}\boldsymbol{\gamma}^{(s)}(\mathfrak{B}(\{A\})^{\triangledown}) &= \mathtt{rn}(\boldsymbol{\gamma}^{(t-s)}(\{A\}^{\triangledown}))\\ &:= \mathrm{T}^{(+)}_{\binom{t}{s}} - \boldsymbol{\gamma}^{(t-s)}(\{A\}^{\triangledown})\cdot\overline{\mathbf{U}}(\tbinom{t}{s})\,,\quad 0\le s\le t\,;\end{aligned}$$

$$T^{(s)}_{\mathfrak{B}(\{A\})^{\triangledown}} = \mathtt{ro}(T^{(t-s)}_{\{A\}^{\triangledown}}) := -T^{(t-s)}_{\{A\}^{\triangledown}}\cdot\overline{\mathbf{U}}(\tbinom{t}{s})\,,\quad 0\le s\le t\,;$$

$$\underbrace{\mathtt{hwt}(\boldsymbol{\gamma}^{(s)}(\mathfrak{B}(\{A\})^{\triangledown}))}_{\#(\mathfrak{B}(\{A\})^{\triangledown}\cap\binom{E_t}{s})} = \underbrace{|\mathfrak{n}(T^{(s)}_{\mathfrak{B}(\{A\})^{\triangledown}})|}_{\#(\mathfrak{B}(\{A\})^{\triangledown}\cap\binom{E_t}{s})}$$
$$= \binom{t}{s} - \binom{t-|A|}{s}, \quad 0 \leq s \leq t;$$

$$\underbrace{\mathtt{hwt}(\boldsymbol{\gamma}^{(t-s)}(\{A\}^{\triangledown}))}_{\#(\{A\}^{\triangledown}\cap\binom{E_t}{t-s})} = \underbrace{|\mathfrak{n}(T^{(t-s)}_{\{A\}^{\triangledown}})|}_{\#(\{A\}^{\triangledown}\cap\binom{E_t}{t-s})} = \binom{t-|A|}{s}, \quad 0 \leq s \leq t.$$

9.4.2.2 The blocker $\mathfrak{B}(\{A\}) = \{\{a\}: a \in A\}$

– The *blocker* of the clutter $\{A\}$ is the clutter

$$\mathfrak{B}(\{A\}) = \{\{a\}: a \in A\}.$$

Thus, $\#\mathfrak{B}(\{A\}) = |A|$, and the members of the blocker $\mathfrak{B}(\{A\})$ are the one-element subsets of the set A.

– We associate with the clutters $\{A\}$ and $\mathfrak{B}(\{A\})$ their characteristic vectors $\boldsymbol{\gamma}(\{A\}) \in \{0,1\}^{2^t}$ and $\boldsymbol{\gamma}(\mathfrak{B}(\{A\})) \in \{0,1\}^{2^t}$, and their characteristic covectors $T_{\{A\}} \in \{1,-1\}^{2^t}$ and $T_{\mathfrak{B}(\{A\})} \in \{1,-1\}^{2^t}$, where

$$\boldsymbol{\gamma}(\{A\}) = \underbrace{(0)}_{\boldsymbol{\gamma}^{(0)}(\{A\})} \bullet \cdots \bullet \underbrace{(0,\ldots,0)}_{\boldsymbol{\gamma}^{(|A|-1)}(\{A\})}$$
$$\bullet\, \boldsymbol{\gamma}^{(|A|)}(\{A\}) \bullet \underbrace{(0,\ldots,0)}_{\boldsymbol{\gamma}^{(|A|+1)}(\{A\})} \bullet \cdots \bullet \underbrace{(0)}_{\boldsymbol{\gamma}^{(t)}(\{A\})},$$

see (9.66), (9.70) and (9.74) in Example 9.8, and

$$\boldsymbol{\gamma}(\mathfrak{B}(\{A\})) = \underbrace{(0)}_{\boldsymbol{\gamma}^{(0)}(\mathfrak{B}(\{A\}))} \bullet \underbrace{\boldsymbol{\chi}(A)}_{\boldsymbol{\gamma}^{(1)}(\mathfrak{B}(\{A\}))}$$
$$\bullet \underbrace{(0,\ldots,0)}_{\boldsymbol{\gamma}^{(2)}(\mathfrak{B}(\{A\}))} \bullet \cdots \bullet \underbrace{(0)}_{\boldsymbol{\gamma}^{(t)}(\mathfrak{B}(\{A\}))},$$

see (9.68), (9.72) and (9.76).

9.4.2.3 More on the increasing families $\{A\}^{\triangledown}$ and $\mathfrak{B}(\{A\})^{\triangledown}$

We can make the following observation:

Remark 9.10. (cf. Remark 9.6). For a nonempty subset $A \subseteq E_t$, we have

$$\begin{aligned}\min\{i \in E_{2^t}\colon T_{\{A\}^\triangledown}(i) = -1\} &:= \min\{i \in E_{2^t}\colon \gamma_i(\{A\}^\triangledown) = 1\}\\ &:= \Gamma^{-1}(A)\,,\\ \max\{i \in E_{2^t}\colon T_{\{A\}^\triangledown}(i) = 1\} &:= \max\{i \in E_{2^t}\colon \gamma_i(\{A\}^\triangledown) = 0\}\\ &= 2^t - \min A\,;\end{aligned}$$

$$\underbrace{\min\{j \in E_{2^t}\colon T_{\mathfrak{B}(\{A\})^\triangledown}(j) = -1\}}_{\min\{j\in E_{2^t}\colon T_{\mathfrak{B}(\{A\})}(j)=-1\}} := \underbrace{\min\{j \in E_{2^t}\colon \gamma_j(\mathfrak{B}(\{A\})^\triangledown) = 1\}}_{\min\{j\in E_{2^t}\colon \gamma_j(\mathfrak{B}(\{A\}))=1\}} = 1 + \min A\,,$$

$$\max\{j \in E_{2^t}\colon T_{\mathfrak{B}(\{A\})^\triangledown}(j) = 1\} := \max\{j \in E_{2^t}\colon \gamma_j(\mathfrak{B}(\{A\})^\triangledown) = 0\} = 1 + 2^t - \Gamma^{-1}(A)\,.$$

- Recall that the partition

$$\{A\}^\triangledown \dot\cup (\mathfrak{B}(\{A\})^{\complement})^\vartriangle = \mathbf{2}^{[t]}$$

implies that

$$\mathfrak{B}(\{A\})^\triangledown = \{D^{\complement}\colon D \in \mathbf{2}^{[t]} - \{A\}^\triangledown\}\,.$$

- Note that

$$\begin{aligned}\boldsymbol{\gamma}(\{A\}^\triangledown) &= \prod_{a\in A}^{*} \boldsymbol{\gamma}(\{\{a\}\}^\triangledown) =: \prod_{a\in A}^{*} \tilde{\mathfrak{a}}(a)\\ &= \prod_{a\in A}^{*} (\mathrm{T}^{(+)}_{2^t} - \tilde{\mathfrak{c}}(a)) = \Big(\prod_{a\in A}^{*} \tilde{\mathfrak{c}}(a)\Big)\cdot \bar{\mathbf{U}}(2^t)\,,\end{aligned}$$

and recall that

$$\boldsymbol{\gamma}(\mathfrak{B}(\{A\})^\triangledown) = \mathtt{rn}(\boldsymbol{\gamma}(\{A\}^\triangledown))\,.$$

Remark 9.11. For a nonempty subset $A \subseteq E_t$, we have:

(i)

$$\boldsymbol{\gamma}(\{A\}^\triangledown) = \prod_{a\in A}^{*} \tilde{\mathfrak{a}}(a)\,.$$

(ii)

$$\boldsymbol{\gamma}(\mathfrak{B}(\{A\})^\triangledown) = \mathrm{T}^{(+)}_{2^t} - \Big(\prod_{a\in A}^{*} \tilde{\mathfrak{a}}(a)\Big)\cdot \bar{\mathbf{U}}(2^t)\,.$$

9.4.3 A Clutter $\mathcal{A} := \{A_1, \ldots, A_\alpha\}$

Let $\mathcal{A} := \{A_1, \ldots, A_\alpha\}$ be a nontrivial clutter on the ground set E_t.

9.4.3.1 The increasing family of blocking sets $\mathfrak{B}(\mathcal{A})^\triangledown$

- See Remark 9.4, and note that

$$\mathcal{A}^\triangledown = \bigcup_{k\in[\alpha]} \{A_k\}^\triangledown = \bigcup_{k\in[\alpha]} \bigcap_{a^k\in A_k} \{\{a^k\}\}^\triangledown ,$$

and

$$\mathfrak{B}(\mathcal{A})^\triangledown = \bigcap_{k\in[\alpha]} \mathfrak{B}(\{A_k\})^\triangledown = \bigcap_{k\in[\alpha]} \bigcup_{a^k\in A_k} \{\{a^k\}\}^\triangledown .$$

- We associate with the increasing families $\mathcal{A}^\triangledown$ and $\mathfrak{B}(\mathcal{A})^\triangledown$ their characteristic vectors $\boldsymbol{\gamma}(\mathcal{A}^\triangledown) \in \{0, 1\}^{2^t}$ and $\boldsymbol{\gamma}(\mathfrak{B}(\mathcal{A})^\triangledown) \in \{0, 1\}^{2^t}$, and their characteristic covectors $T_{\mathcal{A}^\triangledown} \in \{1, -1\}^{2^t}$ and $T_{\mathfrak{B}(\mathcal{A})^\triangledown} \in \{1, -1\}^{2^t}$, where

$$\begin{aligned}\boldsymbol{\gamma}(\mathcal{A}^\triangledown) &= \boldsymbol{\gamma}\Big(\bigcup_{k\in[\alpha]} \bigcap_{a^k\in A_k} \{\{a^k\}\}^\triangledown\Big) \\ &= \underbrace{(0)}_{\boldsymbol{\gamma}^{(0)}(\mathcal{A}^\triangledown)} \cdot \boldsymbol{\gamma}^{(1)}\Big(\bigcup_{k\in[\alpha]} \bigcap_{a^k\in A_k} \{\{a^k\}\}^\triangledown\Big) \cdot \cdots \\ &\quad \cdot \boldsymbol{\gamma}^{(t-1)}\Big(\bigcup_{k\in[\alpha]} \bigcap_{a^k\in A_k} \{\{a^k\}\}^\triangledown\Big) \cdot \underbrace{(1)}_{\boldsymbol{\gamma}^{(t)}(\mathcal{A}^\triangledown)} ,\end{aligned}$$

and

$$\begin{aligned}\boldsymbol{\gamma}(\mathfrak{B}(\mathcal{A})^\triangledown) &= \boldsymbol{\gamma}\Big(\bigcap_{k\in[\alpha]} \bigcup_{a^k\in A_k} \{\{a^k\}\}^\triangledown\Big) \\ &= \underbrace{(0)}_{\boldsymbol{\gamma}^{(0)}(\mathfrak{B}(\mathcal{A})^\triangledown)} \cdot \boldsymbol{\gamma}^{(1)}\Big(\bigcap_{k\in[\alpha]} \bigcup_{a^k\in A_k} \{\{a^k\}\}^\triangledown\Big) \cdot \cdots \\ &\quad \cdot \boldsymbol{\gamma}^{(t-1)}\Big(\bigcap_{k\in[\alpha]} \bigcup_{a^k\in A_k} \{\{a^k\}\}^\triangledown\Big) \cdot \underbrace{(1)}_{\boldsymbol{\gamma}^{(t)}(\mathfrak{B}(\mathcal{A})^\triangledown)} .\end{aligned}$$

9.4.3.2 The blocker $\mathfrak{B}(\mathcal{A})$

- The *blocker* of the clutter $\mathcal{A}$ is the clutter

$$\begin{aligned}\mathfrak{B}(\mathcal{A}) &= \mathbf{min} \bigcap_{k\in[\alpha]} \mathfrak{B}(\{A_k\})^{\triangledown} \\ &= \mathbf{min} \bigcap_{k\in[\alpha]} \{\{a^k\} \colon a^k \in A_k\}^{\triangledown} = \mathbf{min} \bigcap_{k\in[\alpha]} \bigcup_{a^k\in A_k} \{\{a^k\}\}^{\triangledown} .\end{aligned}$$

- We associate with the clutters $\mathcal{A}$ and $\mathfrak{B}(\mathcal{A})$ their characteristic vectors $\boldsymbol{\gamma}(\mathcal{A}) \in \{0,1\}^{2^t}$ and $\boldsymbol{\gamma}(\mathfrak{B}(\mathcal{A})) \in \{0,1\}^{2^t}$, and their characteristic covectors $T_{\mathcal{A}} \in \{1,-1\}^{2^t}$ and $T_{\mathfrak{B}(\mathcal{A})} \in \{1,-1\}^{2^t}$, where

$$\boldsymbol{\gamma}(\mathcal{A}) := \underbrace{(0)}_{\boldsymbol{\gamma}^{(0)}(\mathcal{A})} \,\boldsymbol{.}\, \boldsymbol{\gamma}^{(1)}(\mathcal{A}) \,\boldsymbol{.}\, \cdots \,\boldsymbol{.}\, \boldsymbol{\gamma}^{(t)}(\mathcal{A}) ,$$

and

$$\begin{aligned}\boldsymbol{\gamma}(\mathfrak{B}(\mathcal{A})) = \underbrace{(0)}_{\boldsymbol{\gamma}^{(0)}(\mathfrak{B}(\mathcal{A}))} &\,\boldsymbol{.}\, \boldsymbol{\gamma}^{(1)}(\mathbf{min} \bigcap_{k\in[\alpha]} \bigcup_{a^k\in A_k} \{\{a^k\}\}^{\triangledown}) \\ &\boldsymbol{.}\, \cdots \,\boldsymbol{.}\, \boldsymbol{\gamma}^{(t)}(\mathbf{min} \bigcap_{k\in[\alpha]} \bigcup_{a^k\in A_k} \{\{a^k\}\}^{\triangledown}) .\end{aligned}$$

9.4.3.3 More on the increasing families $\mathcal{A}^{\triangledown}$ and $\mathfrak{B}(\mathcal{A})^{\triangledown}$

- Recall that we have

$$\mathcal{A}^{\triangledown} \,\dot{\cup}\, (\mathfrak{B}(\mathcal{A})^{\complement})^{\vartriangle} = \mathbf{2}^{[t]} ,$$

that is,

$$\mathfrak{B}(\mathcal{A})^{\triangledown} = \{D^{\complement} \colon D \in \mathbf{2}^{[t]} - \mathcal{A}^{\triangledown}\} .$$

- According to Remark 9.11(ii), we have

$$\begin{aligned}\boldsymbol{\gamma}(\mathfrak{B}(\mathcal{A})^{\triangledown}) &= \prod\nolimits^{*}_{i\in[\alpha]} \boldsymbol{\gamma}(\mathfrak{B}(\{A_i\})^{\triangledown}) \\ &= \prod\nolimits^{*}_{i\in[\alpha]} \Big(\mathrm{T}^{(+)}_{2^t} - \Big(\prod\nolimits^{*}_{a^i\in A_i} \tilde{\mathrm{a}}(a^i)\Big) \cdot \overline{\mathbf{U}}(2^t)\Big) \\ &= \Big(\prod\nolimits^{*}_{i\in[\alpha]} \Big(\mathrm{T}^{(+)}_{2^t} - \prod\nolimits^{*}_{a^i\in A_i} \tilde{\mathrm{a}}(a^i)\Big)\Big) \cdot \overline{\mathbf{U}}(2^t) .\end{aligned}$$

Since

$$\boldsymbol{\gamma}(\mathfrak{B}(\mathcal{A})^{\triangledown}) = \mathrm{rn}(\boldsymbol{\gamma}(\mathcal{A}^{\triangledown})) := \mathrm{T}^{(+)}_{2^t} - \boldsymbol{\gamma}(\mathcal{A}^{\triangledown}) \cdot \overline{\mathbf{U}}(2^t) ,$$

by (9.45), we have

$$\Big(\prod_{i\in[\alpha]}^{*}\Big(\mathrm{T}_{2^t}^{(+)}-\prod_{a^i\in A_i}^{*}\tilde{\mathfrak{a}}(a^i)\Big)\Big)\cdot\overline{\mathrm{U}}(2^t)=\mathrm{T}_{2^t}^{(+)}-\boldsymbol{\gamma}(\mathcal{A}^{\triangledown})\cdot\overline{\mathrm{U}}(2^t)\,,$$

that is,

$$\boldsymbol{\gamma}(\mathcal{A}^{\triangledown})\cdot\overline{\mathrm{U}}(2^t)=\mathrm{T}_{2^t}^{(+)}-\Big(\prod_{i\in[\alpha]}^{*}\Big(\mathrm{T}_{2^t}^{(+)}-\prod_{a^i\in A_i}^{*}\tilde{\mathfrak{a}}(a^i)\Big)\Big)\cdot\overline{\mathrm{U}}(2^t)\,.$$

Theorem 9.12. *If* $\mathcal{A} := \{A_1, \ldots, A_\alpha\}$ *is a nontrivial clutter*[16] *on the ground set* E_t, *then we have*:

(i)

$$\boldsymbol{\gamma}(\mathcal{A}^{\triangledown})=\mathrm{T}_{2^t}^{(+)}-\Big(\prod_{i\in[\alpha]}^{*}\Big(\mathrm{T}_{2^t}^{(+)}-\prod_{a^i\in A_i}^{*}\tilde{\mathfrak{a}}(a^i)\Big)\Big)\,. \tag{9.78}$$

(ii)

$$\boldsymbol{\gamma}(\mathfrak{B}(\mathcal{A})^{\triangledown})=\Big(\prod_{i\in[\alpha]}^{*}\Big(\mathrm{T}_{2^t}^{(+)}-\prod_{a^i\in A_i}^{*}\tilde{\mathfrak{a}}(a^i)\Big)\Big)\cdot\overline{\mathrm{U}}(2^t)\,. \tag{9.79}$$

Example 9.13. Suppose $t := 3$, and $E_t = \{1, 2, 3\}$. We have in our hands the characteristic vectors

$$\begin{aligned}
\tilde{\mathfrak{a}}(1) := \tilde{\mathfrak{a}}(1; 2^t) := \boldsymbol{\gamma}(\{\{1\}\}^{\triangledown}) &= (0,\ 1,\ 0,\ 0,\ 1,\ 1,\ 0,\ 1)\\
&\qquad\qquad\qquad \in \{0, 1\}^{2^t}\,,\\
\tilde{\mathfrak{a}}(2) := \tilde{\mathfrak{a}}(2; 2^t) := \boldsymbol{\gamma}(\{\{2\}\}^{\triangledown}) &= (0,\ 0,\ 1,\ 0,\ 1,\ 0,\ 1,\ 1)\,,\\
\tilde{\mathfrak{a}}(3) := \tilde{\mathfrak{a}}(3; 2^t) := \boldsymbol{\gamma}(\{\{3\}\}^{\triangledown}) &= (0,\ 0,\ 0,\ 1,\ 0,\ 1,\ 1,\ 1)\,,
\end{aligned}$$

associated with the principal increasing families that are generated by the clutters $\{\{a\}\}$, for the elements $a \in E_t$ of the ground set.

We are given the clutter $\mathcal{A} := \{A_1, A_2\}$ on the ground set E_t, where $A_1 := \{1, 2\}$ and $A_2 := \{2, 3\}$, and we want to know the characteristic vector $\boldsymbol{\gamma}(\mathfrak{B}(\mathcal{A})^{\triangledown})$ of the increasing family of blocking sets $\mathfrak{B}(\mathcal{A})^{\triangledown}$ of the clutter $\mathcal{A}$.

[16] The assertions of the theorem are certainly correct for an arbitrary nonempty family $\mathcal{A}$ of nonempty subsets of the set E_t.

Turning to Theorem 9.12(ii), we see that

$$\begin{aligned}
\prod^{*}_{a^1\in A_1}\tilde{\mathfrak{a}}(a^1) := \prod^{*}_{a^1\in\{1,2\}}\tilde{\mathfrak{a}}(a^1) &= (0,\ 1,\ 0,\ 0,\ 1,\ 1,\ 0,\ 1)\\
&\quad *(0,\ 0,\ 1,\ 0,\ 1,\ 0,\ 1,\ 1)\\
&= (0,\ 0,\ 0,\ 0,\ 1,\ 0,\ 0,\ 1)\,,\\
\prod^{*}_{a^2\in A_2}\tilde{\mathfrak{a}}(a^2) := \prod^{*}_{a^2\in\{2,3\}}\tilde{\mathfrak{a}}(a^2) &= (0,\ 0,\ 1,\ 0,\ 1,\ 0,\ 1,\ 1)\\
&\quad *(0,\ 0,\ 0,\ 1,\ 0,\ 1,\ 1,\ 1)\\
&= (0,\ 0,\ 0,\ 0,\ 0,\ 0,\ 1,\ 1)\,;\\
\mathrm{T}^{(+)}_{2^t} - \prod^{*}_{a^1\in A_1}\tilde{\mathfrak{a}}(a^1) &= (1,\ 1,\ 1,\ 1,\ 0,\ 1,\ 1,\ 0)\,,\\
\mathrm{T}^{(+)}_{2^t} - \prod^{*}_{a^2\in A_2}\tilde{\mathfrak{a}}(a^2) &= (1,\ 1,\ 1,\ 1,\ 1,\ 1,\ 0,\ 0)\,;\\
\prod^{*}_{i\in[2]}\Big(\mathrm{T}^{(+)}_{2^t} - \prod^{*}_{a^i\in A_i}\tilde{\mathfrak{a}}(a^i)\Big) &= (1,\ 1,\ 1,\ 1,\ 0,\ 1,\ 1,\ 0)\\
&\quad *(1,\ 1,\ 1,\ 1,\ 1,\ 1,\ 0,\ 0)\\
&= (1,\ 1,\ 1,\ 1,\ 0,\ 1,\ 0,\ 0)\,,
\end{aligned}$$

and finally

$$\begin{aligned}
&\boldsymbol{\gamma}(\mathfrak{B}(\mathcal{A})^{\triangledown})\\
&= \Big(\prod^{*}_{i\in[2]}\Big(\mathrm{T}^{(+)}_{2^t} - \prod^{*}_{a^i\in A_i}\tilde{\mathfrak{a}}(a^i)\Big)\Big)\cdot\overline{\mathbf{U}}(2^t)\\
&= (0,\ 0,\ 1,\ 0,\ 1,\ 1,\ 1,\ 1)\,.
\end{aligned}$$

In Example 9.15 on page 252, we will attempt to extract the characteristic vector $\boldsymbol{\gamma}(\mathfrak{B}(\mathcal{A}))$ of the blocker $\mathfrak{B}(\mathcal{A})$ from the above vector $\boldsymbol{\gamma}(\mathfrak{B}(\mathcal{A})^{\triangledown})$.

9.4.3.4 The characteristic vector of the subfamily of inclusion-minimal sets min $\mathcal{F}$ in a family $\mathcal{F}$

Suppose we are given the characteristic vector $\boldsymbol{\gamma}(\mathcal{F})$ of a nonempty family $\mathcal{F}\subset\mathbf{2}^{[t]}$ of subsets of the ground set E_t, such that $\mathcal{F}\not\ni\hat{0}$. We can read off the position numbers of all the inclusion-minimal sets in the family $\mathcal{F}$ in the following straightforward way (see Example 9.15 on page 252):

Algorithm 9.14.

`Input:` The char.-vector $\boldsymbol{\gamma}(\mathcal{F})$ of a family $\mathcal{F} \subset \mathbf{2}^{[t]}$, such that $\emptyset \neq \mathcal{F} \not\ni \hat{0}$.

`Output:` A set M is the set $\operatorname{supp}(\boldsymbol{\gamma}(\mathbf{min}\,\mathcal{F}))$ of position numbers of the members of the clutter $\mathbf{min}\,\mathcal{F}$;
a vector $\boldsymbol{\beta}$ is the char.-vector $\boldsymbol{\gamma}(\mathbf{min}\,\mathcal{F})$ of the clutter $\mathbf{min}\,\mathcal{F}$ (*this data is optional*);
a family $\mathcal{B}$ is the clutter $\mathbf{min}\,\mathcal{F}$ (*this data is optional*).

(0). Define $\boldsymbol{\phi} \in \{0, 1\}^{2^t}$, and store $\boldsymbol{\phi} \leftarrow \boldsymbol{\gamma}(\mathcal{F})$;
define $\boldsymbol{\beta} \in \{0, 1\}^{2^t}$, and store $\boldsymbol{\beta} \leftarrow (0, \ldots, 0)$ (*this action is optional*);
define $\mathcal{B} \subset \mathbf{2}^{[t]}$, and store $\mathcal{B} \leftarrow \emptyset$ (*this action is optional*);
define $M \subset [2^t]$, and store $M \leftarrow \hat{0}$;
define $m \in \mathbb{N}$, and store $m \leftarrow 0$;
define $B \in \mathbf{2}^{[t]}$, and store $B \leftarrow \hat{0}$.

(1). If $|\operatorname{supp}(\boldsymbol{\phi})| = 0$, then go to Step (3),
else go to Step (2).

(2). Store $m \leftarrow \min \operatorname{supp}(\boldsymbol{\phi})$,
and store $M \leftarrow M \,\dot\cup\, \{m\}$,
and store $B \leftarrow \Gamma(m)$,
and store $\mathcal{B} \leftarrow \mathcal{B} \,\dot\cup\, \{B\}$ (*this action is optional*);
store $\boldsymbol{\beta} \leftarrow \boldsymbol{\beta} + \boldsymbol{\sigma}(m)$ (*this action is optional*)
If $|\operatorname{supp}(\boldsymbol{\phi})| = 1$, then go to Step (3),
else store $\boldsymbol{\phi} \leftarrow \boldsymbol{\phi} - \boldsymbol{\phi} * \underbrace{\prod^{*}_{e \in B} \widetilde{\mathfrak{a}}(e)}_{\boldsymbol{\gamma}(\{B\}^{\triangledown})}$.
Go to Step (1).

(3). Stop.

9.4.3.5 More on the blocker $\mathfrak{B}(\mathcal{A})$

If we know (see, e.g., Theorem 9.12(ii)) the characteristic vector $\boldsymbol{\gamma}(\mathcal{F})$ of the increasing family of blocking sets $\mathcal{F} := \mathfrak{B}(\mathcal{A})^{\triangledown}$ of a clutter $\mathcal{A} := \{A_1, \ldots, A_\alpha\}$ on the ground set E_t, then a description of the blocker $\mathbf{min}\,\mathcal{F} := \mathfrak{B}(\mathcal{A})$ can be obtained by an application of Algorithm 9.14 to the vector $\boldsymbol{\gamma}(\mathcal{F})$; see Example 9.15.

Example 9.15. Suppose $t := 3$, and $E_t = \{1, 2, 3\}$. Note that

$$\widetilde{\mathfrak{a}}(2) := \widetilde{\mathfrak{a}}(2; 2^t) := \boldsymbol{\gamma}(\{\{2\}\}^{\triangledown}) = (0, 0, 1, 0, 1, 0, 1, 1) \in \{0, 1\}^{2^t} .$$

We are given the characteristic vector

$$\boldsymbol{\gamma}(\mathcal{F}) := (0, 0, 1, 0, 1, 1, 1, 1) \in \{0, 1\}^{2^t}$$

of the increasing family of blocking sets $\mathcal{F} := \mathfrak{B}(\mathcal{A})^{\triangledown}$ of the clutter $\mathcal{A} := \{\{1, 2\}, \{2, 3\}\}$ on the ground set E_t; see, e.g., Example 9.13 on page 249. In order to find a description of the clutter $\mathbf{min}\,\mathcal{F} := \mathfrak{B}(\mathcal{A})$, let us apply Algorithm 9.14 to the vector $\boldsymbol{\gamma}(\mathcal{F})$:

$$\begin{aligned}
\boldsymbol{\phi} &\leftarrow \boldsymbol{\gamma}(\mathcal{F}) := (0, 0, 1, 0, 1, 1, 1, 1) ;\\
|\operatorname{supp}(\boldsymbol{\phi})| &> 0 ;\\
m &\leftarrow \underbrace{\min \operatorname{supp}((0, 0, 1, 0, 1, 1, 1, 1))}_{3} ,\\
M &\leftarrow \underbrace{\hat{0} \,\dot{\cup}\, \{3\}}_{\{3\}} ,\\
B &\leftarrow \underbrace{\Gamma(3)}_{\{2\}} ,\\
\mathcal{B} &\leftarrow \underbrace{\emptyset \,\dot{\cup}\, \{2\}}_{\{\{2\}\}} ;\\
\boldsymbol{\beta} &\leftarrow \underbrace{(0, 0, 0, 0, 0, 0, 0, 0) + \boldsymbol{\sigma}(3)}_{(0,0,1,0,0,0,0,0)} ;\\
\boldsymbol{\phi} &\leftarrow \underbrace{\begin{matrix}(0,0,1,0,1,1,1,1)\\ \qquad -(0,0,1,0,1,1,1,1)*\widetilde{\mathfrak{a}}(2)\end{matrix}}_{(0,0,0,0,0,1,0,0)} ;
\end{aligned}$$

$$
\begin{aligned}
|\operatorname{supp}(\boldsymbol{\phi})| \quad & > 0\,;\\
m \;\leftarrow\; & \underbrace{\min \operatorname{supp}((0,0,0,0,0,1,0,0))}_{6}\,,\\
M \quad & \leftarrow \underbrace{\{3\} \dot{\cup} \{6\}}_{\{3,6\}}\,,\\
B \quad & \leftarrow \underbrace{\Gamma(6)}_{\{1,3\}}\,,\\
\mathcal{B} \quad & \leftarrow \underbrace{\{\{2\}\} \dot{\cup} \{1,3\}}_{\{\{2\},\{1,3\}\}}\,;\\
\boldsymbol{\beta} \;\leftarrow\; & \underbrace{(0,0,1,0,0,0,0,0) + \boldsymbol{\sigma}(6)}_{(0,0,1,0,0,1,0,0)}\,;\\
|\operatorname{supp}(\boldsymbol{\phi})| \quad & = 1\,;
\end{aligned}
$$

Stop.

We see that the set $\operatorname{supp}(\boldsymbol{\gamma}(\mathbf{min}\,\mathcal{F})) =: M$ of the position numbers of the members of the blocker $\mathfrak{B}(\mathcal{A}) =: \mathbf{min}\,\mathcal{F}$ is the set $\{3, 6\}$.

The characteristic vector $\boldsymbol{\gamma}(\mathbf{min}\,\mathcal{F}) =: \boldsymbol{\beta}$ of the blocker $\mathfrak{B}(\mathcal{A}) =: \mathbf{min}\,\mathcal{F}$ is the vector $(0, 0, 1, 0, 0, 1, 0, 0)$.

The blocker $\mathfrak{B}(\mathcal{A}) =: \mathbf{min}\,\mathcal{F} =: \mathcal{B}$ of the clutter $\mathcal{A} := \{\{1, 2\}, \{2, 3\}\}$ is the clutter $\{\{2\}, \{1, 3\}\}$.

9.5 Decompositions of the Characteristic (Co)vectors of Families

The vertices $R^i \in \{1, -1\}^t$ of the symmetric cycle $\boldsymbol{R}$ in the hypercube graph $\boldsymbol{H}(t, 2)$, given in (9.1)(9.2), are just simply defined components of vertex decompositions in the discrete hypercube $\{1, -1\}^t$.

– In the context of the combinatorics of finite sets, the vertices $R^i \in \{1, -1\}^{2^t}$ of a distinguished *symmetric cycle*

$$\boldsymbol{R} := (R^0, R^1, \ldots, R^{2\cdot 2^t - 1}, R^0)\,,$$

in the hypercube graph $\boldsymbol{H}(2^t, 2)$ on the vertex set $\{1, -1\}^{2^t}$, where

$$
\begin{aligned}
R^0 &:= \mathrm{T}^{(+)}_{2^t}\,,\\
R^s &:= {}_{-[s]}R^0\,, \quad 1 \le s \le 2^t - 1\,,
\end{aligned}
\tag{9.80}
$$

and

$$R^{2^t+k} := -R^k\,, \quad 0 \le k \le 2^t - 1\,, \tag{9.81}$$

have an additional meaning:

Remark 9.16. Let $\boldsymbol{R}$ be the distinguished symmetric cycle in the hypercube graph $\boldsymbol{H}(2^t, 2)$, defined by (9.80)(9.81).

(i) The vertex $R^0 := \mathrm{T}^{(+)}_{2^t} \in \mathrm{V}(\boldsymbol{R})$ is the *characteristic covector* $T_\emptyset$ of the *empty family* $\emptyset$ on the ground set E_t.
The vertex $R^{2^t} := -\mathrm{T}^{(+)}_{2^t} \in \mathrm{V}(\boldsymbol{R})$ is the *characteristic covector* $T_{\mathbf{2}^{[t]}}$ of the *power set* $\mathbf{2}^{[t]}$ of the set E_t.

(ii) If $1 \le i \le 2^t - 1$, then the vertex $R^i \in \mathrm{V}(\boldsymbol{R})$ is the *characteristic covector* $T_{\mathcal{F}}$ of a *decreasing family* $\mathcal{F}$ of subsets of the ground set E_t. In other words, the family $\mathcal{F}$ is a particular abstract simplicial complex, when $1 < i \le 2^t - 1$.
Either the subfamily $\mathbf{max}\,\mathcal{F}$ is an s-uniform clutter, where $s := |\Gamma(\max \mathfrak{n}(T_{\mathcal{F}}))|$, or we have $\{|F|\colon F \in \mathbf{max}\,\mathcal{F}\} = \{s, s-1\}$. Indeed, we have

$$\mathbf{max}\,\mathcal{F} = \underbrace{\left(\mathcal{F} \cap \tbinom{E_t}{s}\right)}_{(\mathbf{max}\,\mathcal{F}) \cap \binom{E_t}{s}} \,\dot{\cup}\, \underbrace{\left(\tbinom{E_t}{s-1} - \left(\mathcal{F} \cap \tbinom{E_t}{s}\right)^{\vartriangle}\right)}_{(\mathbf{max}\,\mathcal{F}) \cap \binom{E_t}{s-1}}\,.$$

(iii) If $2^t + 1 \le i \le 2 \cdot 2^t - 1$, then the vertex $R^i \in \mathrm{V}(\boldsymbol{R})$ is the *characteristic covector* $T_{\mathcal{F}}$ of an *increasing family* $\mathcal{F}$ of subsets of the ground set E_t.
Either the subfamily $\mathbf{min}\,\mathcal{F}$ is an s-uniform clutter, where $s := |\Gamma(\min \mathfrak{n}(T_{\mathcal{F}}))|$, or we have $\{|F|\colon F \in \mathbf{min}\,\mathcal{F}\} = \{s, s+1\}$. We have

$$\mathbf{min}\,\mathcal{F} = \underbrace{\left(\mathcal{F} \cap \tbinom{E_t}{s}\right)}_{(\mathbf{min}\,\mathcal{F}) \cap \binom{E_t}{s}} \,\dot{\cup}\, \underbrace{\left(\tbinom{E_t}{s+1} - \left(\mathcal{F} \cap \tbinom{E_t}{s}\right)^{\triangledown}\right)}_{(\mathbf{min}\,\mathcal{F}) \cap \binom{E_t}{s+1}}\,.$$

If $i = 3 \cdot 2^{t-1}$, then the clutter $\mathbf{min}\,\mathcal{F}$ is *self-dual*.

- A distinguished symmetric cycle $\widetilde{\boldsymbol{R}} := (\widetilde{R}^0, \widetilde{R}^1, \ldots, \widetilde{R}^{2\cdot 2^t-1}, \widetilde{R}^0)$ in the hypercube graph $\widetilde{\boldsymbol{H}}(2^t, 2)$, on its vertex set $\{0,1\}^{2^t}$, is defined[17]

[17] Here $\sigma(e)$ is the eth standard unit vector of the space $\mathbb{R}^{2^t}$.

as follows:

$$\widetilde{R}^0 := (0, \ldots, 0) ,$$
$$\widetilde{R}^s := \sum\nolimits_{e \in [s]} \sigma(e) , \quad 1 \le s \le 2^t - 1 ,$$

and

$$\widetilde{R}^{2^t+k} := \mathrm{T}_{2^t}^{(+)} - \widetilde{R}^k , \quad 0 \le k \le 2^t - 1 .$$

We let $\vec{\mathrm{V}}(\widetilde{\boldsymbol{R}}) := (\widetilde{R}^0, \widetilde{R}^1, \ldots, \widetilde{R}^{2 \cdot 2^t - 1})$ denote the vertex sequence of the cycle $\widetilde{\boldsymbol{R}}$. The vertex set of $\widetilde{\boldsymbol{R}}$ is denoted by $\mathrm{V}(\widetilde{\boldsymbol{R}}) := \{\widetilde{R}^0, \widetilde{R}^1, \ldots, \widetilde{R}^{2 \cdot 2^t - 1}\}$.

- Let $\mathcal{F} \subset \mathbf{2}^{[t]}$ be a family of subsets of the ground set E_t, such that $\emptyset \neq \mathcal{F} \not\ni \hat{0}$. As earlier, we associate with the family $\mathcal{F}$ its characteristic covector $T_{\mathcal{F}} \in \{1, -1\}^{2^t}$, defined by (9.44).

Recall that there exists a unique inclusion-minimal and linearly independent subset

$$\boldsymbol{Q}(T_{\mathcal{F}}, \boldsymbol{R}) \subset \mathrm{V}(\boldsymbol{R}) := \{R^0, R^1, \ldots, R^{2 \cdot 2^t - 1}\}$$

of the vertex set $\mathrm{V}(\boldsymbol{R}) \subset \mathbb{R}^{2^t}$ of the cycle $\boldsymbol{R}$, defined by (9.80)(9.81), of the hypercube graph $\boldsymbol{H}(2^t, 2)$, such that

$$T_{\mathcal{F}} = \sum_{Q \in \boldsymbol{Q}(T_{\mathcal{F}}, \boldsymbol{R})} Q .$$

In other words, there exists a unique row vector $\boldsymbol{x} := \boldsymbol{x}(T_{\mathcal{F}}) := \boldsymbol{x}(T_{\mathcal{F}}, \boldsymbol{R}) = (x_1, \ldots, x_{2^t}) \in \{-1, 0, 1\}^{2^t}$, such that

$$T_{\mathcal{F}} = \sum_{i \in E_{2^t}} x_i \cdot R^{i-1} = \boldsymbol{x}\mathbf{M} , \tag{9.82}$$

where

$$\mathbf{M} := \mathbf{M}(\boldsymbol{R}) := \begin{pmatrix} R^0 \\ R^1 \\ \vdots \\ R^{2^t-1} \end{pmatrix} . \tag{9.83}$$

One can see such a well-structured matrix (in the case $t := 3$) in Example 9.19 on page 262. Thus, we have

$$\boldsymbol{x} = T_{\mathcal{F}} \cdot \mathbf{M}^{-1} ,$$

and

$$\boldsymbol{Q}(T_{\mathcal{F}}, \boldsymbol{R}) := \{x_i \cdot R^{i-1} : x_i \neq 0\} .$$

We use the notation $\mathfrak{q}(T_{\mathcal{F}}) := \mathfrak{q}(T_{\mathcal{F}}, \boldsymbol{R}) := |\boldsymbol{Q}(T_{\mathcal{F}}, \boldsymbol{R})|$ to denote the cardinality of the set $\boldsymbol{Q}(T_{\mathcal{F}}, \boldsymbol{R})$.

- Let us consider the subset

$$\widetilde{\boldsymbol{Q}}(\underbrace{\boldsymbol{\gamma}(\mathcal{F})}_{\frac{1}{2}(\mathrm{T}^{(+)}_{2^t}-T_{\mathcal{F}})}, \widetilde{\boldsymbol{R}}) := \left\{\frac{1}{2}(\mathrm{T}^{(+)}_{2^t} - Q) :\ Q \in \boldsymbol{Q}(T_{\mathcal{F}}, \boldsymbol{R})\right\} \subset \mathrm{V}(\widetilde{\boldsymbol{R}}) ,$$

and let us use the notation $\mathfrak{q}(\boldsymbol{\gamma}(\mathcal{F})) := \mathfrak{q}(\boldsymbol{\gamma}(\mathcal{F}), \widetilde{\boldsymbol{R}}) := |\widetilde{\boldsymbol{Q}}(\boldsymbol{\gamma}(\mathcal{F}), \widetilde{\boldsymbol{R}})|$ to denote its cardinality; by construction, we have $\mathfrak{q}(\boldsymbol{\gamma}(\mathcal{F})) = \mathfrak{q}(T_{\mathcal{F}})$.

In analogy with expression (9.7), we have

$$\boldsymbol{\gamma}(\mathcal{F}) = -\frac{1}{2}\big(\mathfrak{q}(\boldsymbol{\gamma}(\mathcal{F})) - 1\big) \cdot \mathrm{T}^{(+)}_{2^t} + \sum_{\substack{\widetilde{Q}\in\widetilde{\boldsymbol{Q}}(\boldsymbol{\gamma}(\mathcal{F}),\widetilde{\boldsymbol{R}}):\\ \widetilde{Q}\neq(0,\ldots,0)=:\widetilde{R}^0}} \widetilde{Q} . \tag{9.84}$$

- Let $\mathcal{A} \subset \mathbf{2}^{[t]}$ be a nontrivial clutter on the ground set E_t, and let $\mathcal{B} := \mathfrak{B}(\mathcal{A})$ be its blocker. We associate with the families $\mathcal{A}^{\triangledown}$, $\mathcal{B}^{\triangledown}$, $\mathcal{A}$ and $\mathcal{B}$ their characteristic covectors $T_{\mathcal{A}^{\triangledown}}, T_{\mathcal{B}^{\triangledown}}, T_{\mathcal{A}}, T_{\mathcal{B}} \in \{1, -1\}^{2^t}$, and their characteristic vectors $\boldsymbol{\gamma}(\mathcal{A}^{\triangledown}), \boldsymbol{\gamma}(\mathcal{B}^{\triangledown}), \boldsymbol{\gamma}(\mathcal{A}), \boldsymbol{\gamma}(\mathcal{B}) \in \{0, 1\}^{2^t}$. See (co)vectors (9.85)–(9.92) in Example 9.17.

Example 9.17. Suppose $t := 3$, and $E_t = \{1, 2, 3\}$. Let $\boldsymbol{R} := (R^0, R^1, \ldots, R^{2^t-1}, R^0)$ be the distinguished symmetric cycle in the hypercube graph $\boldsymbol{H}(2^t, 2)$ on its vertex set $\{1, -1\}^{2^t}$, defined by (9.80)(9.81).

We are given the *blocking pair* of *clutters* $\mathcal{A} := \{\{1, 2\}, \{2, 3\}\}$ and $\mathcal{B} := \mathfrak{B}(\mathcal{A}) = \{\{1, 3\}, \{2\}\}$ on the ground set E_t.

The families $\mathcal{A}^{\triangledown}$, $\mathcal{B}^{\triangledown}$, $\mathcal{A}$ and $\mathcal{B}$ are described by their characteristic covectors

$$T_{\mathcal{A}^{\triangledown}} := (1,\ 1,\ 1,\ 1, -1,\ 1, -1, -1) \in \{1, -1\}^{2^t} , \tag{9.85}$$

$$T_{\mathcal{B}^{\triangledown}} := (1,\ 1, -1,\ 1, -1, -1, -1, -1) , \tag{9.86}$$

$$T_{\mathcal{A}} := (1,\ 1,\ 1,\ 1, -1,\ 1, -1,\ 1) , \tag{9.87}$$

$$T_{\mathcal{B}} := (1,\ 1, -1,\ 1,\ 1, -1,\ 1,\ 1) , \tag{9.88}$$

and by their characteristic vectors

$$\boldsymbol{\gamma}(\mathcal{A}^{\triangledown}) := (0,\ 0,\ 0,\ 0,\ 1,\ 0,\ 1,\ 1) \in \{0, 1\}^{2^t} , \tag{9.89}$$

$$\boldsymbol{\gamma}(\mathcal{B}^{\triangledown}) := (0,\ 0,\ 1,\ 0,\ 1,\ 1,\ 1,\ 1) , \tag{9.90}$$

$$\boldsymbol{\gamma}(\mathcal{A}) := (0,\ 0,\ 0,\ 0,\ 1,\ 0,\ 1,\ 0) , \tag{9.91}$$

$$\boldsymbol{\gamma}(\mathcal{B}) := (0,\ 0,\ 1,\ 0,\ 0,\ 1,\ 0,\ 0) . \tag{9.92}$$

Turning to decompositions of the form (9.82), we see that

$$\boldsymbol{x}(T_{\mathcal{A}^{\triangledown}}) = (0,\ 0,\ 0,\ 0, -1,\ 1, -1,\ 0) \in \{-1, 0, 1\}^{2^t}, \tag{9.93}$$

$$\boldsymbol{x}(T_{\mathcal{B}^{\triangledown}}) = (0,\ 0, -1,\ 1, -1,\ 0,\ 0,\ 0), \tag{9.94}$$

$$\boldsymbol{x}(T_{\mathcal{A}}) = (1,\ 0,\ 0,\ 0, -1,\ 1, -1,\ 1), \tag{9.95}$$

$$\boldsymbol{x}(T_{\mathcal{B}}) = (1,\ 0, -1,\ 1,\ 0, -1,\ 1,\ 0). \tag{9.96}$$

Thus, we have the decompositions:

$$\begin{aligned}
T_{\mathcal{A}^{\triangledown}} :=&\ (\ 1,\ 1,\ 1,\ 1, -1,\ 1, -1, -1)\\
=&- \underbrace{R^4}_{-R^{12}} + R^5 - \underbrace{R^6}_{-R^{14}}\\
=&-(-1, -1, -1, -1,\ 1,\ 1,\ 1,\ 1)\\
&+(-1, -1, -1, -1, -1,\ 1,\ 1,\ 1)\\
&-(-1, -1, -1, -1, -1, -1,\ 1,\ 1)\\
=&\ R^5 + R^{12} + R^{14}\\
=&\ (-1, -1, -1, -1, -1,\ 1,\ 1,\ 1)\\
&+(\ 1,\ 1,\ 1,\ 1, -1, -1, -1, -1)\\
&+(\ 1,\ 1,\ 1,\ 1,\ 1,\ 1, -1, -1),
\end{aligned}$$

$$\begin{aligned}
T_{\mathcal{B}^{\triangledown}} :=&\ (\ 1,\ 1, -1,\ 1, -1, -1, -1, -1)\\
=&- \underbrace{R^2}_{-R^{10}} + R^3 - \underbrace{R^4}_{-R^{12}}\\
=&-(-1, -1,\ 1,\ 1,\ 1,\ 1,\ 1,\ 1)\\
&+(-1, -1, -1,\ 1,\ 1,\ 1,\ 1,\ 1)\\
&-(-1, -1, -1, -1,\ 1,\ 1,\ 1,\ 1)\\
=&\ R^3 + R^{10} + R^{12}\\
=&\ (-1, -1, -1,\ 1,\ 1,\ 1,\ 1,\ 1)\\
&+(\ 1,\ 1, -1, -1, -1, -1, -1, -1)\\
&+(\ 1,\ 1,\ 1,\ 1, -1, -1, -1, -1),
\end{aligned}$$

$$\begin{aligned}
T_{\mathcal{A}} :=& \ (\ 1,\ \ 1,\ \ 1,\ \ 1,-1,\ \ 1,-1,\ \ 1)\\
=& \ \underbrace{R^0}_{\mathrm{T}^{(+)}_{2t}} - \underbrace{R^4}_{-R^{12}} + R^5 - \underbrace{R^6}_{-R^{14}} + R^7\\
=& \ (\ 1,\ \ 1,\ \ 1,\ \ 1,\ \ 1,\ \ 1,\ \ 1,\ \ 1)\\
&-(-1,-1,-1,-1,\ \ 1,\ \ 1,\ \ 1,\ \ 1)\\
&+(-1,-1,-1,-1,-1,\ \ 1,\ \ 1,\ \ 1)\\
&-(-1,-1,-1,-1,-1,-1,\ \ 1,\ \ 1)\\
&+(-1,-1,-1,-1,-1,-1,-1,\ \ 1)\\
=& \ \underbrace{R^0}_{\mathrm{T}^{(+)}_{2t}} + R^5 + R^7 + R^{12} + R^{14}\\
=& \ (\ 1,\ \ 1,\ \ 1,\ \ 1,\ \ 1,\ \ 1,\ \ 1,\ \ 1)\\
&+(-1,-1,-1,-1,-1,\ \ 1,\ \ 1,\ \ 1)\\
&+(-1,-1,-1,-1,-1,-1,-1,\ \ 1)\\
&+(\ 1,\ \ 1,\ \ 1,\ \ 1,-1,-1,-1,-1)\\
&+(\ 1,\ \ 1,\ \ 1,\ \ 1,\ \ 1,\ \ 1,-1,-1)\,,
\end{aligned}$$

and

$$\begin{aligned}
T_{\mathcal{B}} :=& \ (\ 1,\ \ 1,-1,\ \ 1,\ \ 1,-1,\ \ 1,\ \ 1)\\
=& \ \underbrace{R^0}_{\mathrm{T}^{(+)}_{2t}} - \underbrace{R^2}_{-R^{10}} + R^3 - \underbrace{R^5}_{-R^{13}} + R^6\\
=& \ (\ 1,\ \ 1,\ \ 1,\ \ 1,\ \ 1,\ \ 1,\ \ 1,\ \ 1)\\
&-(-1,-1,\ \ 1,\ \ 1,\ \ 1,\ \ 1,\ \ 1,\ \ 1)\\
&+(-1,-1,-1,\ \ 1,\ \ 1,\ \ 1,\ \ 1,\ \ 1)\\
&-(-1,-1,-1,-1,-1,\ \ 1,\ \ 1,\ \ 1)\\
&+(-1,-1,-1,-1,-1,-1,\ \ 1,\ \ 1)\\
=& \ \underbrace{R^0}_{\mathrm{T}^{(+)}_{2t}} + R^3 + R^6 + R^{10} + R^{13}\\
=& \ (\ 1,\ \ 1,\ \ 1,\ \ 1,\ \ 1,\ \ 1,\ \ 1,\ \ 1)\\
&+(-1,-1,-1,\ \ 1,\ \ 1,\ \ 1,\ \ 1,\ \ 1)\\
&+(-1,-1,-1,-1,-1,-1,\ \ 1,\ \ 1)\\
&+(\ 1,\ \ 1,-1,-1,-1,-1,-1,-1)\\
&+(\ 1,\ \ 1,\ \ 1,\ \ 1,\ \ 1,-1,-1,-1)\,.
\end{aligned}$$

Relations of the form (9.84) imply that

$$
\begin{aligned}
\gamma(\mathcal{A}^{\triangledown}) :=&\ (\ 0,\ 0,\ 0,\ 0,\ 1,\ 0,\ 1,\ 1)\\
=&-\mathrm{T}^{(+)}_{2^t} + \widetilde{R}^{5} + \widetilde{R}^{12} + \widetilde{R}^{14}\\
=&\ (-1,-1,-1,-1,-1,-1,-1,-1)\\
&+(\ 1,\ 1,\ 1,\ 1,\ 1,\ 0,\ 0,\ 0)\\
&+(\ 0,\ 0,\ 0,\ 0,\ 1,\ 1,\ 1,\ 1)\\
&+(\ 0,\ 0,\ 0,\ 0,\ 0,\ 0,\ 1,\ 1)\,,\\
\gamma(\mathcal{B}^{\triangledown}) :=&\ (\ 0,\ 0,\ 1,\ 0,\ 1,\ 1,\ 1,\ 1)\\
=&-\mathrm{T}^{(+)}_{2^t} + \widetilde{R}^{3} + \widetilde{R}^{10} + \widetilde{R}^{12}\\
=&\ (-1,-1,-1,-1,-1,-1,-1,-1)\\
&+(\ 1,\ 1,\ 1,\ 0,\ 0,\ 0,\ 0,\ 0)\\
&+(\ 0,\ 0,\ 1,\ 1,\ 1,\ 1,\ 1,\ 1)\\
&+(\ 0,\ 0,\ 0,\ 0,\ 1,\ 1,\ 1,\ 1)\,,\\
\gamma(\mathcal{A}) :=&\ (\ 0,\ 0,\ 0,\ 0,\ 1,\ 0,\ 1,\ 0)\\
=&-2\mathrm{T}^{(+)}_{2^t} + \underbrace{\widetilde{R}^{0}}_{(0,\ldots,0)} + \widetilde{R}^{5} + \widetilde{R}^{7} + \widetilde{R}^{12} + \widetilde{R}^{14}\\
=&-2\mathrm{T}^{(+)}_{2^t} + \widetilde{R}^{5} + \widetilde{R}^{7} + \widetilde{R}^{12} + \widetilde{R}^{14}\\
=&\ (-2,-2,-2,-2,-2,-2,-2,-2)\\
&+(\ 1,\ 1,\ 1,\ 1,\ 1,\ 0,\ 0,\ 0)\\
&+(\ 1,\ 1,\ 1,\ 1,\ 1,\ 1,\ 1,\ 0)\\
&+(\ 0,\ 0,\ 0,\ 0,\ 1,\ 1,\ 1,\ 1)\\
&+(\ 0,\ 0,\ 0,\ 0,\ 0,\ 0,\ 1,\ 1)\,,
\end{aligned}
$$

and

$$
\begin{aligned}
\gamma(\mathcal{B}) :=&\ (\ 0,\ 0,\ 1,\ 0,\ 0,\ 1,\ 0,\ 0)\\
=&-2\mathrm{T}^{(+)}_{2^t} + \underbrace{\widetilde{R}^{0}}_{(0,\ldots,0)} + \widetilde{R}^{3} + \widetilde{R}^{6} + \widetilde{R}^{10} + \widetilde{R}^{13}\\
=&-2\mathrm{T}^{(+)}_{2^t} + \widetilde{R}^{3} + \widetilde{R}^{6} + \widetilde{R}^{10} + \widetilde{R}^{13}\\
=&\ (-2,-2,-2,-2,-2,-2,-2,-2)\\
&+(\ 1,\ 1,\ 1,\ 0,\ 0,\ 0,\ 0,\ 0)\\
&+(\ 1,\ 1,\ 1,\ 1,\ 1,\ 1,\ 0,\ 0)\\
&+(\ 0,\ 0,\ 1,\ 1,\ 1,\ 1,\ 1,\ 1)\\
&+(\ 0,\ 0,\ 0,\ 0,\ 0,\ 1,\ 1,\ 1)\,.
\end{aligned}
$$

- Let $\overline{\mathbf{T}}(2^t)$ denote the forward shift matrix of order 2^t. Corollary 8.3(i) and Proposition 8.2(iv), both restated in dimensionality 2^t, suggest the following:

Theorem 9.18. *Let $\boldsymbol{R}$ be the symmetric cycle in the hypercube graph $\boldsymbol{H}(2^t, 2)$, on its vertex set $\{1, -1\}^{2^t}$, defined by (9.80)(9.81).*

Let $\mathcal{A} \subset \mathbf{2}^{[t]}$ be a nontrivial clutter on the ground set E_t, and let $\mathcal{B} := \mathfrak{B}(\mathcal{A})$ be its blocker. Since the characteristic covectors of the increasing families $\mathcal{A}^{\triangledown}$ and $\mathcal{B}^{\triangledown}$ obey the relation

$$T_{\mathcal{B}^{\triangledown}} = \mathtt{ro}(T_{\mathcal{A}^{\triangledown}})\ ,$$

we have:

(i)

$$\mathfrak{q}(T_{\mathcal{B}^{\triangledown}}) := |\boldsymbol{Q}(T_{\mathcal{B}^{\triangledown}}, \boldsymbol{R})| = |\boldsymbol{Q}(T_{\mathcal{A}^{\triangledown}}, \boldsymbol{R})| =: \mathfrak{q}(T_{\mathcal{A}^{\triangledown}})\ ,$$

and

$$\boldsymbol{x}(T_{\mathcal{B}^{\triangledown}}) = \boldsymbol{x}(T_{\mathcal{A}^{\triangledown}}) \cdot \overline{\mathbf{U}}(2^t) \cdot \overline{\mathbf{T}}(2^t)\ .$$

(ii) *Suppose that the subset $\mathfrak{n}(T_{\mathcal{A}^{\triangledown}}) = \operatorname{supp}(\boldsymbol{\gamma}(\mathcal{A}^{\triangledown})) \subset E_{2^t}$ is a disjoint union*

$$[i_1, j_1] \dot\cup [i_2, j_2] \dot\cup \cdots \dot\cup [i_{\varrho-1}, j_{\varrho-1}] \dot\cup [i_\varrho, j_\varrho]$$

of inclusion-maximal intervals such that

$$j_1+2 \le i_2,\ \ j_2+2 \le i_3,\ \ \ldots,\ \ j_{\varrho-2}+2 \le i_{\varrho-1},\ \ j_{\varrho-1}+2 \le i_\varrho\ ,$$

for some ϱ. We have

$$\mathfrak{q}(T_{\mathcal{B}^{\triangledown}}) = \mathfrak{q}(T_{\mathcal{A}^{\triangledown}}) = 2\varrho - 1\ ;$$

$$\boldsymbol{x}(T_{\mathcal{A}^{\triangledown}}) = \sum_{1\le k\le \varrho-1} \boldsymbol{\sigma}(j_k + 1) - \sum_{1\le \ell\le \varrho} \boldsymbol{\sigma}(i_\ell)\ ,$$

and

$$\boldsymbol{x}(T_{\mathcal{B}^{\triangledown}}) = \sum_{1\le k\le \varrho-1} \boldsymbol{\sigma}(2^t - j_k + 1) - \sum_{1\le \ell\le \varrho} \boldsymbol{\sigma}(2^t - i_\ell + 2)\ .$$

Cf. covectors (9.85)(9.86) and '$\boldsymbol{x}$-vectors' (9.93)(9.94) in Example 9.17.

9.5.1 A Clutter $\{\{a\}\}$

As earlier in Section 9.4.1, let $\{\{a\}\}$ be a clutter on the ground set E_t, whose only member is a *one-element* subset $\{a\} \subset E_t$.
- Let us associate with the characteristic covector $\mathfrak{a}(a) := T_{\{\{a\}\}^\triangledown}$ of the principal increasing family $\{\{a\}\}^\triangledown$ the row vector $\boldsymbol{x}(\mathfrak{a}(a)) := \boldsymbol{x}(\mathfrak{a}(a), \boldsymbol{R}) \in \{-1, 0, 1\}^{2^t}$, described in (9.82), where $\boldsymbol{R}$ is the distinguished symmetric cycle in the hypercube graph $\boldsymbol{H}(2^t, 2)$, defined by (9.80)(9.81). Recall that

$$\boldsymbol{x}(\mathfrak{a}(a)) = \mathfrak{a}(a) \cdot \mathbf{M}^{-1}\,, \tag{9.97}$$

where the matrix $\mathbf{M}$ is defined by (9.83), and

$$\boldsymbol{Q}(\mathfrak{a}(a), \boldsymbol{R}) := \{x_i \cdot R^{i-1} : x_i \neq 0\}\,, \quad \text{and} \quad \mathfrak{a}(a) = \sum_{Q \in \boldsymbol{Q}(\mathfrak{a}(a), \boldsymbol{R})} Q\,;$$

see Example 9.19.

Example 9.19. Suppose $t := 3$, and $E_t = \{1, 2, 3\}$.
The characteristic covectors associated with the principal increasing families that are generated by the clutters $\{\{a\}\}$, for the elements $a \in E_t$ of the ground set, are as follows:

$$\begin{aligned}
\mathfrak{a}(1) := \mathfrak{a}(1; 2^t) := T_{\{\{1\}\}^\triangledown} &= (1, -1,\ \ 1,\ \ 1, -1, -1,\ \ 1, -1) \\
&\qquad\qquad\qquad\qquad \in \{1, -1\}^{2^t}\,, \\
\mathfrak{a}(2) := \mathfrak{a}(2; 2^t) := T_{\{\{2\}\}^\triangledown} &= (1,\ \ 1, -1,\ \ 1, -1,\ \ 1, -1, -1)\,, \\
\mathfrak{a}(3) := \mathfrak{a}(3; 2^t) := T_{\{\{3\}\}^\triangledown} &= (1,\ \ 1,\ \ 1, -1,\ \ 1, -1, -1, -1)\,.
\end{aligned}$$

The corresponding $\boldsymbol{x}$-vectors, given in (9.97), for the symmetric cycle $\boldsymbol{R}$ in the hypercube graph $\boldsymbol{H}(2^t, 2)$, defined by (9.80)(9.81), are:

$$\begin{aligned}
\boldsymbol{x}(\mathfrak{a}(1)) := \boldsymbol{x}(\mathfrak{a}(1), \boldsymbol{R}) &= (0, -1,\ \ 1,\ \ 0, -1,\ \ 0,\ \ 1, -1) \\
&\qquad\qquad\qquad\qquad \in \{-1, 0, 1\}^{2^t}\,, \\
\boldsymbol{x}(\mathfrak{a}(2)) := \boldsymbol{x}(\mathfrak{a}(2), \boldsymbol{R}) &= (0,\ \ 0, -1,\ \ 1, -1,\ \ 1, -1,\ \ 0)\,, \\
\boldsymbol{x}(\mathfrak{a}(3)) := \boldsymbol{x}(\mathfrak{a}(3), \boldsymbol{R}) &= (0,\ \ 0,\ \ 0, -1,\ \ 1, -1,\ \ 0,\ \ 0)\,.
\end{aligned}$$

Indeed, we see that

$$\boldsymbol{x}(\mathfrak{a}(2)) \cdot \mathbf{M} = \begin{pmatrix}0 & 0 & -1 & 1 & -1 & 1 & -1 & 0\end{pmatrix}$$
$$\times \begin{pmatrix} 1 & 1 & 1 & 1 & 1 & 1 & 1 & 1\\ -1 & 1 & 1 & 1 & 1 & 1 & 1 & 1\\ -1 & -1 & 1 & 1 & 1 & 1 & 1 & 1\\ -1 & -1 & -1 & 1 & 1 & 1 & 1 & 1\\ -1 & -1 & -1 & -1 & 1 & 1 & 1 & 1\\ -1 & -1 & -1 & -1 & -1 & 1 & 1 & 1\\ -1 & -1 & -1 & -1 & -1 & -1 & 1 & 1\\ -1 & -1 & -1 & -1 & -1 & -1 & -1 & 1 \end{pmatrix}$$
$$= \begin{pmatrix}1 & 1 & -1 & 1 & -1 & 1 & -1 & -1\end{pmatrix} =: \mathfrak{a}(2)\,.$$

- For the row vector $\boldsymbol{y}(1+a) := \boldsymbol{y}(1+a; 2^t) \in \{-1, 0, 1\}^{2^t}$, defined by

$$\boldsymbol{y}(1+a) := \boldsymbol{x}({}_{-\{1+a\}}\mathrm{T}^{(+)}_{2^t}) =: \boldsymbol{x}(T_{\{\{a\}\}})\,,$$

we have (see Section 5.2):

$$\boldsymbol{y}(1+a) = \boldsymbol{\sigma}(1) - \boldsymbol{\sigma}(1+a) + \boldsymbol{\sigma}(2+a)\,.$$

In other words,

$$\boldsymbol{Q}(T_{\{\{a\}\}}, \boldsymbol{R}) = \Bigl\{ \underbrace{R^0}_{\mathrm{T}^{(+)}_{2^t}},\ R^{1+a},\ R^{2^t+a} \Bigr\}\,,$$

and

$$T_{\mathfrak{B}(\{\{a\}\})} = T_{\{\{a\}\}} = \mathrm{T}^{(+)}_{2^t} + R^{1+a} + R^{2^t+a}\,.$$

Equivalently,

$$\widetilde{\boldsymbol{Q}}(\boldsymbol{\gamma}(\{\{a\}\}), \widetilde{\boldsymbol{R}}) = \Bigl\{ \underbrace{\widetilde{R}^0}_{(0,\ldots,0)},\ \widetilde{R}^{1+a},\ \widetilde{R}^{2^t+a} \Bigr\}\,,$$

and

$$\begin{aligned}\boldsymbol{\gamma}(\mathfrak{B}(\{\{a\}\})) = \boldsymbol{\gamma}(\{\{a\}\}) &= -\frac{1}{2}(3-1)\cdot \mathrm{T}^{(+)}_{2^t} + \widetilde{R}^{1+a} + \widetilde{R}^{2^t+a}\\ &= -\mathrm{T}^{(+)}_{2^t} + \widetilde{R}^{1+a} + \widetilde{R}^{2^t+a}\,.\end{aligned}$$

9.5.2 A Clutter $\{A\}$

Let $\{A\}$ be a clutter on the ground set E_t, whose only member is a *nonempty subset* $A \subseteq E_t$, as earlier in Section 9.4.2.

- Dealing with the distinguished symmetric cycle $\boldsymbol{R}$ in the hypercube graph $\boldsymbol{H}(2^t, 2)$, defined by (9.80)(9.81), with its matrix $\mathbf{M}$ given in (9.83), and with $\mathbf{x}$-vectors described in (9.82), for the row vector

$$\boldsymbol{y}(\Gamma^{-1}(A)) := \mathbf{x}({}_{-\{\Gamma^{-1}(A)\}}\mathrm{T}^{(+)}_{2^t}) =: \mathbf{x}(T_{\{A\}}) \\ = T_{\{A\}} \cdot \mathbf{M}^{-1} \in \{-1, 0, 1\}^{2^t}\,, \quad (9.98)$$

we have (see Section 5.2):

$$\boldsymbol{y}(\Gamma^{-1}(A)) = \begin{cases} \sigma(1) - \sigma(\Gamma^{-1}(A)) + \sigma(1 + \Gamma^{-1}(A))\,, & \\ & \text{if } A \neq E_t, \\ -\sigma(2^t)\,, & \text{if } A = E_t. \end{cases}$$

In other words,

$$\boldsymbol{Q}(T_{\{A\}}, \boldsymbol{R}) = \begin{cases} \{\underbrace{R^0}_{\mathrm{T}^{(+)}_{2^t}}, R^{\Gamma^{-1}(A)}, R^{2^t+\Gamma^{-1}(A)-1}\}\,, & \text{if } A \neq E_t, \\ \{R^{2\cdot 2^t-1}\}\,, & \text{if } A = E_t, \end{cases}$$

and

$$T_{\{A\}} = \begin{cases} \mathrm{T}^{(+)}_{2^t} + R^{\Gamma^{-1}(A)} + R^{2^t+\Gamma^{-1}(A)-1}\,, & \text{if } A \neq E_t, \\ R^{2\cdot 2^t-1}\,, & \text{if } A = E_t. \end{cases}$$

Equivalently,

$$\widetilde{\boldsymbol{Q}}(\boldsymbol{\gamma}(\{A\}), \widetilde{\boldsymbol{R}}) = \begin{cases} \{\underbrace{\widetilde{R}^0}_{(0,\ldots,0)}, \widetilde{R}^{\Gamma^{-1}(A)}, \widetilde{R}^{2^t+\Gamma^{-1}(A)-1}\}\,, & \text{if } A \neq E_t, \\ \{\widetilde{R}^{2\cdot 2^t-1}\}\,, & \text{if } A = E_t, \end{cases}$$

and

$$\boldsymbol{\gamma}(\{A\}) = \begin{cases} -\tfrac{1}{2}(3-1)\cdot \mathrm{T}^{(+)}_{2^t} + \widetilde{R}^{\Gamma^{-1}(A)} + \widetilde{R}^{2^t+\Gamma^{-1}(A)-1} & \\ = -\mathrm{T}^{(+)}_{2^t} + \widetilde{R}^{\Gamma^{-1}(A)} + \widetilde{R}^{2^t+\Gamma^{-1}(A)-1}\,, & \text{if } A \neq E_t, \\ \widetilde{R}^{2\cdot 2^t-1}\,, & \text{if } A = E_t. \end{cases}$$

- Recall that $\boldsymbol{\gamma}(\{A\}^{\triangledown}) = \prod^{*}_{a\in A} \tilde{\mathfrak{a}}(a)$, and $\boldsymbol{\gamma}(\mathfrak{B}(\{A\})^{\triangledown}) = \mathtt{rn}(\boldsymbol{\gamma}(\{A\}^{\triangledown}))$.

Remark 9.20. (cf. Remark 9.11). For a nonempty subset $A \subseteq E_t$, we have

(i)

$$\boldsymbol{\gamma}(\{A\}^{\triangledown}) = \prod_{a \in A}^{*} \Big(\frac{1}{2}\big(\mathrm{T}_{2^t}^{(+)} - \boldsymbol{x}(\mathfrak{a}(a)) \cdot \mathbf{M}\big)\Big) .$$

(ii)

$$\boldsymbol{\gamma}(\mathfrak{B}(\{A\})^{\triangledown}) = \mathrm{T}_{2^t}^{(+)} - \Big(\prod_{a \in A}^{*} \Big(\frac{1}{2}\big(\mathrm{T}_{2^t}^{(+)} - \boldsymbol{x}(\mathfrak{a}(a)) \cdot \mathbf{M}\big)\Big)\Big) \cdot \overline{\mathrm{U}}(2^t) .$$

– Since the *blocker* of the clutter $\{A\}$ is the clutter $\mathfrak{B}(\{A\}) = \{\{a\} : a \in A\}$, and $\boldsymbol{\gamma}(\mathfrak{B}(\{A\})) = \sum_{a \in A} \boldsymbol{\gamma}(\{\{a\}\})$, we have

$$\boldsymbol{\gamma}(\mathfrak{B}(\{A\})) = \sum_{a \in A} (-\mathrm{T}_{2^t}^{(+)} + \widetilde{R}^{1+a} + \widetilde{R}^{2^t+a}) ,$$

that is,

$$\boldsymbol{\gamma}(\mathfrak{B}(\{A\})) = -|A| \cdot \mathrm{T}_{2^t}^{(+)} + \sum_{a \in A} (\widetilde{R}^{1+a} + \widetilde{R}^{2^t+a}) .$$

9.5.3 A Clutter $\mathcal{A} := \{A_1, \ldots, A_\alpha\}$

As earlier in Section 9.4.3, let $\mathcal{A} := \{A_1, \ldots, A_\alpha\}$ be a nontrivial clutter on the ground set E_t.

– In analogy with Remark 5.6, dealing with the symmetric cycle $\boldsymbol{R}$ in the hypercube graph $\boldsymbol{H}(2^t, 2)$, defined by (9.80)(9.81), with its matrix $\mathbf{M}$ given in (9.83), with $\boldsymbol{x}$-vectors described in (9.82), and with '$\boldsymbol{y}$-vectors' given in (9.98), we have

$$\boldsymbol{x}(T_{\mathcal{A}}) = (1 - \#\mathcal{A}) \cdot \boldsymbol{\sigma}(1) + \sum_{A \in \mathcal{A}} \boldsymbol{y}(\Gamma^{-1}(A)) ,$$

that is,

$$\boldsymbol{x}(T_{\mathcal{A}}) = \begin{cases} \boldsymbol{\sigma}(1) + \sum_{A \in \mathcal{A}} (-\boldsymbol{\sigma}(\Gamma^{-1}(A)) + \boldsymbol{\sigma}(1 + \Gamma^{-1}(A))) , & \text{if } \mathcal{A} \neq \{E_t\}, \\ -\boldsymbol{\sigma}(2^t) , & \text{if } \mathcal{A} = \{E_t\}, \end{cases}$$

or

$$T_{\mathcal{A}} = \begin{cases} \mathrm{T}_{2^t}^{(+)} + \sum_{A \in \mathcal{A}} (R^{\Gamma^{-1}(A)} + R^{2^t + \Gamma^{-1}(A) - 1}) , & \text{if } \mathcal{A} \neq \{E_t\}, \\ R^{2 \cdot 2^t - 1} , & \text{if } \mathcal{A} = \{E_t\}. \end{cases}$$

We also have

$$\gamma(\mathcal{A}) = \begin{cases} -(\#\mathcal{A}) \cdot \mathrm{T}_{2^t}^{(+)} + \sum_{A\in\mathcal{A}}(\widetilde{R}^{\Gamma^{-1}(A)} + \widetilde{R}^{2^t+\Gamma^{-1}(A)-1}), & \text{if } \mathcal{A} \neq \{E_t\}, \\ \widetilde{R}^{2\cdot 2^t-1}, & \text{if } \mathcal{A} = \{E_t\}. \end{cases}$$

- Theorem 9.12 can be accompanied with the following statement:

Corollary 9.21. *If $\mathcal{A} := \{A_1, \ldots, A_\alpha\}$ is a nontrivial clutter on the ground set E_t, then we have:*

(i)

$$\gamma(\mathcal{A}^{\triangledown}) = \mathrm{T}_{2^t}^{(+)} - \Big(\prod_{i\in[\alpha]}^{*}\Big(\mathrm{T}_{2^t}^{(+)} - \prod_{a^i\in A_i}^{*}\big(\tfrac{1}{2}(\mathrm{T}_{2^t}^{(+)} - \mathbf{x}(\mathfrak{a}(a^i))\cdot\mathbf{M})\big)\Big)\Big).$$

(ii)

$$\gamma(\mathfrak{B}(\mathcal{A})^{\triangledown}) = \Big(\prod_{i\in[\alpha]}^{*}\Big(\mathrm{T}_{2^t}^{(+)} - \prod_{a^i\in A_i}^{*}\big(\tfrac{1}{2}(\mathrm{T}_{2^t}^{(+)} - \mathbf{x}(\mathfrak{a}(a^i))\cdot\mathbf{M})\big)\Big)\Big)\cdot\overline{\mathbf{U}}(2^t).$$

Chapter 10

Vertex Decompositions and Subtope Decompositions in Hypercube Graphs

If T' and T'' are *adjacent vertices* in the hypercube graph $\boldsymbol{H}(t, 2)$ on its vertex set $\{1, -1\}^t$ (i.e., the *Hamming distance* between the *tuples* T' and T'' is 1 or, equivalently, the standard *scalar product* $\langle T', T'' \rangle$ of the *vectors* T' and T'' of the space $\mathbb{R}^t$ is $t - 2$), then we label the edge $\{T', T''\}$ of the graph $\boldsymbol{H}(t, 2)$ by the row tuple

$$S := \frac{1}{2}(T' + T'') \in \{1, -1, 0\}^t\,, \tag{10.1}$$

called the common *subtope* of the vertices T' and T''. The subtope (10.1) can be interpreted as the *midpoint* of a straight *line segment* in $\mathbb{R}^t$, with the *endpoints* T' and T''. If we let $\mathbb{S}^{t-1}(r)$ denote the $(t-1)$-dimensional *sphere* of radius r in $\mathbb{R}^t$, centered at the origin, then S is the *point of tangency* of the *tangent line* to the sphere $\mathbb{S}^{t-1}(\sqrt{t-1})$ passing through the *points* T' and T'' that lie on the sphere $\mathbb{S}^{t-1}(\sqrt{t})$.

Formula (10.1) implies that the subtope S of the hypercube graph $\boldsymbol{H}(t, 2)$ is a *subtope* of the oriented matroid $\mathcal{H} := (E_t, \{1, -1\}^t)$. Indeed, it can be regarded as the *meet* $S := T' \wedge T''$ of the *topes* T' and T'' in the *big face lattice* of $\mathcal{H}$.

Symmetric Cycles
Andrey O. Matveev

ISBN 978-981-4968-81-2 (Hardcover), 978-1-003-43832-8 (eBook)
www.jennystanford.com

An edge $\{\widetilde{T}', \widetilde{T}''\}$ of the *hypercube graph* $\widetilde{\boldsymbol{H}}(t, 2)$ on the vertex set $\{0, 1\}^t$, where $\widetilde{T}' := \frac{1}{2}(\mathrm{T}^{(+)} - T')$ and $\widetilde{T}'' := \frac{1}{2}(\mathrm{T}^{(+)} - T'')$, could be labeled by the row tuple $\widetilde{S} := \frac{1}{2}(\widetilde{T}' + \widetilde{T}'') = \frac{1}{2}(\mathrm{T}^{(+)} - S) \in \{0, 1, \frac{1}{2}\}^t$.

Recall that a *symmetric cycle* $\boldsymbol{D} := (D^0, D^1, \ldots, D^{2t-1}, D^0)$ in the hypercube graph $\boldsymbol{H}(t, 2)$ is defined to be its cycle with the vertex sequence

$$\vec{\mathrm{V}}(\boldsymbol{D}) := (D^0, D^1, \ldots, D^{2t-1}), \tag{10.2}$$

such that

$$D^{k+t} = -D^k, \quad 0 \le k \le t-1. \tag{10.3}$$

The sequence $\vec{\mathrm{V}}(\boldsymbol{D})$ is an ordered *maximal positive basis* of the space $\mathbb{R}^t$; see Subsection 2.1.4. In Section 10.1 of this chapter we will see that if the cardinality t of the ground set E_t is *even*, then the sequence of subtopes, of the form (10.1), associated with the edges of the cycle $\boldsymbol{D}$ is also an ordered *maximal positive basis* of $\mathbb{R}^t$.

In Section 10.2 we describe related (de)composition constructions for vertices and subtopes of the hypercube graph $\boldsymbol{H}(t, 2)$, and we present a detailed example to illustrate them.

In addition, in Section 10.3 we consider vertex decompositions in hypercube graphs with respect to the edges of their distinguished symmetric cycles.

10.1 Symmetric Cycles in Hypercube Graphs: Vertices, Edges and Subtopes

Given a symmetric cycle $\boldsymbol{D} := (D^0, D^1, \ldots, D^{2t-1}, D^0)$ in the hypercube graph $\boldsymbol{H}(t, 2)$, with its vertex sequence $\vec{\mathrm{V}}(\boldsymbol{D})$ described in (10.2)(10.3), we denote by $\vec{\mathcal{E}}(\boldsymbol{D})$ the edge sequence of the cycle:

$$\vec{\mathcal{E}}(\boldsymbol{D}) := (\{D^0, D^1\}, \{D^1, D^2\}, \ldots, \{D^{2t-2}, D^{2t-1}\}, \{D^{2t-1}, D^0\}).$$

Let $\mathcal{S} \subset \{1, -1, 0\}^t$ denote the set of all subtopes of the hypercube graph $\boldsymbol{H}(t, 2)$. Since the quantity $|\mathcal{S}|$ by definition is the number of edges of the graph $\boldsymbol{H}(t, 2)$, we have $|\mathcal{S}| = 2^{t-1}t$. By means of the map

$$\vec{\mathcal{E}}(\boldsymbol{D}) \to \mathcal{S}\,,$$
$$\{D', D''\} \mapsto \frac{1}{2}(D' + D'') =: D' \wedge D''\,,$$

we associate with the edge sequence $\vec{\mathcal{E}}(\boldsymbol{D})$ of the cycle $\boldsymbol{D}$ the corresponding sequence of subtopes $\vec{\mathrm{S}}(\boldsymbol{D})$:

$$\begin{aligned}\vec{\mathrm{S}}(\boldsymbol{D}) : &= \big(S^0 := D^0 \wedge D^1,\ S^1 := D^1 \wedge D^2,\ \ldots,\\ S^{2t-2} : &= D^{2t-2} \wedge D^{2t-1},\ S^{2t-1} := D^{2t-1} \wedge D^0\big)\,,\end{aligned} \tag{10.4}$$

where we have

$$S^{k+t} = -S^k\,,\quad 0 \le k \le t-1\,.$$

Recall that the subsequence $(D^0, D^1, \ldots, D^{t-1})$ of the vertex sequence $\vec{\mathrm{V}}(\boldsymbol{D})$ of the cycle (10.2)(10.3) is an ordered *basis* of the space $\mathbb{R}^t$. Looking at the matrix expression

$$\underbrace{\begin{pmatrix} S^0\\ S^1\\ S^2\\ \vdots\\ S^{t-3}\\ S^{t-2}\\ S^{t-1}\end{pmatrix}}_{\mathbf{W}:=\mathbf{W}(\boldsymbol{D})} = \frac{1}{2}\underbrace{\begin{pmatrix} 1\,1\,0\cdots 0\,0\,0\\ 0\,1\,1\cdots 0\,0\,0\\ \vdots\ \vdots\ \vdots\ \cdots\ \vdots\ \vdots\ \vdots\\ 0\,0\,0\cdots 1\,1\,0\\ 0\,0\,0\cdots 0\,1\,1\\ -1\,0\,0\cdots 0\,0\,1\end{pmatrix}}_{\boldsymbol{N}(t)\in\mathbb{R}^{t\times t}} \cdot \underbrace{\begin{pmatrix} D^0\\ D^1\\ D^2\\ \vdots\\ D^{t-3}\\ D^{t-2}\\ D^{t-1}\end{pmatrix}}_{\mathbf{M}:=\mathbf{M}(\boldsymbol{D})}\,, \tag{10.5}$$

we see that this relation

$$\mathbf{W} = \frac{1}{2}\boldsymbol{N}(t) \cdot \mathbf{M}$$

implies that

$$\operatorname{rank}\mathbf{W} = \operatorname{rank}\boldsymbol{N}(t) = \begin{cases} t-1\,, & \text{if } t \text{ is odd}\,,\\ t\,, & \text{if } t \text{ is even}\,.\end{cases}$$

Thus, if t is *even*, then the sequence of subtopes $(S^0, S^1, \ldots, S^{t-1})$ is an ordered *basis* of the space $\mathbb{R}^t$, the sequence $\vec{\mathrm{S}}(\boldsymbol{D})$ defined by (10.4) is an ordered *maximal positive basis* of $\mathbb{R}^t$, and we have

$$\mathbf{M} =: \boldsymbol{P}(t) \cdot \mathbf{W}\,,$$

where the (i, j)th entries of the *Toeplitz matrix* (see, e.g., Ref. [23])

$$\boldsymbol{P}(t) := 2\boldsymbol{N}(t)^{-1} \in \mathbb{R}^{t\times t}$$

are

$$\begin{cases} (-1)^{i+j}\,, & \text{if } i \le j\,, \\ (-1)^{i+j+1}\,, & \text{if } i > j\,, \end{cases} \tag{10.6}$$

that is, we have

$$\underbrace{\begin{pmatrix} D^0 \\ D^1 \\ D^2 \\ \vdots \\ D^{t-3} \\ D^{t-2} \\ D^{t-1} \end{pmatrix}}_{\mathbf{M}} = \underbrace{\begin{pmatrix} 1 & -1 & 1 & \cdots & -1 & 1 & -1 \\ 1 & 1 & -1 & \cdots & 1 & -1 & 1 \\ \vdots & \vdots & \vdots & \cdots & \vdots & \vdots & \vdots \\ 1 & -1 & 1 & \cdots & 1 & -1 & 1 \\ -1 & 1 & -1 & \cdots & 1 & 1 & -1 \\ 1 & -1 & 1 & \cdots & -1 & 1 & 1 \end{pmatrix}}_{\boldsymbol{P}(t)} \cdot \underbrace{\begin{pmatrix} S^0 \\ S^1 \\ S^2 \\ \vdots \\ S^{t-3} \\ S^{t-2} \\ S^{t-1} \end{pmatrix}}_{\mathbf{W}} .$$

For example, the matrices $\boldsymbol{P}(4)$ and $\boldsymbol{P}(6)$ are

$$\boldsymbol{P}(4) = \begin{pmatrix} 1 & -1 & 1 & -1 \\ 1 & 1 & -1 & 1 \\ -1 & 1 & 1 & -1 \\ 1 & -1 & 1 & 1 \end{pmatrix}, \text{ and } \boldsymbol{P}(6) = \begin{pmatrix} 1 & -1 & 1 & -1 & 1 & -1 \\ 1 & 1 & -1 & 1 & -1 & 1 \\ -1 & 1 & 1 & -1 & 1 & -1 \\ 1 & -1 & 1 & 1 & -1 & 1 \\ -1 & 1 & -1 & 1 & 1 & -1 \\ 1 & -1 & 1 & -1 & 1 & 1 \end{pmatrix}.$$

Since the entries of matrices $\boldsymbol{P}(t)$ belong to the set $\{1, -1\}$, the rows of these matrices can be viewed as vertices of the *hypercube graph* $\boldsymbol{H}(t, 2)$ on the vertex set $\{1, -1\}^t$:

Remark 10.1. If t is *even*, then the rows $\boldsymbol{P}(t)^i =: P^{i-1}, 1 \le i \le t$, of the *nonsingular* matrix

$$\boldsymbol{P}(t) := \begin{pmatrix} P^0 \\ P^1 \\ P^2 \\ \vdots \\ P^{t-3} \\ P^{t-2} \\ P^{t-1} \end{pmatrix} \in \{1, -1\}^{t \times t} \tag{10.7}$$

with entries (10.6) constitute a *distinguished* sequence of certain t vertices of the hypercube graph $\boldsymbol{H}(t, 2)$ for which, on the one hand, we have

$$P(t) = \mathbf{M}(\boldsymbol{D}) \cdot \mathbf{W}(\boldsymbol{D})^{-1}\,, \tag{10.8}$$

for *any* symmetric cycle $\boldsymbol{D}$ in the graph $\boldsymbol{H}(t, 2)$.

On the other hand, for each *row* $\boldsymbol{P}(t)^i =: P^{i-1}$, $1 \leq i \leq t$, of the matrix (10.7) with entries (10.6), regarded as a *vertex* of the hypercube graph $\boldsymbol{H}(t, 2)$, there exists a *unique* row vector $\boldsymbol{x}^{i-1} := \boldsymbol{x}^{i-1}(P^{i-1}) := \boldsymbol{x}^{i-1}(P^{i-1}, \boldsymbol{D}) \in \{-1, 0, 1\}^t$, with $|\operatorname{supp}(\boldsymbol{x}^{i-1})|$ *odd*, such that

$$P^{i-1} = \boldsymbol{x}^{i-1} \cdot \mathbf{M}(\boldsymbol{D})\,;$$

see Subsection 2.1.4. Thus, we have

$$\boldsymbol{P}(t) = \underbrace{\begin{pmatrix} \boldsymbol{x}^0 \\ \boldsymbol{x}^1 \\ \boldsymbol{x}^2 \\ \vdots \\ \boldsymbol{x}^{t-3} \\ \boldsymbol{x}^{t-2} \\ \boldsymbol{x}^{t-1} \end{pmatrix}}_{\mathbf{X}(\boldsymbol{P}(t),\, \boldsymbol{D})} \cdot \mathbf{M}(\boldsymbol{D})\,. \tag{10.9}$$

Relations (10.8) and (10.9) yield

$$\mathbf{M}(\boldsymbol{D}) \cdot \mathbf{W}(\boldsymbol{D})^{-1} = \mathbf{X}(\boldsymbol{P}(t), \boldsymbol{D}) \cdot \mathbf{M}(\boldsymbol{D})\,,$$

for any symmetric cycle $\boldsymbol{D}$ of the hypercube graph $\boldsymbol{H}(t, 2)$.

10.2 Vertex Decompositions and Subtope Decompositions in Hypercube Graphs with Respect to the Edges of Their Symmetric Cycles

In this section we discuss (de)composition constructions related to the edge sequences of symmetric cycles in hypercube graphs. The results and their proofs are accompanied by a detailed example.

Proposition 10.2. *Let* $\boldsymbol{D} := (D^0, D^1, \ldots, D^{2t-1}, D^0)$ *be a symmetric cycle in the hypercube graph* $\boldsymbol{H}(t, 2)$*, where* t *is* even. *Let* $\vec{\mathrm{S}}(\boldsymbol{D})$ *be the corresponding sequence of subtopes defined by* (10.4).

(i) *Given a vertex* $T \in \{1, -1\}^t$ *of the graph* $\boldsymbol{H}(t, 2)$, *there exist a* unique *subset of subtopes* $\overline{\boldsymbol{Q}}(T, \vec{\mathrm{S}}(\boldsymbol{D})) \subset \vec{\mathrm{S}}(\boldsymbol{D})$ *and a* unique *set of* integers $\{\lambda_{Q'}:\ Q' \in \overline{\boldsymbol{Q}}(T, \vec{\mathrm{S}}(\boldsymbol{D}))\}$ *such that*

$$\sum_{Q' \in \overline{\boldsymbol{Q}}(T, \vec{\mathrm{S}}(\boldsymbol{D}))} \lambda_{Q'} \cdot Q' = T \ ,$$

where

$$|\ \overline{\boldsymbol{Q}}(T, \vec{\mathrm{S}}(\boldsymbol{D}))\ | = t \ ,$$

and

$$1 \leq \lambda_{Q'} \leq t - 1 \ , \quad \textit{and} \quad \lambda_{Q'} \ \textit{are all} \ \text{odd} \ .$$

(ii) *If* $S \in \mathcal{S}$ *is a* subtope *of the graph* $\boldsymbol{H}(t, 2)$, *then there exist a* unique inclusion-minimal *subset of subtopes* $\overline{\boldsymbol{Q}}(S, \vec{\mathrm{S}}(\boldsymbol{D})) \subset \vec{\mathrm{S}}(\boldsymbol{D})$ *and a* unique *set of* integers $\{\lambda_{Q'}:\ Q' \in \overline{\boldsymbol{Q}}(S, \vec{\mathrm{S}}(\boldsymbol{D}))\}$ *such that*

$$\sum_{Q' \in \overline{\boldsymbol{Q}}(S, \vec{\mathrm{S}}(\boldsymbol{D}))} \lambda_{Q'} \cdot Q' = S \ ,$$

where

$$1 \leq \lambda_{Q'} \leq t - 1 \ .$$

Proof. (i) See Example 10.3(i).
Let $\boldsymbol{x} := \boldsymbol{x}(T) := \boldsymbol{x}(T, \boldsymbol{D}) = (x_1, \ldots, x_t) \in \{-1, 0, 1\}^t$ be the unique row vector such that

$$T = \boldsymbol{x} \cdot \mathbf{M}(\boldsymbol{D}) \ ,$$

where $|\operatorname{supp}(\boldsymbol{x})|$ is *odd*; see Section 2.1.4. In other words,

$$T = \boldsymbol{x} \cdot \boldsymbol{P}(t)\mathbf{W}(\boldsymbol{D}) =: \overline{\boldsymbol{x}} \cdot \mathbf{W}(\boldsymbol{D}) \ , \tag{10.10}$$

where the vector $\overline{\boldsymbol{x}} := \overline{\boldsymbol{x}}(T) := \overline{\boldsymbol{x}}(T, \vec{\mathrm{S}}(\boldsymbol{D})) = (\overline{x}_1, \ldots, \overline{x}_t) \in \mathbb{Z}^t$ is defined by

$$\overline{\boldsymbol{x}} := \boldsymbol{x} \cdot \boldsymbol{P}(t) \ . \tag{10.11}$$

Since the entries of the matrix $\boldsymbol{P}(t)$ belong to the set $\{1, -1\}$, the components of $\overline{\boldsymbol{x}}$ are all *odd* integers and, as a consequence,

$$|\operatorname{supp}(\overline{\boldsymbol{x}})| = t \ .$$

We have

$$\overline{\boldsymbol{Q}}(T, \vec{\mathrm{S}}(\boldsymbol{D})) = \{\mathrm{sign}(\overline{x}_i) \cdot S^{i-1} \colon 1 \le i \le t\}\,,$$
$$Q' \in \{-S^{j-1}, S^{j-1}\} \implies \lambda_{Q'} = |\overline{x}_j|\,, \quad 1 \le j \le t\,.$$

(ii) See Example 10.3(ii).

Let T' and T'' be the two vertices of the graph $\boldsymbol{H}(t, 2)$ such that $S = \frac{1}{2}(T' + T'')$. If $\boldsymbol{x}_{T'} := \boldsymbol{x}(T') \in \{-1, 0, 1\}^t$ and $\boldsymbol{x}_{T''} := \boldsymbol{x}(T'') \in \{-1, 0, 1\}^t$ are the unique row vectors such that $T' = \boldsymbol{x}_{T'} \cdot \mathbf{M}(\boldsymbol{D})$ and $T'' = \boldsymbol{x}_{T''} \cdot \mathbf{M}(\boldsymbol{D})$, then we have

$$S = \frac{1}{2}\left(\boldsymbol{x}_{T'} + \boldsymbol{x}_{T''}\right) \cdot \mathbf{M}(\boldsymbol{D})$$
$$= \frac{1}{2}\left(\boldsymbol{x}_{T'} + \boldsymbol{x}_{T''}\right) \cdot \boldsymbol{P}(t)\mathbf{W}(\boldsymbol{D})\,,$$

or

$$S = \overline{\boldsymbol{x}} \cdot \mathbf{W}(\boldsymbol{D})\,,$$

where the vector $\overline{\boldsymbol{x}} := \overline{\boldsymbol{x}}(S) := \overline{\boldsymbol{x}}(S, \vec{\mathrm{S}}(\boldsymbol{D})) = (\overline{x}_1, \ldots, \overline{x}_t) \in \mathbb{Z}^t$ is the vector

$$\overline{\boldsymbol{x}} := \frac{1}{2}\left(\overline{\boldsymbol{x}}_{T'} + \overline{\boldsymbol{x}}_{T''}\right) = \frac{1}{2}\left(\boldsymbol{x}_{T'} + \boldsymbol{x}_{T''}\right) \cdot \boldsymbol{P}(t)\,.$$

Thus,

$$\overline{\boldsymbol{Q}}(S, \vec{\mathrm{S}}(\boldsymbol{D})) = \{\mathrm{sign}(\overline{x}_{T',i} + \overline{x}_{T'',i}) \cdot S^{i-1} \colon$$
$$\overline{x}_{T'',i} \neq -\overline{x}_{T',i},\ 1 \le i \le t\}\,,$$

$$Q' \in \{-S^{j-1}, S^{j-1}\}$$
$$\implies \lambda_{Q'} = \frac{1}{2}|\overline{x}_{T',j} + \overline{x}_{T'',j}|\,, \quad 1 \le j \le t\,. \quad \square$$

Example 10.3. Suppose $t := 6$. Consider a *symmetric cycle* $\boldsymbol{D}$ of the hypercube graph $\boldsymbol{H}(t, 2)$, depicted in Figure 10.1. The corresponding matrices $\mathbf{M}$ and $\mathbf{W}$ that describe *vertices* and *subtopes*

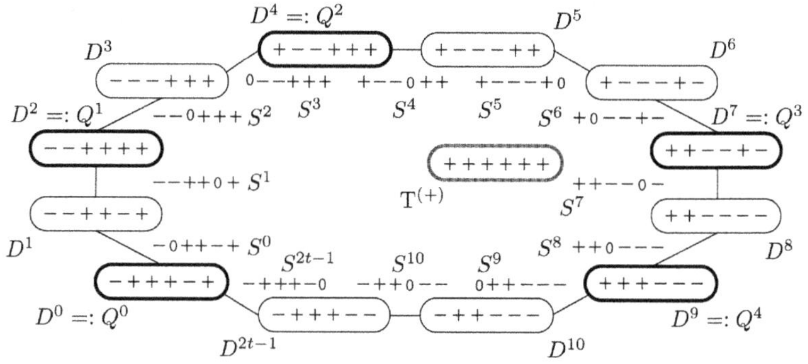

Figure 10.1 A *symmetric cycle* $\boldsymbol{D} := (D^0, D^1, \ldots, D^{2t-1}, D^0)$ in the *hypercube graph* $\boldsymbol{H}((t := 6), 2)$ on its vertex set $\{1, -1\}^t$.
The edges of the cycle $\boldsymbol{D}$ are labeled by the corresponding *subtopes*; for example, the edge $\{D^0, D^1\}$ is labeled by the subtope $S^0 := (-1, 0, 1, 1, -1, 1)$.
For the vertex $\mathrm{T}^{(+)} := (1, 1, 1, 1, 1, 1)$ of the graph $\boldsymbol{H}(t, 2)$, on the one hand, we have $\mathrm{T}^{(+)} = Q^0 + Q^1 + Q^2 + Q^3 + Q^4$ for a *unique inclusion-minimal* subset $\boldsymbol{Q}(\mathrm{T}^{(+)}, \boldsymbol{D}) =: \{Q^0, \ldots, Q^4\}$ of five *vertices* in the vertex sequence $\vec{\mathrm{V}}(\boldsymbol{D})$ of the cycle $\boldsymbol{D}$, where $Q^0 := D^0$, $Q^1 := D^2$, $Q^2 := D^4$, $Q^3 := D^7$ and $Q^4 := D^9$; see Section 2.1.
On the other hand, $\mathrm{T}^{(+)} = S^1 + S^2 + 5S^4 + 3S^6 + 3S^9 + 5S^{2t-1}$ for the *unique* set $\overline{\boldsymbol{Q}}(\mathrm{T}^{(+)}, \vec{\mathrm{S}}(\boldsymbol{D}))$ of $t := 6$ *subtopes* S^1, S^2, S^4, S^6, S^9 and S^{2t-1} labeling edges of the cycle $\boldsymbol{D}$, for which the corresponding *integer* coefficients are all *positive* and *odd*; see Proposition 10.2(i).

of the cycle $\boldsymbol{D}$ in the expression (10.5) are as follows:

$$\mathbf{M} := \mathbf{M}(\boldsymbol{D}) = \begin{pmatrix} D^0 \\ D^1 \\ D^2 \\ D^3 \\ D^4 \\ D^5 \end{pmatrix} = \begin{pmatrix} -1 & 1 & 1 & 1 & -1 & 1 \\ -1 & -1 & 1 & 1 & -1 & 1 \\ -1 & -1 & 1 & 1 & 1 & 1 \\ -1 & -1 & -1 & 1 & 1 & 1 \\ 1 & -1 & -1 & 1 & 1 & 1 \\ 1 & -1 & -1 & -1 & 1 & 1 \end{pmatrix},$$

$$\mathbf{W} := \mathbf{W}(\boldsymbol{D}) = \begin{pmatrix} S^0 \\ S^1 \\ S^2 \\ S^3 \\ S^4 \\ S^5 \end{pmatrix} = \begin{pmatrix} -1 & 0 & 1 & 1 & -1 & 1 \\ -1 & -1 & 1 & 1 & 0 & 1 \\ -1 & -1 & 0 & 1 & 1 & 1 \\ 0 & -1 & -1 & 1 & 1 & 1 \\ 1 & -1 & -1 & 0 & 1 & 1 \\ 1 & -1 & -1 & -1 & 1 & 0 \end{pmatrix}.$$

(i) Here we illustrate the proof of Proposition 10.2(i).

We would like to find the decomposition of the positive vertex $\mathrm{T}^{(+)} := (1, 1, 1, 1, 1, 1)$ with respect to the set $\vec{\mathrm{S}}(\boldsymbol{D})$ of subtopes (10.4) associated with the edges of the cycle $\boldsymbol{D}$; see Figure 10.1.

We have

$$\boldsymbol{x} := \boldsymbol{x}(\mathrm{T}^{(+)}, \boldsymbol{D}) = (1, -1, 1, -1, 1, 0)\,,$$

and

$$\overline{\boldsymbol{x}} := \overline{\boldsymbol{x}}(\mathrm{T}^{(+)}, \vec{\mathrm{S}}(\boldsymbol{D})) = \boldsymbol{x} \cdot \boldsymbol{P}(t)$$

$$= (1, -1, 1, -1, 1, 0) \cdot \begin{pmatrix} 1 & -1 & 1 & -1 & 1 & -1 \\ 1 & 1 & -1 & 1 & -1 & 1 \\ -1 & 1 & 1 & -1 & 1 & -1 \\ 1 & -1 & 1 & 1 & -1 & 1 \\ -1 & 1 & -1 & 1 & 1 & -1 \\ 1 & -1 & 1 & -1 & 1 & 1 \end{pmatrix}$$

$$= (-3, 1, 1, -3, 5, -5)\,.$$

Thus, we have

$$\begin{aligned} \overline{\boldsymbol{Q}}(\mathrm{T}^{(+)}, \vec{\mathrm{S}}(\boldsymbol{D})) &= \{\mathrm{sign}(\overline{x}_i) \cdot S^{i-1} \colon 1 \le i \le t\} \\ &= \{\underbrace{-S^0}_{S^6}, S^1, S^2, \underbrace{-S^3}_{S^9}, S^4, \underbrace{-S^5}_{S^{2t-1}}\} \\ &= \{S^1, S^2, S^4, S^6, S^9, S^{2t-1}\}\,, \end{aligned}$$

that is,

$$\begin{aligned} \mathrm{T}^{(+)} &:= (1, 1, 1, 1, 1, 1) \\ &= |\overline{x}_2| \cdot S^1 + |\overline{x}_3| \cdot S^2 + |\overline{x}_5| \cdot S^4 \\ &\quad + |\overline{x}_1| \cdot S^6 + |\overline{x}_4| \cdot S^9 + |\overline{x}_6| \cdot S^{2t-1} \\ &= S^1 + S^2 + 5S^4 + 3S^6 + 3S^9 + 5S^{2t-1} \\ &= (-1, -1, 1, 1, 0, 1) + (-1, -1, 0, 1, 1, 1) \\ &\quad + 5 \cdot (1, -1, -1, 0, 1, 1) + 3 \cdot (1, 0, -1, -1, 1, -1) \\ &\quad + 3 \cdot (0, 1, 1, -1, -1, -1) + 5 \cdot (-1, 1, 1, 1, -1, 0)\,, \end{aligned}$$

for the *unique* set $\overline{\boldsymbol{Q}}(\mathrm{T}^{(+)}, \vec{\mathrm{S}}(\boldsymbol{D})) = \{S^1, S^2, S^4, S^6, S^9, S^{2t-1}\}$ of $t := 6$ subtopes associated with edges of the cycle $\boldsymbol{D}$ for which the corresponding *integer* coefficients are all *positive* and *odd*.

(ii) Let us now illustrate the proof of Proposition 10.2(ii). Consider the subtope

$$S := (-1, -1, 0, 1, -1, -1)$$

that labels an edge $\{T', T''\}$ of the hypercube graph $\boldsymbol{H}(t, 2)$, where

$$T' := (-1, -1, 1, 1, -1, -1)\,, \text{ and } T'' := (-1, -1, -1, 1, -1, -1)\,.$$

We have

$$\begin{aligned} S &= \underbrace{\frac{1}{2}\left(\overline{\boldsymbol{x}}_{T'} + \overline{\boldsymbol{x}}_{T''}\right)}_{\overline{\boldsymbol{x}}(S)} \cdot \mathbf{W}(\boldsymbol{D}) \\ &= \underbrace{\frac{1}{2}\left(\boldsymbol{x}_{T'} + \boldsymbol{x}_{T''}\right) \cdot \boldsymbol{P}(t)}_{\overline{\boldsymbol{x}}(S)} \mathbf{W}(\boldsymbol{D})\,. \end{aligned}$$

Since

$$\boldsymbol{x}_{T'} := \boldsymbol{x}(T') = (-1,\ \ 1,\ \ 0,\ \ 0,\ \ 0, -1)\,,$$

and

$$\boldsymbol{x}_{T''} := \boldsymbol{x}(T'') = (-1,\ \ 1, -1,\ \ 1,\ \ 0, -1)\,,$$

we have

$$\begin{aligned} \overline{\boldsymbol{x}}(S) &= \frac{1}{2}\big((-1, 1, 0, 0, 0, -1) + (-1, 1, -1, 1, 0, -1)\big) \\ &\quad \times \begin{pmatrix} 1 & -1 & 1 & -1 & 1 & -1 \\ 1 & 1 & -1 & 1 & -1 & 1 \\ -1 & 1 & 1 & -1 & 1 & -1 \\ 1 & -1 & 1 & 1 & -1 & 1 \\ -1 & 1 & -1 & 1 & 1 & -1 \\ 1 & -1 & 1 & -1 & 1 & 1 \end{pmatrix} \\ &= \left(-1, 1, -\frac{1}{2}, \frac{1}{2}, 0, -1\right) \cdot \begin{pmatrix} 1 & -1 & 1 & -1 & 1 & -1 \\ 1 & 1 & -1 & 1 & -1 & 1 \\ -1 & 1 & 1 & -1 & 1 & -1 \\ 1 & -1 & 1 & 1 & -1 & 1 \\ -1 & 1 & -1 & 1 & 1 & -1 \\ 1 & -1 & 1 & -1 & 1 & 1 \end{pmatrix} \\ &= (0, 2, -3, 4, -4, 2)\,. \end{aligned}$$

We see that

$$S := (-1, -1, 0, 1, -1, -1) = 2S^1 - 3 \cdot \underbrace{S^2}_{-S^8} + 4S^3 - 4 \cdot \underbrace{S^4}_{-S^{10}} + 2S^5$$

or, in other words,

$$\begin{aligned} S &:= (-1, -1, 0, 1, -1, -1) \\ &= 2S^1 + 4S^3 + 2S^5 + 3S^8 + 4S^{10} \\ &= 2 \cdot (-1, -1, 1, 1, 0, 1) + 4 \cdot (0, -1, -1, 1, 1, 1) \\ &\quad + 2 \cdot (1, -1, -1, -1, 1, 0) + 3 \cdot (1, 1, 0, -1, -1, -1) \\ &\quad + 4 \cdot (-1, 1, 1, 0, -1, -1)\,, \end{aligned}$$

for the *unique inclusion-minimal* set $\overline{\boldsymbol{Q}}(S, \vec{\mathrm{S}}(\boldsymbol{D})) = \{S^1, S^3, S^5, S^8, S^{10}\}$ of subtopes associated with edges of the cycle $\boldsymbol{D}$ for which the corresponding *integer* coefficients are all *positive.*

10.3 Vertex Decompositions in Hypercube Graphs with Respect to the Edges of Their Distinguished Symmetric Cycles

Let $\boldsymbol{R} := (R^0, R^1, \ldots, R^{2t-1}, R^0)$ be our familiar distinguished symmetric cycle in the *hypercube graph* $\boldsymbol{H}(t, 2)$, where t is *even,* defined as follows:

$$\begin{aligned} R^0 &:= \mathrm{T}^{(+)}\,, \\ R^s &:= {}_{-[s]}R^0\,, \quad 1 \le s \le t-1\,, \end{aligned} \tag{10.12}$$

and

$$R^{k+t} := -R^k\,, \quad 0 \le k \le t-1\,. \tag{10.13}$$

Recall that no matter how large the dimension t of the *discrete hypercube* $\{1, -1\}^t$ is, the four assertions of Proposition 5.9 allow us to find the *linear algebraic* decompositions

$$T = \sum_{Q \in \boldsymbol{Q}(T, \boldsymbol{R})} Q$$

of vertices T of the graph $\boldsymbol{H}(t, 2)$ by means of the *inclusion-minimal* subsequences $\boldsymbol{Q}(T, \boldsymbol{R}) \subset \vec{\mathrm{V}}(\boldsymbol{R})$ of the vertex sequence $\vec{\mathrm{V}}(\boldsymbol{R}) :=$

$(R^0, R^1, \ldots, R^{2t-1})$ of the cycle $\boldsymbol{R}$ in an *explicit* and *computation-free* way.

We now present a (subtope) companion to Proposition 5.9. As earlier, $\boldsymbol{P}(t)^s$, $1 \leq s \leq t$, denotes the sth row P^{s-1} of the matrix $\boldsymbol{P}(t)$ given in (10.7), with entries (10.6).

Proposition 10.4. *Let $\boldsymbol{R}$ be the distinguished symmetric cycle in the hypercube graph $\boldsymbol{H}(t, 2)$ on its vertex set $\{1, -1\}^t$, defined by* (10.12)(10.13), *where t is* even.

Let A be a nonempty subset of the ground set E_t, viewed as a disjoint *union*

$$A = [i_1, j_1] \,\dot{\cup}\, [i_2, j_2] \,\dot{\cup}\, \cdots \,\dot{\cup}\, [i_\varrho, j_\varrho]$$

of inclusion-maximal intervals of E_t, such that

$$j_1 + 2 \leq i_2, \quad j_2 + 2 \leq i_3, \quad \ldots, \quad j_{\varrho-1} + 2 \leq i_\varrho \,,$$

for some $\varrho := \varrho(A)$.

Consider the vector $\overline{\boldsymbol{x}}({}_{-A}\mathrm{T}^{(+)}, \vec{\mathrm{S}}(\boldsymbol{R})) \in \mathbb{Z}^t$ defined by

$$ {}_{-A}\mathrm{T}^{(+)} =: \overline{\boldsymbol{x}}({}_{-A}\mathrm{T}^{(+)}, \vec{\mathrm{S}}(\boldsymbol{R})) \cdot \boldsymbol{W}(\boldsymbol{R}) \,,$$

cf. (10.10) *and* (10.11).

(i) *If $\{1, t\} \cap A = \{1\}$, then*

$$\overline{\boldsymbol{x}}({}_{-A}\mathrm{T}^{(+)}, \vec{\mathrm{S}}(\boldsymbol{R})) = \sum_{1 \leq k \leq \varrho} \boldsymbol{P}(t)^{j_k+1} \;-\; \sum_{2 \leq \ell \leq \varrho} \boldsymbol{P}(t)^{i_\ell} \,,$$

that is, for a component $\overline{x}_e$ of this vector, where $e \in E_t$, we have

$$\overline{x}_e({}_{-A}\mathrm{T}^{(+)}, \vec{\mathrm{S}}(\boldsymbol{R})) = \sum_{1 \leq k \leq \varrho} \begin{cases} (-1)^{e+j_k+1} \,, & \text{if } j_k < e \,, \\ (-1)^{e+j_k} \,, & \text{if } j_k \geq e \end{cases}$$

$$- \sum_{2 \leq \ell \leq \varrho} \begin{cases} (-1)^{e+i_\ell} \,, & \text{if } i_\ell \leq e \,, \\ (-1)^{e+i_\ell+1} \,, & \text{if } i_\ell > e \,. \end{cases}$$

(ii) *If $\{1, t\} \cap A = \{1, t\}$, then*

$$\overline{\boldsymbol{x}}({}_{-A}\mathrm{T}^{(+)}, \vec{\mathrm{S}}(\boldsymbol{R})) = -\boldsymbol{P}(t)^1 + \sum_{1 \leq k \leq \varrho-1} \boldsymbol{P}(t)^{j_k+1} \;-\; \sum_{2 \leq \ell \leq \varrho} \boldsymbol{P}(t)^{i_\ell} \,,$$

that is, for a component $\overline{x}_e$ of this vector we have

$$\overline{x}_e({}_{-A}\mathrm{T}^{(+)}, \vec{\mathrm{S}}(\boldsymbol{R})) = (-1)^e + \sum_{1\le k\le \varrho-1} \begin{cases} (-1)^{e+j_k+1}, & \text{if } j_k < e, \\ (-1)^{e+j_k}, & \text{if } j_k \ge e \end{cases} - \sum_{2\le \ell\le \varrho} \begin{cases} (-1)^{e+i_\ell}, & \text{if } i_\ell \le e, \\ (-1)^{e+i_\ell+1}, & \text{if } i_\ell > e. \end{cases}$$

(iii) *If* $|\{1, t\} \cap A| = 0$, *then*

$$\overline{\boldsymbol{x}}({}_{-A}\mathrm{T}^{(+)}, \vec{\mathrm{S}}(\boldsymbol{R})) = \boldsymbol{P}(t)^1 + \sum_{1\le k\le \varrho} \boldsymbol{P}(t)^{j_k+1} - \sum_{1\le \ell\le \varrho} \boldsymbol{P}(t)^{i_\ell},$$

that is, for a component $\overline{x}_e$ of this vector we have

$$\overline{x}_e({}_{-A}\mathrm{T}^{(+)}, \vec{\mathrm{S}}(\boldsymbol{R})) = (-1)^{e+1} + \sum_{1\le k\le \varrho} \begin{cases} (-1)^{e+j_k+1}, & \text{if } j_k < e, \\ (-1)^{e+j_k}, & \text{if } j_k \ge e \end{cases} - \sum_{1\le \ell\le \varrho} \begin{cases} (-1)^{e+i_\ell}, & \text{if } i_\ell \le e, \\ (-1)^{e+i_\ell+1}, & \text{if } i_\ell > e. \end{cases}$$

(iv) *If* $\{1, t\} \cap A = \{t\}$, *then*

$$\overline{\boldsymbol{x}}({}_{-A}\mathrm{T}^{(+)}, \vec{\mathrm{S}}(\boldsymbol{R})) = \sum_{1\le k\le \varrho-1} \boldsymbol{P}(t)^{j_k+1} - \sum_{1\le \ell\le \varrho} \boldsymbol{P}(t)^{i_\ell},$$

that is, for a component $\overline{x}_e$ of this vector we have

$$\overline{x}_e({}_{-A}\mathrm{T}^{(+)}, \vec{\mathrm{S}}(\boldsymbol{R})) = \sum_{1\le k\le \varrho-1} \begin{cases} (-1)^{e+j_k+1}, & \text{if } j_k < e, \\ (-1)^{e+j_k}, & \text{if } j_k \ge e \end{cases} - \sum_{1\le \ell\le \varrho} \begin{cases} (-1)^{e+i_\ell}, & \text{if } i_\ell \le e, \\ (-1)^{e+i_\ell+1}, & \text{if } i_\ell > e. \end{cases}$$

In particular, we have

$$1 \le j < t \implies \overline{\boldsymbol{x}}({}_{-[j]}\mathrm{T}^{(+)}, \vec{\mathrm{S}}(\boldsymbol{R})) = \boldsymbol{P}(t)^{j+1};$$

$$\overline{\boldsymbol{x}}(\mathrm{T}^{(-)}, \vec{\mathrm{S}}(\boldsymbol{R})) = -\overline{\boldsymbol{x}}(\mathrm{T}^{(+)}, \vec{\mathrm{S}}(\boldsymbol{R})) = -\boldsymbol{P}(t)^1;$$

$$1 < i < j < t \implies \overline{\boldsymbol{x}}({}_{-[i,j]}\mathrm{T}^{(+)}, \vec{\mathrm{S}}(\boldsymbol{R})) = \boldsymbol{P}(t)^1 - \boldsymbol{P}(t)^i + \boldsymbol{P}(t)^{j+1} ;$$

$$1 < i \le t \implies \overline{\boldsymbol{x}}({}_{-[i,t]}\mathrm{T}^{(+)}, \vec{\mathrm{S}}(\boldsymbol{R})) = -\boldsymbol{P}(t)^i .$$

If $s \in E_t$, then for a row vector $\overline{\boldsymbol{y}}(s) := \overline{\boldsymbol{y}}(s;t)$ defined by

$$\overline{\boldsymbol{y}}(s) := \overline{\boldsymbol{x}}({}_{-s}\mathrm{T}^{(+)}, \vec{\mathrm{S}}(\boldsymbol{R})) ,$$

we have

$$\overline{\boldsymbol{y}}(s) = \begin{cases} \boldsymbol{P}(t)^2 , & \text{if } s = 1, \\ \boldsymbol{P}(t)^1 - \boldsymbol{P}(t)^s + \boldsymbol{P}(t)^{s+1} , & \text{if } 1 < s < t, \\ -\boldsymbol{P}(t)^t , & \text{if } s = t. \end{cases}$$

Appendix A

Enumeration of Smirnov Words over Three-letter and Four-letter Alphabets

The words without consecutive equal letters (*waves*, *normal words*), called the *Smirnov words* after the work [109], are investigated, applied and enumerated, e.g., in Refs. [110, 111, 112], in Examples 7.45, 8.14 and 8.16 of Ref. [113], in Refs. [114, 115, 116, 117], in Examples III.24 and IV.10 of Ref. [118], in Refs. [119, 120, 121], in Section 2.4 and in Exercise 3.5.1 of Ref. [27], on pages 164–166 of Ref. [122], in Refs. [123, 124, 125, 126, 127, 128, 129, 130], in Examples 2.2.10 and 13.3.5 of Ref. [131], in Section 4.8 of Ref. [132], in Refs. [24, 133, 134, 135, 136, 137, 138, 139, 140]. The importance of these words can easily be explained (see page 205 in Ref. [118]):

> Start from a Smirnov word and substitute for any letter a_j that appears in it an arbitrary nonempty sequence of letters a_j. When this operation is done at all places of a Smirnov word, it gives rise to an unconstrained word. Conversely, any word can be associated to a unique Smirnov word by collapsing into single letters maximal groups of contiguous equal letters.

Symmetric Cycles
Andrey O. Matveev

ISBN 978-981-4968-81-2 (Hardcover), 978-1-003-43832-8 (eBook)
www.jennystanford.com

A.1 Ternary Smirnov Words

Let (θ, α, β) be an ordered three-letter alphabet, and let 'u', 'v' and 'w' be formal variables which mark the letters θ, α and β, respectively. Let $e_{\cdot}(\cdot\cdot)$ denote elementary symmetric polynomials. The ordinary trivariate generating function of the set of ternary Smirnov words is

$$\frac{1}{1-\left(\frac{u}{1+u}+\frac{v}{1+v}+\frac{w}{1+w}\right)}=\frac{(1+u)(1+v)(1+w)}{1-e_2(u,v,w)-2e_3(u,v,w)}\,; \quad \text{(A.1)}$$

see Example III.24 in Ref. [118], and Section 2.4.16 of Ref. [27], on the general multivariate generating function of the Smirnov words.

In the theoretical framework of the breakthrough article [24], let us consider the system of generating functions

$$\begin{cases} f_\theta(u,v,w) = u + u\,f_\alpha(u,v,w) + u\,f_\beta(u,v,w)\,, \\ f_\alpha(u,v,w) = \qquad v\,f_\theta(u,v,w) + v\,f_\beta(u,v,w)\,, \\ f_\beta(u,v,w) = \qquad w\,f_\theta(u,v,w) + w\,f_\alpha(u,v,w) \end{cases}$$

rewritten, for short, as

$$\begin{cases} f_\theta = u + u\cdot(f_\alpha + f_\beta)\,, \\ f_\alpha = \qquad v\cdot(f_\theta + f_\beta)\,, \\ f_\beta = \qquad w\cdot(f_\theta + f_\alpha)\,. \end{cases} \quad \text{(A.2)}$$

For a letter $\mathfrak{s} \in (\theta, \alpha, \beta)$, the generating function $f_\mathfrak{s} := f_\mathfrak{s}(u,v,w)$ is meant to count the ternary Smirnov words starting with the letter θ and ending with the letter $\mathfrak{s}$.

The solutions to the system (A.2) are

$$f_\theta = \frac{u(1-e_2(v,w))}{1-e_2(u,v,w)-2e_3(u,v,w)}\,,$$

$$f_\alpha == \frac{uv(1+w)}{1-e_2(u,v,w)-2e_3(u,v,w)}$$

and

$$f_\beta = \frac{uw(1+v)}{1-e_2(u,v,w)-2e_3(u,v,w)}\,,$$

cf. (A.1). Thus, we have

$$f_\theta = \sum_{k=1}^{\infty} \left(\frac{\mathrm{v} + \mathrm{w} + 2\mathrm{vw}}{1 - \mathrm{vw}} \right)^{k-1} \mathrm{u}^k \tag{A.3}$$

$$= \underbrace{(1 - \mathrm{e}_2(\mathrm{v}, \mathrm{w})) \sum_{k=1}^{\infty} \frac{(\mathrm{e}_1(\mathrm{v}, \mathrm{w}) + 2\mathrm{e}_2(\mathrm{v}, \mathrm{w}))^{k-1}}{(1 - \mathrm{e}_2(\mathrm{v}, \mathrm{w}))^k} \mathrm{u}^k}_{\text{cf. (A.11)}}, \tag{A.4}$$

$$f_\alpha = \underbrace{\mathrm{v}(1 + \mathrm{w}) \sum_{k=1}^{\infty} \frac{(\mathrm{e}_1(\mathrm{v}, \mathrm{w}) + 2\mathrm{e}_2(\mathrm{v}, \mathrm{w}))^{k-1}}{(1 - \mathrm{e}_2(\mathrm{v}, \mathrm{w}))^k} \mathrm{u}^k}_{\text{cf. (A.13)}} \tag{A.5}$$

and

$$f_\beta = \mathrm{w}(1 + \mathrm{v}) \sum_{k=1}^{\infty} \frac{(\mathrm{v} + \mathrm{w} + 2\mathrm{vw})^{k-1}}{(1 - \mathrm{vw})^k} \mathrm{u}^k \tag{A.6}$$

$$= \mathrm{w}(1 + \mathrm{v}) \sum_{k=1}^{\infty} \frac{(\mathrm{e}_1(\mathrm{v}, \mathrm{w}) + 2\mathrm{e}_2(\mathrm{v}, \mathrm{w}))^{k-1}}{(1 - \mathrm{e}_2(\mathrm{v}, \mathrm{w}))^k} \mathrm{u}^k .$$

Remark A.1. In the framework of Ref. [24], the numbers $\mathfrak{T}(\theta, \mathfrak{s}; k, i, j)$ of distinct ternary Smirnov words that start with the letter θ, end with a letter $\mathfrak{s} \in \{\theta, \beta\}$, and contain k letters θ, i letters α, and j letters β, can be read off from the power series representations of the generating functions $f_\mathfrak{s}$, given in (A.3) and (A.6), as the coefficients $[\mathrm{u}^k \mathrm{v}^i \mathrm{w}^j] f_\mathfrak{s}$ of $\mathrm{u}^k \mathrm{v}^i \mathrm{w}^j$:

(i)

$$\mathfrak{T}(\theta, \theta; k, i, j) = \begin{cases} \binom{k-1}{\frac{k+i-j-1}{2}} \binom{\frac{k+i+j-3}{2}}{k-2}, & \text{if } k+i+j \text{ odd}, \\ (k+i-j) \cdot \binom{k-1}{\frac{k+i-j}{2}} \binom{\frac{k+i+j}{2}-2}{k-2}, & \\ & \text{if } k+i+j \text{ even}. \end{cases} \tag{A.7}$$

(ii)

$$\mathfrak{T}(\theta,\beta;k,i,j)=$$
$$\begin{cases} (k+j-i)\cdot\binom{k-1}{\frac{k+j-i-1}{2}}\binom{\frac{k+i+j-3}{2}}{k-1}\ , & \text{if } k+i+j \text{ odd}\ ,\\ \binom{k-1}{\frac{k+j-i}{2}-1}\binom{\frac{k+i+j}{2}-1}{k-1}+(k+j-i)\cdot\binom{k-1}{\frac{k+j-i}{2}}\binom{\frac{k+i+j}{2}-2}{k-1}\ , & \\ \qquad\qquad\qquad \text{if } k+i+j \text{ even}\ , \end{cases}$$
$$=\mathfrak{T}(\beta,\theta;k,i,j)\,. \qquad \text{(A.8)}$$

A.2 Smirnov Words over a Four-letter Alphabet

The ordinary quadrivariate generating function of the set of Smirnov words over the ordered alphabet $(\theta, \alpha, \beta, \gamma)$ with its four letters marked by the formal variables 'u', 'v', 'w' and 'x', is

$$\frac{1}{1-\left(\frac{\mathrm{u}}{1+\mathrm{u}}+\frac{\mathrm{v}}{1+\mathrm{v}}+\frac{\mathrm{w}}{1+\mathrm{w}}+\frac{\mathrm{x}}{1+\mathrm{x}}\right)}$$
$$=\frac{(1+\mathrm{u})(1+\mathrm{v})(1+\mathrm{w})(1+\mathrm{x})}{1-\mathrm{e}_2(\mathrm{u},\mathrm{v},\mathrm{w},\mathrm{x})-2\mathrm{e}_3(\mathrm{u},\mathrm{v},\mathrm{w},\mathrm{x})-3\mathrm{e}_4(\mathrm{u},\mathrm{v},\mathrm{w},\mathrm{x})}\,.$$

For letters $\mathfrak{s} \in (\theta, \alpha, \beta, \gamma)$, the generating functions $g_{\mathfrak{s}} := g_{\mathfrak{s}}(\mathrm{u}, \mathrm{v}, \mathrm{w}, \mathrm{x})$ defined by the system

$$\begin{cases} g_\theta = \mathrm{u}+\mathrm{u}\cdot(g_\alpha+g_\beta+g_\gamma)\ ,\\ g_\alpha = \qquad \mathrm{v}\cdot(g_\theta+g_\beta+g_\gamma)\ ,\\ g_\beta = \qquad \mathrm{w}\cdot(g_\theta+g_\alpha+g_\gamma)\ ,\\ g_\gamma = \qquad \mathrm{x}\cdot(g_\theta+g_\alpha+g_\beta)\ , \end{cases} \qquad \text{(A.9)}$$

are intended for enumerating the Smirnov words starting with the letter θ and ending with the letter $\mathfrak{s}$.

The solutions to the system (A.9) are

$$g_\theta=\frac{\mathrm{u}(1-\mathrm{e}_2(\mathrm{v},\mathrm{w},\mathrm{x}))}{1-\mathrm{e}_2(\mathrm{u},\mathrm{v},\mathrm{w},\mathrm{x})-2\mathrm{e}_3(\mathrm{u},\mathrm{v},\mathrm{w},\mathrm{x})-3\mathrm{e}_4(\mathrm{u},\mathrm{v},\mathrm{w},\mathrm{x})}$$
$$=(1-(\mathrm{vw}+\mathrm{vx}+\mathrm{wx}))$$
$$\times\sum_{k=1}^{\infty}\frac{(\mathrm{v}+\mathrm{w}+\mathrm{x}+2(\mathrm{vw}+\mathrm{vx}+\mathrm{wx})+3\mathrm{vwx})^{k-1}}{(1-(\mathrm{vw}+\mathrm{vx}+\mathrm{wx}+2\mathrm{vwx}))^k}\mathrm{u}^k \qquad \text{(A.10)}$$

$$= (1 - e_2(v, w, x)) \underbrace{\times \sum_{k=1}^{\infty} \frac{(e_1(v, w, x) + 2e_2(v, w, x) + 3e_3(v, w, x))^{k-1}}{(1 - e_2(v, w, x) - 2e_3(v, w, x))^k} u^k}_{\text{cf. (A.4)}}, \tag{A.11}$$

$$g_\alpha = \frac{uv(1+w)(1+x)}{1 - e_2(u, v, w, x) - 2e_3(u, v, w, x) - 3e_4(u, v, w, x)}$$

$$= v(1+w)(1+x) \times \sum_{k=1}^{\infty} \frac{(v + w + x + 2(vw + vx + wx) + 3vwx)^{k-1}}{(1 - (vw + vx + wx + 2vwx))^k} u^k \tag{A.12}$$

$$= v(1+w)(1+x) \underbrace{\times \sum_{k=1}^{\infty} \frac{(e_1(v, w, x) + 2e_2(v, w, x) + 3e_3(v, w, x))^{k-1}}{(1 - e_2(v, w, x) - 2e_3(v, w, x))^k} u^k}_{\text{cf. (A.5)}}, \tag{A.13}$$

$$g_\beta = \frac{uw(1+v)(1+x)}{1 - e_2(u, v, w, x) - 2e_3(u, v, w, x) - 3e_4(u, v, w, x)}$$

$$= w(1+v)(1+x) \times \sum_{k=1}^{\infty} \frac{(e_1(v, w, x) + 2e_2(v, w, x) + 3e_3(v, w, x))^{k-1}}{(1 - e_2(v, w, x) - 2e_3(v, w, x))^k} u^k \tag{A.14}$$

and

$$g_\gamma = \frac{ux(1+v)(1+w)}{1 - e_2(u, v, w, x) - 2e_3(u, v, w, x) - 3e_4(u, v, w, x)}$$

$$= x(1+v)(1+w) \times \sum_{k=1}^{\infty} \frac{(e_1(v, w, x) + 2e_2(v, w, x) + 3e_3(v, w, x))^{k-1}}{(1 - e_2(v, w, x) - 2e_3(v, w, x))^k} u^k . \tag{A.15}$$

Remark A.2. In the framework of Ref. [24], the numbers $\mathfrak{F}(\theta, \mathfrak{s}; k, i, j, h)$ of distinct Smirnov words, over the four-letter alphabet

$(\theta, \alpha, \beta, \gamma)$ and with the Parikh vector (k, i, j, h), that start with the letter θ and end with a letter $\mathfrak{s} \in \{\theta, \alpha\}$, can be read off, in one way or another, from the power series representations of the generating functions $g_{\mathfrak{s}}$, given in (A.10) and (A.12), as the coefficients $[\mathrm{u}^k \mathrm{v}^i \mathrm{w}^j \mathrm{x}^h] g_{\mathfrak{s}}$ of $\mathrm{u}^k \mathrm{v}^i \mathrm{w}^j \mathrm{x}^h$. For example, we have:

(i)

$$\begin{aligned}
\mathfrak{F}(\theta, \theta; k, i, j, h) = & \\
& \sum_{\substack{0 \le \mathrm{p} \le k-1, \\ 0 \le \mathrm{r} \le \lfloor \frac{1}{2}(i+j+h-k+1) \rfloor}} \ \sum_{\substack{\mathrm{p} \le \mathrm{s} \le k-1, \\ \mathrm{r} \le \mathrm{t} \le \lfloor \frac{1}{2}(i+j+h-k+1) \rfloor}} \binom{k+\mathrm{t}-1}{\mathrm{p},\ \mathrm{r},\ \mathrm{s}-\mathrm{p},\ \mathrm{t}-\mathrm{r},\ k-\mathrm{s}-1} \\
& \times 2^{-i-j-h+3k-2\mathrm{s}-2\mathrm{r}+4\mathrm{t}-3} \cdot 3^{i+j+h-2k+\mathrm{s}+\mathrm{r}-3\mathrm{t}+2} \\
& \times \Bigg(\binom{k-\mathrm{s}-1}{-i-j-h+3k-2\mathrm{s}-\mathrm{r}+3\mathrm{t}-3} \binom{-i-j-h+3k-2\mathrm{s}+3\mathrm{t}-3}{-i+k+\mathrm{p}-\mathrm{s}+\mathrm{t}-1} \\
& \times \binom{-j-h+2k-2\mathrm{p}+2\mathrm{t}-2}{-j+k-\mathrm{p}+\mathrm{t}-1} - \frac{4}{9} \cdot \binom{k-\mathrm{s}-1}{-i-j-h+3k-2\mathrm{s}-\mathrm{r}+3\mathrm{t}-1} \\
& \times \binom{-i-j-h+3k-2\mathrm{s}+3\mathrm{t}}{-i+k+\mathrm{p}-\mathrm{s}+\mathrm{t}} \binom{-j-h+2k-2\mathrm{p}+2\mathrm{t}}{-j+k-\mathrm{p}+\mathrm{t}} \Bigg) . \qquad \text{(A.16)}
\end{aligned}$$

(ii)

$$\begin{aligned}
\mathfrak{F}(\theta, \alpha; k, i, j, h) = & \\
& \sum_{\substack{0 \le \mathrm{p} \le k-1, \\ 0 \le \mathrm{r} \le \lfloor \frac{1}{2}(i+j+h-k) \rfloor}} \ \sum_{\substack{\mathrm{p} \le \mathrm{s} \le k-1, \\ \mathrm{r} \le \mathrm{t} \le \lfloor \frac{1}{2}(i+j+h-k) \rfloor}} \binom{k+\mathrm{t}-1}{\mathrm{p},\ \mathrm{r},\ \mathrm{s}-\mathrm{p},\ \mathrm{t}-\mathrm{r},\ k-\mathrm{s}-1} \\
& \times 2^{-i-j-h+3k-2\mathrm{s}-2\mathrm{r}+4\mathrm{t}-2} \cdot 3^{i+j+h-2k+\mathrm{s}+\mathrm{r}-3\mathrm{t}+1} \\
& \times \Bigg(\binom{k-\mathrm{s}-1}{-i-j-h+3k-2\mathrm{s}-\mathrm{r}+3\mathrm{t}-2} \binom{-i-j-h+3k-2\mathrm{s}+3\mathrm{t}-2}{-i+k+\mathrm{p}-\mathrm{s}+\mathrm{t}} \\
& \times \binom{-j-h+2k-2\mathrm{p}+2\mathrm{t}-2}{-j+k-\mathrm{p}+\mathrm{t}-1} + \frac{2}{3} \cdot \Bigg(\binom{k-\mathrm{s}-1}{-i-j-h+3k-2\mathrm{s}-\mathrm{r}+3\mathrm{t}-1} \\
& \times \binom{-i-j-h+3k-2\mathrm{s}+3\mathrm{t}-1}{-i+k+\mathrm{p}-\mathrm{s}+\mathrm{t}} + \frac{2}{3} \cdot \binom{k-\mathrm{s}-1}{-i-j-h+3k-2\mathrm{s}-\mathrm{r}+3\mathrm{t}} \\
& \times \binom{-i-j-h+3k-2\mathrm{s}+3\mathrm{t}}{-i+k+\mathrm{p}-\mathrm{s}+\mathrm{t}} \Bigg) \cdot \binom{-j-h+2k-2\mathrm{p}+2\mathrm{t}}{-j+k-\mathrm{p}+\mathrm{t}} \Bigg) \\
= & \ \mathfrak{F}(\alpha, \theta; k, i, j, h) . \qquad \text{(A.17)}
\end{aligned}$$

Appendix B

The Increasing Families of Sets Generated by Self-dual Clutters

For convenience, we begin by recalling some terminology and notation. As earlier, we denote by E_t, where $t \geq 3$, the set of integers $[t] := [1, \ldots, t] := \{1, \ldots, t\}$. Recall that a finite collection of sets $\mathcal{F}$ is called a *family*. The family $\mathbf{2}^{[t]}$ of *all* subsets of the set E_t is called the *power set* of E_t. The *empty set* is denoted by $\hat{0}$, and the *empty family* containing *no sets* is denoted by $\emptyset$. We denote by $|\cdot|$ the cardinalities of sets, while the numbers of sets in families are denoted by $\#\cdot$. Sometimes we say that the cardinality of a set A is the *size* of A.

If $\mathcal{F}$ is a set family such that $\emptyset \neq \mathcal{F} \neq \{\hat{0}\}$, then the union $\mathrm{V}(\mathcal{F}) := \bigcup_{F \in \mathcal{F}} F$ is called the *vertex set* of $\mathcal{F}$. *Any* finite set S such that $S \supseteq \mathrm{V}(\mathcal{F})$, *fixed* for someone's research purposes, is called the *ground set* of the family $\mathcal{F}$.

Given a nonempty family $\mathcal{F} \subseteq \mathbf{2}^{[t]}$ on its ground set E_t, we let $\mathcal{F}^{\complement} := \{F^{\complement}\colon\ F \in \mathcal{F}\}$ denote the family of their *complements*, where $F^{\complement} := E_t - F$. We also associate with the family $\mathcal{F}$ the family

$$\mathcal{F}^* := \{G^{\complement} : G \in \mathbf{2}^{[t]} - \mathcal{F}\}\ .$$

A family of sets $\mathcal{A}$, such that $\emptyset \neq \mathcal{A} \neq \{\hat{0}\}$, is called a *nontrivial clutter* if for any indices $i \neq j$ of members of the family $\{A_1, \ldots, A_\alpha\} =: \mathcal{A}$ we have $A_i \not\subseteq A_j$.

Symmetric Cycles
Andrey O. Matveev

ISBN 978-981-4968-81-2 (Hardcover), 978-1-003-43832-8 (eBook)
www.jennystanford.com

For a subset $A \subseteq E_t$, the *principal increasing family* of sets $\{A\}^{\triangledown} \subseteq \mathbf{2}^{[t]}$, *generated* by the *one-member* clutter $\{A\}$ on its *ground set* E_t, is defined by $\{A\}^{\triangledown} := \{B \subseteq E_t\colon\ B \supseteq A\}$. In particular, we have $\{\hat{0}\}^{\triangledown} = \mathbf{2}^{[t]}$, and $\{E_t\}^{\triangledown} = \{E_t\}$.

Given a nonempty clutter $\mathcal{A} := \{A_1, \ldots, A_\alpha\} \subset \mathbf{2}^{[t]}$, the *increasing family* of sets $\mathcal{A}^{\triangledown} \subseteq \mathbf{2}^{[t]}$, *generated* by $\mathcal{A}$ on its *ground set* E_t, is defined as the union $\mathcal{A}^{\triangledown} := \bigcup_{A \in \mathcal{A}} \{A\}^{\triangledown}$ of the principal increasing families $\{A\}^{\triangledown}$.

A subset $B \subseteq E_t$ is a *blocking set* of a nontrivial clutter $\mathcal{A} \subset \mathbf{2}^{[t]}$ if it holds $|B \cap A| > 0$, for each member $A \in \mathcal{A}$. The *blocker* $\mathfrak{B}(\mathcal{A})$ of the clutter $\mathcal{A}$ is defined to be the family of all *inclusion-minimal blocking sets* of $\mathcal{A}$; see, e.g., the monographs [5, 18, 27, 36, 37, 38, 39, 40, 41, 42, 44, 45, 46, 47, 48, 49, 50, 51, 52, 53, 54]. Additional references can be found in Section 9.1 on page 226. The increasing family $\mathfrak{B}(\mathcal{A})^{\triangledown}$ is by definition the family of *all* blocking sets of the clutter $\mathcal{A}$.

Recall that for a nontrivial clutter $\mathcal{A}$ on its ground set E_t we have

$$\mathfrak{B}(\mathcal{A})^{\triangledown} = (\mathcal{A}^{\triangledown})^{*} .$$

Clutters $\mathcal{A}$ with the property

$$\mathfrak{B}(\mathcal{A}) = \mathcal{A} ,$$

or, equivalently,

$$\mathfrak{B}(\mathcal{A})^{\triangledown} = \mathcal{A}^{\triangledown} ,$$

are called *self-dual* or *identically self-blocking*; see, e.g., Section 5.7 of Ref. [5], Section 2.1 of Ref. [36], Chapter 9 of Ref. [48], and Refs. [104, 105] on such clutters.

A clutter $\mathcal{A}$ on its ground set E_t is self-dual if and only if we have

$$(\mathcal{A}^{\triangledown})^{*} = \mathcal{A}^{\triangledown} , \qquad \text{(B.1)}$$

and although the (lack of) self-duality of clutters does not depend structurally on the cardinalities of their vertex sets and ground sets, relation (B.1) suggests the following criterion:

Remark B.1. (see Corollary 5.28(i) of Ref. [5]) A clutter $\mathcal{A} \subset \mathbf{2}^{[t]}$ on its ground set E_t is *self-dual* if and only if

$$\#\mathcal{A}^{\triangledown} = 2^{t-1} . \qquad \text{(B.2)}$$

In Theorem B.11 we consider condition (B.2) from a perspective of constraints that are satisfied by the numbers of k-sets in the increasing families $\mathcal{A}^{\triangledown}$ generated by self-dual clutters $\mathcal{A} \subset \mathbf{2}^{[t]}$ on their ground set E_t of even cardinality t.

B.1 Long f- and h-Vectors of Set Families

Let us associate with each family $\mathcal{F} \subseteq \mathbf{2}^{[t]}$ on the ground set E_t its *long f-vector*

$$\boldsymbol{f}(\mathcal{F};t) := \big(f_0(\mathcal{F};t),\ f_1(\mathcal{F};t), \ldots,\ f_t(\mathcal{F};t)\big) \in \mathbb{N}^{t+1},$$

where

$$f_k(\mathcal{F};t) := \#\{F \in \mathcal{F} : |F| = k\}, \quad 0 \le k \le t.$$

If 'x' is a formal variable, then the *long h-vector*

$$\boldsymbol{h}(\mathcal{F};t) := \big(h_0(\mathcal{F};t),\ h_1(\mathcal{F};t), \ldots,\ h_t(\mathcal{F};t)\big) \in \mathbb{Z}^{t+1}$$

of the family $\mathcal{F}$ is defined by means of the relation

$$\sum_{i=0}^{t} h_i(\mathcal{F};t) \cdot \mathrm{x}^{t-i} := \sum_{i=0}^{t} f_i(\mathcal{F};t) \cdot (\mathrm{x}-1)^{t-i}.$$

Example B.2. (cf. Example B.6). For any element $a \in E_t$, the principal increasing family $\{\{a\}\}^{\triangledown} \subset \mathbf{2}^{[t]}$, generated by the *self-dual* clutter $\{\{a\}\}$ on its ground set E_t, is described by the vectors

$$\boldsymbol{f}(\{\{a\}\}^{\triangledown};t) = (0,\ \tbinom{t-1}{1-1},\ \tbinom{t-1}{2-1}, \ldots,\ \tbinom{t-1}{t-1})$$

and

$$\boldsymbol{h}(\{\{a\}\}^{\triangledown};t) = (0,\quad 1,\quad 0, \ldots, 0).$$

Example B.3. (i) (a) Given the self-dual clutter

$$\mathcal{A} := \{\{1,2\},\{1,3\},\{2,3\}\} \tag{B.3}$$

that generates on its ground set $E_3 = \mathrm{V}(\mathcal{A})$ the increasing family

$$\mathcal{A}^{\triangledown} = \{\{1,2\},\{1,3\},\{2,3\},E_3\},$$

we have

$$\boldsymbol{f}(\mathcal{A}^{\triangledown};3) = (0,\quad 0,\quad 3,\quad 1),$$
$$\boldsymbol{h}(\mathcal{A}^{\triangledown};3) = (0,\quad 0,\quad 3,-2).$$

(b) If we choose the set $E_4 \supsetneqq \mathrm{V}(\mathcal{A})$ as the ground set of the self-dual clutter (B.3), instead of the set E_3, then $\mathcal{A}$ generates on E_4 the increasing family

$$\mathcal{A}^{\triangledown} = \{\{1, 2\}, \{1, 3\}, \{2, 3\}, \{1, 2, 3\}, \{1, 2, 4\}, \{1, 3, 4\}, \{2, 3, 4\}, E_4\}\,,$$

and we have

$$\begin{aligned} \boldsymbol{f}(\mathcal{A}^{\triangledown}; 4) &= (0, \quad 0, \quad 3, \quad 4, \quad 1)\,, \\ \boldsymbol{h}(\mathcal{A}^{\triangledown}; 4) &= (0, \quad 0, \quad 3, -2, \quad 0)\,. \end{aligned}$$

(ii) (a) For the self-dual clutter

$$\mathcal{A} := \{\{2\}\} \tag{B.4}$$

that generates on its ground set $E_3 \supsetneqq \mathrm{V}(\mathcal{A})$ the principal increasing family

$$\mathcal{A}^{\triangledown} = \{\{2\}, \{1, 2\}, \{2, 3\}, E_3\}\,,$$

we have

$$\begin{aligned} \boldsymbol{f}(\mathcal{A}^{\triangledown}; 3) &= (0, \quad 1, \quad 2, \quad 1)\,, \\ \boldsymbol{h}(\mathcal{A}^{\triangledown}; 3) &= (0, \quad 1, \quad 0, \quad 0)\,. \end{aligned}$$

(b) For the self-dual clutter (B.4) that generates on its ground set $E_4 \supsetneqq \mathrm{V}(\mathcal{A})$ the principal increasing family

$$\mathcal{A}^{\triangledown} = \{\{2\}, \{1, 2\}, \{2, 3\}, \{2, 4\}, \{1, 2, 3\}, \{1, 2, 4\}, \{2, 3, 4\}, E_4\}\,,$$

we have

$$\begin{aligned} \boldsymbol{f}(\mathcal{A}^{\triangledown}; 4) &= (0, \quad 1, \quad 3, \quad 3, \quad 1)\,, \\ \boldsymbol{h}(\mathcal{A}^{\triangledown}; 4) &= (0, \quad 1, \quad 0, \quad 0, \quad 0)\,. \end{aligned}$$

(iii) The self-dual clutter

$$\mathcal{A} := \{\{1, 2, 3, 4\}, \{1, 5\}, \{2, 5\}, \{3, 5\}, \{4, 5\}\}\,,$$

that generates on its ground set $E_5 = \mathrm{V}(\mathcal{A})$ the increasing family

$$\begin{aligned} \mathcal{A}^{\triangledown} = \{&\{1, 5\}, \{2, 5\}, \{3, 5\}, \{4, 5\}, \{1, 2, 5\}, \{1, 3, 5\}, \{1, 4, 5\}, \\ &\{2, 3, 5\}, \{2, 4, 5\}, \{3, 4, 5\}, \{1, 2, 3, 4\}, \{1, 2, 3, 5\}, \\ &\{1, 2, 4, 5\}, \{1, 3, 4, 5\}, \{2, 3, 4, 5\}, E_5\}\,, \end{aligned}$$

is described by the vectors

$$\begin{aligned} \boldsymbol{f}(\mathcal{A}^{\triangledown}; 5) &= (0, \quad 0, \quad 4, \quad 6, \quad 5, \quad 1)\,, \\ \boldsymbol{h}(\mathcal{A}^{\triangledown}; 5) &= (0, \quad 0, \quad 4, -6, \quad 5, -2)\,. \end{aligned}$$

Remark B.4. Given a family $\mathcal{F} \subseteq \mathbf{2}^{[t]}$ on its ground set E_t, we have:

(i) (see Proposition 2.1(iii)(a) of Ref. [5]).

$$h_\ell(\mathcal{F};t) = (-1)^\ell \sum_{k=0}^{\ell} (-1)^k \binom{t-k}{t-\ell} f_k(\mathcal{F};t) \,, \quad 0 \le \ell \le t\,; \tag{B.5}$$

$$f_\ell(\mathcal{F};t) = \sum_{k=0}^{\ell} \binom{t-k}{t-\ell} h_k(\mathcal{F};t) \,, \qquad 0 \le \ell \le t\,. \tag{B.6}$$

(ii) (see Proposition 2.1(iii)(b) of Ref. [5]).

$$\begin{aligned} h_0(\mathcal{F};t) &= f_0(\mathcal{F};t)\,;\\ h_1(\mathcal{F};t) &= f_1(\mathcal{F};t) - t f_0(\mathcal{F};t)\,;\\ h_{t-1}(\mathcal{F};t) &= (-1)^{t-1} \sum_{k=0}^{t-1} (-1)^k (t-k)\, f_k(\mathcal{F};t)\,;\\ h_t(\mathcal{F};t) &= (-1)^t \sum_{k=0}^{t} (-1)^k f_k(\mathcal{F};t)\,; \end{aligned} \tag{B.7}$$

$$\sum_{k=0}^{t} h_k(\mathcal{F};t) = f_t(\mathcal{F};t)\,.$$

(iii) (see Proposition 2.1(iii)(c) of Ref. [5]).

$$\sum_{k=0}^{t} f_k(\mathcal{F};t) = \sum_{k=0}^{t} 2^{t-k}\, h_k(\mathcal{F};t) = \#\mathcal{F}\,.$$

(iv)

$$\boldsymbol{h}(\mathcal{F};t) + \boldsymbol{h}(\mathbf{2}^{[t]} - \mathcal{F};t) = (1,0,0,\ldots,0) = \boldsymbol{h}(\mathbf{2}^{[t]};t)\,.$$

Being *redundant* analogues of *standard* h-vectors of abstract simplicial complexes (see Refs. [19, 97, 141, 142, 143, 144]), *long h-vectors* (see, e.g., page 170 of Ref. [145], and page 265 of Ref. [146]) demonstrate their usefulness below in relations (B.8), where they describe an enumerative connection between the families $\mathcal{F}$ and $\mathcal{F}^*$.

Remark B.5. Given a family $\mathcal{F} \subseteq \mathbf{2}^{[t]}$ on its ground set E_t, we have:

(i)

$$\#\mathcal{F}^* + \#\mathcal{F} = 2^t \, .$$

More precisely,

$$f_\ell(\mathcal{F}^*;t) + f_{t-\ell}(\mathcal{F};t) = \tbinom{t}{\ell} \, , \quad 0 \le \ell \le t \, .$$

(ii) (see Proposition 2.1(iii)(d) of Ref. [5]).

$$h_\ell(\mathcal{F}^*;t) + (-1)^\ell \sum_{k=\ell}^{t} \tbinom{k}{\ell} \, h_k(\mathcal{F};t) = \delta_{\ell,0} \, , \quad 0 \le \ell \le t \, , \tag{B.8}$$

where $\delta_{\ell,0} := 1$ if $\ell = 0$, and $\delta_{\ell,0} := 0$ otherwise. Note that

$$h_t(\mathcal{F}^*;t) = (-1)^{t+1} h_t(\mathcal{F};t) \, .$$

B.2 Abstract Simplicial Complexes Δ, Such That $\Delta^* = \Delta$, on Their Ground Sets E_t of Even Cardinality t

Although we are interested in the enumerative properties of specific increasing families of sets, in this section we turn to specific *abstract simplicial complexes* (i.e., '*decreasing*' families of '*faces*') because the numbers of faces of size k in complexes are known to satisfy the classical *Kruskal–Katona–Schützenberger* (*KKS*) *inequalities*; see, e.g., Section 10.3 of Ref. [143], and Refs. [19, 33, 36, 48, 54, 97, 107, 141, 144, 147, 148, 149, 150, 151].

Let $\Delta \subseteq \mathbf{2}^{[t]}$, where $\emptyset \neq \Delta \neq \{\hat{0}\}$, be an abstract simplicial complex on its ground set E_t, with its *vertex set* $\mathrm{V}(\Delta) \subseteq E_t$. By definition of complex, the following implications hold: $(B \in \Delta,$ $A \subset B) \Longrightarrow A \in \Delta$. The *inclusion-maximal* faces of Δ are called its *facets*. Sometimes one says that a complex with *one* facet (i.e., the power set of a set) is a *simplex*. The long f-vector of the complex Δ has the form

$$\boldsymbol{f}(\Delta;t) := \big(1, \underbrace{f_1(\Delta;t)}_{:=|\mathrm{V}(\Delta)|>0}, \ldots, \underbrace{f_{d(\Delta)}(\Delta;t)}_{>0}, 0, \ldots, 0\big) \, ,$$

where $d(\Delta) := \max\{k \in [t] \colon f_k(\Delta;t) > 0\}$.

Example B.6. (cf. Example B.2). For any element $a \in E_t$, the *simplex* $\Delta := \{F : F \subseteq (E_t - \{a\})\} = \Delta^*$, whose facet is the subset $(E_t - \{a\}) \subset E_t$ of size $(t-1)$, is described by the vectors

$$\boldsymbol{f}(\Delta; t) = (\tbinom{t-1}{0}, \ldots, \tbinom{t-1}{t-2}, \tbinom{t-1}{t-1}, 0)$$

and

$$\boldsymbol{h}(\Delta; t) = (1, -1, \quad 0, \ldots, 0) .$$

Remark B.7. An abstract simplicial complex Δ with *vertex set* $\mathrm{V}(\Delta)$, such that Δ *coincides* with its *Alexander dual complex* $\Delta^{\vee}$, defined by

$$\Delta^{\vee} := \{\mathrm{V}(\Delta) - F : F \in \mathbf{2}^{\mathrm{V}(\Delta)} - \Delta\}$$

when $\mathrm{V}(\Delta^{\vee}) = \mathrm{V}(\Delta)$, is known as an *Alexander self-dual complex.* Note that we have

$$\Delta = \Delta^{\vee} \iff \#\Delta = 2^{|\mathrm{V}(\Delta)|-1} .$$

See, e.g., Refs. [54, 97, 98], and Ref. [99], on *combinatorial Alexander duality.*

Remark B.8. Suppose that $\Delta \subset \mathbf{2}^{[t]}$ is an abstract simplicial complex, such that $\Delta^* = \Delta$, on its ground set E_t. Then we have:

(i)

$$\#\Delta = \sum_{k=0}^{t} f_k(\Delta; t) = \sum_{k=0}^{d(\Delta)} f_k(\Delta; t) = \sum_{k=0}^{t} 2^{t-k} h_k(\Delta; t) = 2^{t-1} .$$

(ii)

$$f_\ell(\Delta; t) + f_{t-\ell}(\Delta; t) = \tbinom{t}{\ell} , \quad 0 \le \ell \le t . \qquad \text{(B.9)}$$

In particular, if t is *even*, then

$$f_{t/2}(\Delta; t) = \frac{1}{2}\tbinom{t}{t/2} .$$

Example B.9. (i) The abstract simplicial complex

$$\Delta := \{\hat{0}, \{1\}, \{2\}, \{3\}, \{4\}, \{1,2\}, \{1,3\}, \{2,3\}\},$$

such that $\Delta^* = \Delta$, on its ground set $E_4 = \mathrm{V}(\Delta)$, is described by the vectors

$$\begin{aligned} \boldsymbol{f}(\Delta;4) &= (1, \ \ 4, \ \ 3, \ \ 0, \ \ 0), \\ \boldsymbol{h}(\Delta;4) &= (1, \ \ 0, -3, \ \ 2, \ \ 0). \end{aligned}$$

The complex Δ is an Alexander self-dual complex, that is, $\Delta = \Delta^{\vee}$, because $8 = \#\Delta = 2^{|\mathrm{V}(\Delta)|-1} = 2^{4-1}$.

(ii) The simplex

$$\Delta := \{\hat{0}, \{2\}, \{3\}, \{4\}, \{2,3\}, \{2,4\}, \{3,4\}, \{2,3,4\}\},$$

such that $\Delta^* = \Delta$, on its ground set $E_4 \supsetneqq \mathrm{V}(\Delta)$, is described by the vectors

$$\begin{aligned} \boldsymbol{f}(\Delta;4) &= (1, \ \ 3, \ \ 3, \ \ 1, \ \ 0), \\ \boldsymbol{h}(\Delta;4) &= (1, -1, \ \ 0, \ \ 0, \ \ 0). \end{aligned}$$

Now suppose again that $\Delta \subset \mathbf{2}^{[t]}$ is an abstract simplicial complex, such that $\Delta^* = \Delta$, on its ground set E_t of *even* cardinality t. We would like to apply the KKS inequalities to the complex Δ.

- Since $\frac{1}{2}\binom{t}{t/2} = \binom{t-1}{t/2}$, we have the trivial $(t/2)$-*binomial expansion*

$$f_{t/2}(\Delta;t) = \binom{\alpha_{t/2}(t/2)}{t/2}$$

of the middle component of the long f-vector $\boldsymbol{f}(\Delta;t)$, where

$$\alpha_{t/2}(t/2) = t-1.$$

Then relations (B.9) and the KKS inequalities (lower and upper bound versions; see Corollaries 10.1 and 10.2 of Ref. [143]) imply that

$$\begin{aligned} &\binom{t}{(t/2)-1} - f_{(t/2)+1}(\Delta;t) \\ &\qquad = f_{(t/2)-1}(\Delta;t) \geq \binom{\alpha_{t/2}(t/2)}{(t/2)-1} = \binom{t-1}{(t/2)-1} = \binom{t-1}{t/2}, \end{aligned}$$

$$\begin{aligned} &\binom{t}{(t/2)+1} - f_{(t/2)-1}(\Delta;t) \\ &\qquad = f_{(t/2)+1}(\Delta;t) \leq \binom{\alpha_{t/2}(t/2)}{(t/2)+1} = \binom{t-1}{(t/2)+1} = \binom{t-1}{(t/2)-2}. \end{aligned}$$

Further, note that if our complex Δ had the minimum possible number of its faces of size $((t/2)-1)$, namely $\binom{t-1}{(t/2)-1}$ faces, then the following implication would hold:

$$f_{(t/2)-1}(\Delta;t) := \binom{t-1}{(t/2)-1} = \binom{t-1}{t/2}$$
$$\Longrightarrow \quad f_{(t/2)+1}(\Delta;t) = \binom{t}{(t/2)+1} - \binom{t-1}{(t/2)-1}$$
$$= \binom{t}{(t/2)-1} - \binom{t-1}{(t/2)-1} = \binom{t-1}{(t/2)-2}\,.$$

Note also that $\binom{t}{(t/2)-1} - \binom{t-1}{(t/2)-1} = \binom{t-1}{(t/2)-2} = \binom{t-1}{(t/2)+1}$.

- If $((t/2)-1) > 1$, then suppose that our complex Δ indeed has the *minimum possible* number of faces of size $((t/2)-1)$, that is, $f_{(t/2)-1}(\Delta;t) := \binom{t-1}{(t/2)-1}$. In this case we have the $((t/2)-1)$-*binomial expansion*

$$f_{(t/2)-1}(\Delta;t) = \binom{\alpha_{(t/2)-1}((t/2)-1)}{(t/2)-1}$$

of the component $f_{(t/2)-1}(\Delta;t)$ of the vector $\boldsymbol{f}(\Delta;t)$, where

$$\alpha_{(t/2)-1}((t/2)-1) = t-1\,.$$

Relations (B.9) and the KKS inequalities (lower bound version; see Corollary 10.1 of Ref. [143]) imply that

$$\binom{t}{(t/2)-2} - f_{(t/2)+2}(\Delta;t)$$
$$= f_{(t/2)-2}(\Delta;t) \geq \binom{\alpha_{(t/2)-1}((t/2)-1)}{(t/2)-2} = \binom{t-1}{(t/2)-2}\,.$$

On the other hand, suppose that the complex Δ has the *maximum possible* number of faces of size $((t/2)+1)$, that is, $f_{(t/2)+1}(\Delta;t) := \binom{t-1}{(t/2)+1}$. We have the $((t/2)+1)$-*binomial expansion*

$$f_{(t/2)+1}(\Delta;t) = \binom{\alpha_{(t/2)+1}((t/2)+1)}{(t/2)+1}$$

of the component $f_{(t/2)+1}(\Delta;t)$ of the vector $\boldsymbol{f}(\Delta;t)$, where

$$\alpha_{(t/2)+1}((t/2)+1) = t-1\,.$$

Now relations (B.9) and the KKS inequalities (upper bound version; see Corollary 10.2 of Ref. [143]) imply that

$$\binom{t}{(t/2)+2} - f_{(t/2)-2}(\Delta;t)$$
$$= f_{(t/2)+2}(\Delta;t) \leq \binom{\alpha_{(t/2)+1}((t/2)+1)}{(t/2)+2} = \binom{t-1}{(t/2)+2}\,.$$

Note that $\binom{t}{(t/2)-2} - \binom{t-1}{(t/2)-2} = \binom{t-1}{(t/2)-3} = \binom{t-1}{(t/2)+2}$.

– Proceeding by induction, we arrive at the following counterpart of Theorem B.11:

Lemma B.10. (*see also Example* B.6). *Let* $\Delta \subset \mathbf{2}^{[t]}$ *be an abstract simplicial complex on its ground set* E_t *of* even *cardinality t. If*

$$\Delta^* = \Delta \,,$$

then we have

$$f_0(\Delta;t) = 1\,, \quad \textit{and} \quad f_t(\Delta;t) = 0\,;$$
$$f_{t/2}(\Delta;t) = \frac{1}{2}\tbinom{t}{t/2} = \tbinom{t-1}{t/2}\,;$$
$$f_{(t/2)-i}(\Delta;t) \geq \tbinom{t-1}{(t/2)-i} = \tbinom{t-1}{(t/2)+i-1}\,, \quad 1 \leq i \leq (t/2)-1\,;$$
$$f_{(t/2)+j}(\Delta;t) \leq \tbinom{t-1}{(t/2)+j} = \tbinom{t-1}{(t/2)-j-1}\,, \quad 1 \leq j \leq (t/2)-1\,;$$

$$f_{(t/2)-k}(\Delta;t) + f_{(t/2)+k}(\Delta;t)$$
$$= \tbinom{t}{(t/2)-k} = \tbinom{t}{(t/2)+k}\,, \quad 1 \leq k \leq (t/2)-1\,.$$

B.3 The Increasing Families of Sets Generated by Self-Dual Clutters on Their Ground Sets E_t of Even Cardinality *t*

Let $\mathcal{F} \subset \mathbf{2}^{[t]}$ be a family on its ground set E_t. The implication

$$\mathcal{F}^* = \mathcal{F} \iff (\mathbf{2}^{[t]} - \mathcal{F})^* = \mathbf{2}^{[t]} - \mathcal{F}\,,$$

allows us to obtain from Lemma B.10 the following result:

Theorem B.11. (*see also Example* B.2). *Let* $\mathcal{A} \subset \mathbf{2}^{[t]}$ *be a clutter on its ground set* E_t *of* even *cardinality t. If the clutter* $\mathcal{A}$ *is* self-dual, *that is,*

$$\mathfrak{B}(\mathcal{A}) = \mathcal{A}\,,$$

then we have

$$f_0(\mathcal{A}^\triangledown;t) = 0\,, \quad \textit{and} \quad f_t(\mathcal{A}^\triangledown;t) = 1\,;$$
$$f_{t/2}(\mathcal{A}^\triangledown;t) = \frac{1}{2}\tbinom{t}{t/2} = \tbinom{t-1}{t/2}\,;$$
$$f_{(t/2)-i}(\mathcal{A}^\triangledown;t) \leq \tbinom{t-1}{(t/2)-i-1} = \tbinom{t-1}{(t/2)+i}\,, \quad 1 \leq i \leq (t/2)-1\,;$$
$$f_{(t/2)+j}(\mathcal{A}^\triangledown;t) \geq \tbinom{t-1}{(t/2)+j-1} = \tbinom{t-1}{(t/2)-j}\,, \quad 1 \leq j \leq (t/2)-1\,;$$

$$f_{(t/2)-k}(\mathcal{A}^{\triangledown};t) + f_{(t/2)+k}(\mathcal{A}^{\triangledown};t) = \binom{t}{(t/2)-k} = \binom{t}{(t/2)+k}, \quad 1 \le k \le (t/2)-1 .$$

B.4 Remarks on Set Families $\mathcal{F}$ Such That $\mathcal{F}^* = \mathcal{F}$

Remark B.12. Let $\mathcal{F} \subset \mathbf{2}^{[t]}$ be a family, such that $\mathcal{F}^* = \mathcal{F}$, on its ground set E_t.

(i) We have

$$\#\mathcal{F} = 2^{t-1} .$$

More precisely,

$$f_\ell(\mathcal{F};t) + f_{t-\ell}(\mathcal{F};t) = \binom{t}{\ell}, \quad 0 \le \ell \le t . \qquad \text{(B.10)}$$

(ii) Relations (B.10) and (B.6) imply that

$$\sum_{k=0}^{\ell} \binom{t-k}{t-\ell} h_k(\mathcal{F};t) + \sum_{j=0}^{t-\ell} \binom{t-j}{\ell} h_j(\mathcal{F};t) = \binom{t}{\ell}, \quad 0 \le \ell \le t .$$

Given a family $\mathcal{F} \subset \mathbf{2}^{[t]}$, such that $\mathcal{F}^* = \mathcal{F}$, on its ground set E_t, from (B.5) we have for any ℓ, where $0 \le \ell \le t$:

$$h_\ell(\mathcal{F};t) = (-1)^\ell \sum_{k=0}^{\ell} (-1)^k \binom{t-k}{t-\ell} f_k(\mathcal{F};t)$$
$$= (-1)^\ell \sum_{k=0}^{\ell} (-1)^k \binom{t-k}{t-\ell} \left(\binom{t}{k} - f_{t-k}(\mathcal{F};t) \right) .$$

Remark B.13. Let $\mathcal{F} \subset \mathbf{2}^{[t]}$ be a family, such that $\mathcal{F}^* = \mathcal{F}$, on its ground set E_t.

(i) We have

$$h_\ell(\mathcal{F};t) = \delta_{\ell,0} - (-1)^{t-\ell} \sum_{j=t-\ell}^{t} (-1)^j \binom{j}{t-\ell} f_j(\mathcal{F};t), \quad 0 \le \ell \le t; \qquad \text{(B.11)}$$

$$f_\ell(\mathcal{F};t) = \binom{t}{\ell} - \sum_{j=0}^{t-\ell} \binom{t-j}{\ell} h_j(\mathcal{F};t), \quad 0 \le \ell \le t . \qquad \text{(B.12)}$$

(ii) Relations (B.11) and (B.5) yield

$$(-1)^{\ell}\sum_{k=0}^{\ell}(-1)^{k}\tbinom{t-k}{t-\ell}\, f_k(\mathcal{F};t)$$
$$+(-1)^{t-\ell}\sum_{j=t-\ell}^{t}(-1)^{j}\tbinom{j}{t-\ell}\, f_j(\mathcal{F};t)=\delta_{\ell,0}\,,\quad 0\le\ell\le t\,.$$

Remark B.14. Let $\mathcal{F}\subset\mathbf{2}^{[t]}$ be a family, such that $\mathcal{F}^{*}=\mathcal{F}$, on its ground set E_t of *odd* cardinality t. From (B.7) we have

$$\begin{aligned}h_t(\mathcal{F};t)&=(-1)^{t}\sum_{k=0}^{t}(-1)^{k}f_k(\mathcal{F};t)\\&=-\sum_{k=0}^{\lfloor t/2\rfloor}(-1)^{k}\Big(f_k(\mathcal{F};t)-(\tbinom{t}{k}-f_k(\mathcal{F};t))\Big)\\&=\tbinom{t-1}{(t-1)/2}-2\sum_{k=0}^{(t-1)/2}(-1)^{k}f_k(\mathcal{F};t)\,,\end{aligned}$$

and

$$\begin{aligned}h_t(\mathcal{F};t)&=(-1)^{t}\sum_{j=0}^{t}(-1)^{j}f_j(\mathcal{F};t)\\&=-\sum_{j=\lceil t/2\rceil}^{t}(-1)^{j}\Big(f_j(\mathcal{F};t)-(\tbinom{t}{j}-f_j(\mathcal{F};t))\Big)\\&=-\tbinom{t-1}{(t-1)/2}-2\sum_{j=(t+1)/2}^{t}(-1)^{j}f_j(\mathcal{F};t)\,.\end{aligned}$$

As a consequence, we see that

$$\sum_{k=0}^{(t-1)/2}(-1)^{k}f_k(\mathcal{F};t)=\tbinom{t-1}{(t-1)/2}+\sum_{j=(t+1)/2}^{t}(-1)^{j}f_j(\mathcal{F};t)\,.$$

Remark B.15. Let $\mathcal{F}\subset\mathbf{2}^{[t]}$ be a family, such that $\mathcal{F}^{*}=\mathcal{F}$, on its ground set E_t.

(i) From (B.8) we have

$$h_\ell(\mathcal{F};t)=\delta_{\ell,0}+(-1)^{\ell+1}\sum_{k=\ell}^{t}\tbinom{k}{\ell}\,h_k(\mathcal{F};t)\,,\quad 0\le\ell\le t\,. \tag{B.13}$$

If t is *even*, then we see that

$$h_t(\mathcal{F};t) = 0\,. \tag{B.14}$$

(ii) For *even* indices ℓ, where $2 \le \ell \le t$, relations (B.13) imply that

$$\sum_{k=\ell}^{t} \tbinom{k}{\ell-1}\, h_k(\mathcal{F};t) = 0\,, \tag{B.15}$$

that is,

$$\ell\, h_\ell(\mathcal{F};t) + \sum_{k=\ell+1}^{t} \tbinom{k}{\ell-1}\, h_k(\mathcal{F};t) = 0\,;$$

we also have

$$2\, h_\ell(\mathcal{F};t) + \sum_{k=\ell+1}^{t} \tbinom{k}{\ell}\, h_k(\mathcal{F};t) = 0\,.$$

If t is *odd*, then we have

$$(t-1)\, h_{t-1}(\mathcal{F};t) + \tbinom{t}{2}\, h_t(\mathcal{F};t) = 0\,,$$

that is,

$$2\, h_{t-1}(\mathcal{F};t) + t\, h_t(\mathcal{F};t) = 0\,.$$

If t is *even*, then we have

$$\frac{t}{2}\, h_{t/2}(\mathcal{F};t) + \sum_{k=(t/2)+1}^{t-1} \tbinom{k}{(t/2)-1}\, h_k(\mathcal{F};t) = 0\,,$$

and

$$h_{t/2}(\mathcal{F};t) + \frac{1}{2} \sum_{k=(t/2)+1}^{t-1} \tbinom{k}{t/2}\, h_k(\mathcal{F};t) = 0\,. \tag{B.16}$$

Remark B.16. Let $\mathcal{F} \subset \mathbf{2}^{[t]}$ be a family, such that $\mathcal{F}^* = \mathcal{F}$, on its ground set E_t.

If t is *even*, then relations (B.14) and (B.15) in the case $\ell := 2$ yield

$$h_t(\mathcal{F};t) = 0\,, \quad \text{and} \quad \sum_{k=2}^{t-1} k\, h_k(\mathcal{F};t) = 0\,.$$

Similarly, if t is *odd*, then from (B.15) we have

$$\sum_{k=2}^{t} k\, h_k(\mathcal{F};t) = 0\,.$$

Remark B.17. Let $\mathcal{F} \subset \mathbf{2}^{[t]}$ be a family, such that $\mathcal{F}^* = \mathcal{F}$, on its ground set E_t of *even* cardinality t.

(i) On the one hand, since

$$f_{t/2}(\mathcal{F};t) = \frac{1}{2}\binom{t}{t/2}, \tag{B.17}$$

relations (B.6) imply that

$$\begin{aligned} &\sum_{k=0}^{t/2} \binom{t-k}{t/2} h_k(\mathcal{F};t) = f_{t/2}(\mathcal{F};t) \\ &= h_{t/2}(\mathcal{F};t) + \sum_{k=0}^{(t/2)-1} \binom{t-k}{t/2} h_k(\mathcal{F};t) = \frac{1}{2}\binom{t}{t/2}. \end{aligned} \tag{B.18}$$

On the other hand, (B.17) and relations (B.12) imply that

$$\binom{t}{t/2} - \sum_{j=0}^{t/2} \binom{t-j}{t/2} h_{t/2}(\mathcal{F};t) = f_{t/2}(\mathcal{F};t) = \frac{1}{2}\binom{t}{t/2},$$

that is,

$$\sum_{j=0}^{t/2} \binom{t-j}{t/2} h_{t/2}(\mathcal{F};t) = \frac{1}{2}\binom{t}{t/2}.$$

(ii) Relations (B.16) and (B.18) imply that

$$\sum_{k=0}^{(t/2)-1} \binom{t-k}{t/2} h_k(\mathcal{F};t) - \frac{1}{2} \sum_{j=(t/2)+1}^{t-1} \binom{j}{t/2} h_j(\mathcal{F};t) = \frac{1}{2}\binom{t}{t/2},$$

and

$$\begin{aligned} h_{t/2}(\mathcal{F};t) = \frac{1}{4}\binom{t}{t/2} - \frac{1}{2} \sum_{k=0}^{(t/2)-1} \binom{t-k}{t/2} h_k(\mathcal{F};t) \\ -\frac{1}{4} \sum_{j=(t/2)+1}^{t-1} \binom{j}{t/2} h_j(\mathcal{F};t). \end{aligned}$$

Bibliography

1. *Björner A., Las Vergnas M., Sturmfels B., White N., Ziegler G.M.* Oriented matroids. Second edition. Encyclopedia of Mathematics, 46. Cambridge: Cambridge University Press, 1999.
2. *Horadam K.J.* Hadamard matrices and their applications. Princeton, NJ: Princeton University Press, 2007.
3. *Seberry J.* Orthogonal designs. Hadamard matrices, quadratic forms and algebras. Revised and updated edition of the 1979 original. Cham: Springer, 2017.
4. *Seberry J., Yamada M.* Hadamard matrices. Constructions using number theory and algebra. Hoboken, NJ: John Wiley & Sons, Inc., 2020.
5. *Matveev A.O.* Pattern recognition on oriented matroids. Berlin: De Gruyter, 2017.
6. *Audet C., Hare W.* Derivative-free and blackbox optimization. With a foreword by *J.E. Dennis Jr.* Springer Series in Operations Research and Financial Engineering. Cham: Springer, 2017.
7. *Stanley R.P.* Enumerative combinatorics. Volume 1. Second edition. Cambridge Studies in Advanced Mathematics, 49. Cambridge: Cambridge University Press, 2012.
8. *OEIS*, The on-line encyclopedia of integer sequences, 2022.
9. *Eğecioğlu Ö., Iršič V.* Fibonacci-run graphs I: Basic properties. Discrete Applied Mathematics, 2021, 295, pp. 70–84.
10. *Hsu W.-J.* Fibonacci cubes: A new interconnection topology. IEEE Transactions on Parallel and Distributed Systems, 1993, 4, no. 1, pp. 3–12.
11. *Hsu W.-J., Page C.V., Liu J.-S.* Fibonacci cubes: A class of self-similar graphs. Fibonacci Quarterly, 1993, 31, no. 1, pp. 65–72.
12. *Klavžar S.* Structure of Fibonacci cubes: A survey. Journal of Combinatorial Optimization, 2013, 25, no. 4, pp. 505–522.
13. *Munarini E., Perelli Cippo C., Zagaglia Salvi N.* On the Lucas cubes. Fibonacci Quarterly, 2001, 39, no. 1, pp. 12–21.

14. *Ilić A., Klavžar S., Rho Y.* Generalized Fibonacci cubes. Discrete Mathematics, 2012, 312, no. 1, pp. 2–11.
15. *Eğecioğlu Ö., Saygi E., Saygi Z.* *k*-Fibonacci cubes: A family of subgraphs of Fibonacci cubes. International Journal of Foundations of Computer Science, 2020, 31, no. 5, pp. 639–661.
16. *Eğecioğlu Ö., Iršič V.* Fibonacci-run graphs II: Degree sequences. Discrete Applied Mathematics, 2021, 300, pp. 56–71.
17. *Klavžar S., Mollard M.* Daisy cubes and distance cube polynomial. European Journal of Combinatorics, 2019, 80, pp. 214–223.
18. *Gainanov D.N.* Graphs for pattern recognition. Infeasible systems of linear inequalities. Berlin: De Gruyter, 2016.
19. *Ziegler G.M.* Lectures on polytopes. Revised edition. Graduate Texts in Mathematics, 152. Berlin: Springer-Verlag, 1998.
20. *Horn R.A., Johnson C.R.* Matrix analysis. Second edition. Cambridge: Cambridge University Press, 2013.
21. *Bóna M.* (*ed.*) Handbook of enumerative combinatorics. Discrete Mathematics and its Applications (Boca Raton). Boca Raton, FL: CRC Press, 2015.
22. *Heubach S., Mansour T.* Combinatorics of compositions and words. Discrete Mathematics and its Applications (Boca Raton). Boca Raton, FL: CRC Press, 2010.
23. *Nikolski N.* Toeplitz matrices and operators. Cambridge Studies in Advanced Mathematics, 182. *D. Gibbons, G. Gibbons* (trans.). Cambridge: Cambridge University Press, 2020.
24. *Prodinger H.* Ternary Smirnov words and generating functions. Integers, 2018, 18, Paper A69.
25. *Kaufmann A.* Introduction à la combinatorique en vue des applications. Paris: Dunod, 1968. [in French]
26. *Comtet L.* Advanced combinatorics. The art of finite and infinite expansions. Revised and enlarged edition. Dordrecht: D. Reidel Publishing Co., 1974.
27. *Goulden I.P., Jackson D.M.* Combinatorial enumeration. With a foreword by *G.-C. Rota*. Reprint of the 1983 original. Mineola, NY: Dover Publications, Inc., 2004.
28. *Kaplansky I.* Solution of the "Problème des ménages". Bulletin of the American Mathematical Society, 1943, 49, pp. 784–785. Reprinted in: *I. Gessel, G.-C. Rota* (eds.). Classic papers in combinatorics. Reprint of the 1987 original. Modern Birkhäuser Classics. Boston, MA: Birkhäuser Boston, Inc., 2009, pp. 122–123.

29. *Riordan J.* Combinatorial identities. Reprint of the 1968 original. Huntington, N.Y.: Robert E. Krieger Publishing Co., 1979.
30. *Ryser H.J.* Combinatorial mathematics. The Carus Mathematical Monographs, No. 14. New York: The Mathematical Association of America, 1963.
31. *Mansour T., Sun Y.* On the number of combinations without certain separations. European Journal of Combinatorics, 2008, 29, pp. 1200–1206.
32. *Graham R.L., Knuth D.E., Patashnik O.* Concrete mathematics. A foundation for computer science. Second edition. Reading, MA: Addison-Wesley, 1994.
33. *Godsil C., Meagher K.* Erdős–Ko–Rado theorems: Algebraic approaches. Cambridge Studies in Advanced Mathematics, 149. Cambridge: Cambridge University Press, 2016.
34. *MacWilliams F.J., Sloane N.J.A.* The theory of error-correcting codes. North-Holland Mathematical Library, 16. Amsterdam: North-Holland Publishing Co., 1977.
35. *Edmonds J.* Matroid partition. 50 years of integer programming 1958–2008. From the early years to the state-of-the-art. 12th Combinatorial Optimization Workshop AUSSOIS 2008 held in Aussois, January 7–11, 2008. *M. Jünger, T.M. Liebling, D. Naddef, G.L. Nemhauser, W.R. Pulleyblank, G. Reinelt, G. Rinaldi* and *L.A. Wolsey* (eds.). Berlin: Springer-Verlag, 2010, pp. 199–217.
36. *Berge C.* Hypergraphs. Combinatorics of finite sets. Translated from the French. North-Holland Mathematical Library, 45. Amsterdam: North-Holland Publishing Co., 1989.
37. *Bretto A.* Hypergraph theory. An introduction. Mathematical Engineering. Cham: Springer, 2013.
38. *Conforti M., Cornuéjols G., Zambelli G.* Integer programming. Cham: Springer, 2014.
39. *Cornuéjols G.* Combinatorial optimization. Packing and covering. CBMS–NSF Regional Conference Series in Applied Mathematics, 74. Philadelphia, PA: Society for Industrial and Applied Mathematics (SIAM), 2001.
40. *Crama Y., Hammer P.L.* Boolean functions. Theory, algorithms, and applications. Encyclopedia of Mathematics and its Applications, 142. Cambridge: Cambridge University Press, 2011.
41. *Cygan M., Fomin F.V., Kowalik Ł., Lokshtanov D., Marx D., Pilipczuk M., Pilipczuk M., Saurabh S.* Parameterized algorithms. Cham: Springer, 2015.

42. *Fomin F.V., Lokshtanov D., Saurabh S., Zehavi M.* Kernelization. Theory of parameterized preprocessing. Cambridge: Cambridge University Press, 2019.
43. *Garey M.R., Johnson D.S.* Computers and intractability. A guide to the theory of NP-completeness. A Series of Books in the Mathematical Sciences. San Francisco, CA: W.H. Freeman and Co., 1979.
44. *Grötschel M., Lovász L., Schrijver A.* Geometric algorithms and combinatorial optimization. Second edition. Algorithms and Combinatorics, 2. Berlin: Springer-Verlag, 1993.
45. *Haynes T.W., Hedetniemi S.T., Henning M.A.* (eds.). Structures of domination in graphs. Developments in Mathematics, 66. Cham: Springer, 2021.
46. *Henning M.A., Yeo A.* Total domination in graphs. Springer Monographs in Mathematics. New York: Springer, 2013.
47. *Henning M.A., Yeo A.* Transversals in linear uniform hypergraphs. Developments in Mathematics, 63. Cham: Springer, 2020.
48. *Jukna S.* Extremal combinatorics. With applications in computer science. Second edition. Texts in Theoretical Computer Science. An EATCS Series. Heidelberg: Springer, 2011.
49. *Mirsky L.* Transversal theory. An account of some aspects of combinatorial mathematics. Mathematics in Science and Engineering, 75. New York: Academic Press, 1971.
50. *Nagamochi H., Ibaraki T.* Algorithmic aspects of graph connectivity. Encyclopedia of Mathematics and its Applications, 123. Cambridge: Cambridge University Press, 2008.
51. *Nemhauser G.L., Wolsey L.A.* Integer and combinatorial optimization. Reprint of the 1988 original. Wiley-Interscience Series in Discrete Mathematics and Optimization. New York: John Wiley & Sons, Inc., 1999.
52. *Scheinerman E.R., Ullman D.H.* Fractional graph theory. A rational approach to the theory of graphs. With a foreword by *Claude Berge*. Reprint of the 1997 original. Mineola, NY: Dover Publications, Inc., 2011.
53. *Schrijver A.* Combinatorial optimization. Polyhedra and efficiency. Vol. C. Disjoint paths, hypergraphs. Chapters 70–83. Algorithms and Combinatorics, 24,C. Berlin: Springer-Verlag, 2003.
54. *Villarreal R.H.* Monomial algebras. Second edition. Monographs and Research Notes in Mathematics. Boca Raton, FL: CRC Press, 2015.

55. *Adaricheva K., Nation J.B.* Discovery of the *D*-basis in binary tables based on hypergraph dualization. Theoretical Computer Science, 2017, 658, pp. 307–315.

56. *Aharoni R., Howard D.* Cross-intersecting pairs of hypergraphs. Journal of Combinatorial Theory Ser. A, 2017, 148, pp. 15–26.

57. *Alon N.* Transversal numbers of uniform hypergraphs. Graphs and Combinatorics, 1990, 6, pp. 1–4.

58. *Amburg I., Kleinberg J., Benson A.R.* Planted hitting set recovery in hypergraphs. Journal of Physics: Complexity, 2021, 2, no. 3 (14 pp.).

59. *Araújo J., Bougeret M., Campos V.A., Sau I.* Parameterized complexity of computing maximum minimal blocking and hitting sets. Algorithmica, 2023, 85, pp. 444–491.

60. *Bailey J., Stuckey P.J.* Discovery of minimal unsatisfiable subsets of constraints using hitting set dualization. Practical Aspects of Declarative Languages. 7th International Symposium PADL 2005 held in Long Beach, CA, USA, January 10–11, 2005. *M. Hermenegildo* and *D. Cabeza* (eds.). Lecture Notes in Computer Science, 3350. Berlin: Springer-Verlag, 2005, pp. 174–186.

61. *Barát J.* Intersecting and 2-intersecting hypergraphs with maximal covering number: The Erdős–Lovász theme revisited. Journal of Combinatorial Designs, 2021, 29, no. 3, pp. 193–209.

62. *van Bevern R., Smirnov P.V.* Optimal-size problem kernels for *d*-Hitting Set in linear time and space. Information Processing Letters, 2020, 163 (9 pp.).

63. *Björner A., Hultman A.* A note on blockers in posets. Annals of Combinatorics, 2004, 8, no. 2, pp. 123–131.

64. *Bläsius T., Friedrich T., Lischeid J., Meeks K., Schirneck M.* Efficiently enumerating hitting sets of hypergraphs arising in data profiling. Meeting on Algorithm Engineering and Experiments ALENEX 2019 held in San Diego, CA, USA, January 7–8, 2019. *S. Kobourov* and *H. Meyerhenke* (eds.), pp. 130–143.

65. *Bläsius T., Friedrich T., Schirneck M.* The complexity of dependency detection and discovery in relational databases. Theoretical Computer Science, 2022, 900, pp. 79–96.

66. *Boros E., Makino K.* A fast and simple parallel algorithm for the monotone duality problem. International Colloquium on Automata, Languages and Programming ICALP, Rhodes. Part I. *S. Albers, A. Marchetti-Spaccamela, Y. Matias, S.E. Nikoletseas* and *W. Thomas*

(eds.). Lecture Notes in Computer Science, 5555. Berlin: Springer, 2009, pp. 183–194.

67. *Bousquet N.* Hitting sets: VC-dimension and Multicut. Ph.D. Thesis. Université Montpellier II, 2013.
68. *Bringmann K., Kozma L., Moran S., Narayanaswamy N.S.* Hitting Set in hypergraphs of low VC-dimension. 24th Annual European Symposium on Algorithms ESA 2016 held in Aarhus, Denmark, August 22–24, 2016. *P. Sankowski* and *C. Zaroliagis* (eds.). LIPIcs 57 ESA 2016, Article no. 23, pp. 23:1–23:18.
69. *Bujtás Cs., Henning M.A., Tuza Zs.* Transversals and domination in uniform hypergraphs. European Journal of Combinatorics, 2012, 33, pp. 62–71.
70. *Bujtás Cs., Rote G., Tuza Zs.* Optimal strategies in fractional games: Vertex cover and domination. Preprint [arXiv:2105.03890], 2021.
71. *Chen X.*, Hu X., *Zang W.* Dual integrality in combinatorial optimization. Handbook of Combinatorial Optimization. Second edition, *P.M. Pardalos, D.-Z. Du* and *R.L. Graham* (eds.). SpringerReference. New York: Springer, 2013, pp. 995–1063.
72. *Chvátal V., McDiarmid C.* Small transversals in hypergraphs. Combinatorica, 1992, 12, pp. 19–26.
73. *Cochefert M., Couturier J.-F., Gaspers S., Kratsch D.* Faster algorithms to enumerate hypergraph transversals. 12th Latin American Symposium LATIN 2016: Theoretical Informatics held in Ensenada, Mexico, April 11–15, 2016. *E. Kranakis, G. Navarro* and *E. Chávez* (eds.). Lecture Notes in Computer Science, 9644. Berlin: Springer-Verlag, 2016, pp. 306–318.
74. *Cordovil R., Fukuda K., Moreira M.L.* Clutters and matroids. Discrete Mathematics, 1991, 89, no. 2, pp. 161–171.
75. *Damaschke P.* Parameterized algorithms for double hypergraph dualization with rank limitation and maximum minimal vertex cover. Discrete Optimization, 2011, 8, no. 1, pp. 18–24.
76. *Defrain O., Nourine L., Uno T.* On the dualization in distributive lattices and related problems. Discrete Applied Mathematics, 2021, 300, pp. 85–96.
77. *Eiter T., Gottlob G.* Identifying the minimal transversals of a hypergraph and related problems. SIAM Journal on Computing, 1995, 24, no. 6, pp. 1278–1304.
78. *Eiter T., Gottlob G., Makino K.* New results on monotone dualization and generating hypergraph transversals. SIAM Journal on Computing, 2003, 32, no. 2, pp. 514–537.

79. *Eiter T., Makino K., Gottlob G.* Computational aspects of monotone dualization: A brief survey. Discrete Applied Mathematics, 2008, 156, pp. 2035–2049.
80. *Elbassioni K.M.* On the complexity of monotone dualization and generating minimal hypergraph transversals. Discrete Applied Mathematics, 2008, 156, no. 11, pp. 2109–2123.
81. *Florian M., Henningsen S., Ndolo C., Scheuermann B.* The sum of its parts: Analysis of Federated Byzantine Agreement Systems. Preprint [arXiv:2002.08101], 2020.
82. *Fredman M.L., Khachiyan L.* On the complexity of dualization of monotone disjunctive normal forms. Journal of Algorithms, 1996, 21, no. 3, pp. 618–628.
83. *Füredi Z.* Matchings and covers in hypergraphs. Graphs and Combinatorics, 1988, 4, no. 2, pp. 115–206.
84. *Gainer-Dewar A., Vera-Licona P.* The minimal hitting set generation problem: Algorithms and computation. SIAM Journal on Discrete Mathematics, 2017, 31, no. 1, pp. 63–100.
85. *Hébert C., Bretto A., Crémilleux B.* A data mining formalization to improve hypergraph minimal transversal computation. Fundamenta Informaticae, 2007, 80, no. 4, pp. 415–433.
86. *Kanté M.M., Limouzy V., Mary A., Nourine L.*. On the enumeration of minimal dominating sets and related notions. SIAM Journal on Discrete Mathematics, 2014, 28, no. 4, pp. 1916–1929.
87. *Karp R.M.* Reducibility among combinatorial problems. *R.E. Miller, J.W. Thatcher* and *J.D. Bohlinger* (eds.). Complexity of Computer Computations. The IBM Research Symposia Series, IBM Thomas J. Watson Research Center, 1972. Yorktown Heights, N.Y.: Plenum, pp. 85–103.
88. *Kavvadias D.J., Stavropoulos E.C.* An efficient algorithm for the transversal hypergraph generation. Journal of Graph Algorithms and Applications, 2005, no. 2, pp. 239–264.
89. *Khachiyan L., Boros E., Elbassioni K., Gurvich V.* An efficient implementation of a quasi-polynomial algorithm for generating hypergraph transversals and its application in joint generation. Discrete Applied Mathematics, 2006, 154, pp. 2350–2372.
90. *Liffiton M.H., Sakallah K.A.* Algorithms for computing minimal unsatisfiable subsets of constraints. Journal of Automated Reasoning, 2008, 40, no. 1, pp. 1–33.
91. *Lin L., Jiang Y.* The computation of hitting sets: Review and new algorithms. Information Processing Letters, 2003, 86, pp. 177–184.

92. *Murakami K., Uno T.* Efficient algorithms for dualizing large-scale hypergraphs. Discrete Applied Mathematics, 2014, 170, pp. 83–94.
93. *Nourine L., Petit J.-M.* Beyond hypergraph dualization. Encyclopedia of Algorithms. Second edition, *M.-Y. Kao* (ed.). SpringerReference. New York: Springer, 2016, pp. 189–192.
94. *Pill I., Quaritsch T.* Optimizations for the Boolean approach to computing minimal hitting sets. 20th European Conference on Artificial Intelligence ECAI 2012 held in Montpellier, August 27–31, 2012. *L. De Raedt, C. Bessiere, D. Dubois, P. Doherty, P. Frasconi, F. Heintz* and *P. Lucas* (eds.). Amsterdam: IOS Press, 2012, pp. 648–653.
95. *Reiter R.* A theory of diagnosis from first principles. Artificial Intelligence, 1987, 32, no. 1, pp. 57–95.
96. *Yolov N.* Blocker size via matching minors. Discrete Mathematics, 2018, 341, no. 8, pp. 2237–2242.
97. *Herzog J., Hibi T.* Monomial ideals. Graduate Texts in Mathematics, 260. London: Springer-Verlag London, Ltd., 2011.
98. *Miller E., Sturmfels B.* Combinatorial commutative algebra. Graduate Texts in Mathematics, 227. New York: Springer-Verlag, 2005.
99. *Björner A., Tancer M.* Note: Combinatorial Alexander duality: A short and elementary proof. Discrete and Computational Geometry, 2009, 42, no. 4, pp. 586–593.
100. *Edmonds J., Fulkerson D.R.* Bottleneck extrema. Journal of Combinatorial Theory, 1970, 8, pp. 299–306.
101. *Isbell J.R.* A class of simple games. Duke Mathematical Journal, 1958, 25, pp. 423–439.
102. *Lawler E.L.* Covering problems: Duality relations and a new method of solution. SIAM Journal on Applied Mathematics, 1966, 14, pp. 1115–1132.
103. *Lehman A.* A solution of the Shannon switching game. Journal of the Society for Industrial and Applied Mathematics, 1964, 12, pp. 687–725.
104. *Abdi A., Cornuéjols G., Lee D.* Identically self-blocking clutters. 20th International Conference Integer Programming and Combinatorial Optimization IPCO 2019 held in Ann Arbor, MI, USA, May 22–24, 2019. *A. Lodi* and *V. Nagarajan* (eds.). Lecture Notes in Computer Science, 11480. Cham: Springer, 2019, pp. 1–12.
105. *Abdi A., Pashkovich K.* Delta minors, delta free clutters, and entanglement. SIAM Journal on Discrete Mathematics, 2018, 32, no. 3, pp. 1750–1774.

106. *Seymour P.D.* The forbidden minors of binary clutters. Journal of the London Mathematical Society (2), 1975/76, 12, no. 3, pp. 356–360.

107. *Knuth D.E.* The art of computer programming. Vol. 4A. Combinatorial algorithms. Part 1. Upper Saddle River, NJ: Addison–Wesley, 2011.

108. *Grabisch M.* Set functions, games and capacities in decision making. With a foreword by *D. Bouyssou*. Theory and Decision Library C. Game Theory, Social Choice, Decision Theory, and Optimization, 46. Cham: Springer, 2016.

109. *Smirnov N.V., Sarmanov O.V., Zaharov V.K.* A local limit theorem for the number of transitions in a Markov chain and its applications. Dokl. Akad. Nauk SSSR, 1966, 167, pp. 1238–1241. [in Russian]

110. *Athanasiadis C.A.* The local h-polynomial of the edgewise subdivision of the simplex. Bulletin of the Hellenic Mathematical Society, 2016, 60, pp. 11–19.

111. *Avidon M., Mabry R., Sisson P.* Enumerating row arrangements of three species. Mathematics Magazine, 2001, 74, no. 2, pp. 130–134.

112. *Carlitz L.* Enumeration of sequences by rises and falls: A refinement of the Simon Newcomb problem. Duke Mathematical Journal, 1972, 39, pp. 267–280.

113. *Dobrushkin V.A.* Methods in algorithmic analysis. Chapman & Hall / CRC Computer and Information Science Series. Boca Raton, FL: Chapman & Hall / CRC, 2010.

114. *Dollhopf J., Goulden I.P., Greene C.* Words avoiding a reflexive acyclic relation. Electronic Journal of Combinatorics, 2006, 11, no. 2, Paper 28.

115. *Eifler L.Q., Reid K.B., Jr., Roselle D.P.* Sequences with adjacent elements unequal. Aequationes Mathematicae, 1971, 6, pp. 256–262.

116. *Ellzey B., Wachs M.L.* On enumerators of Smirnov words by descents and cyclic descents. Journal of Combinatorics, 2020, 11, no. 3, pp. 413–456.

117. *Farmer F.D.* Cellular homology for posets. Mathematica Japonica, 1978/79, 23, no. 6, pp. 607–613.

118. *Flajolet P., Sedgewick R.* Analytic combinatorics. Cambridge: Cambridge University Press, 2009.

119. *Freiberg U., Heuberger C., Prodinger H.* Application of Smirnov words to waiting time distributions of runs. Electronic Journal of Combinatorics, 2017, 24, no. 3, Paper 3.55.

120. *Gafni A.* Longest run of equal parts in a random integer composition. Discrete Mathematics, 2015, 338, no. 2, pp. 236–247.

121. *Gessel I.M.* Generating functions and enumeration of sequences. Ph.D. Thesis. Massachusetts Institute of Technology, 1977.

122. *Honsberger R.* From Erdős to Kiev. Problems of Olympiad caliber. The Dolciani Mathematical Expositions, 17. Washington, DC: Mathematical Association of America, 1996.

123. *Kolchin V.F., Chistyakov V.P.* Combinatorial problems of probability theory. Journal of Soviet Mathematics, 1975, 4, no. 3, pp. 217–243.

124. *Konvalinka M., Tewari V.* Smirnov trees. Electronic Journal of Combinatorics, 2019, 26, no. 3, Paper 3.22.

125. *Koshy T., Grimaldi R.* Ternary words and Jacobsthal numbers. The Fibonacci Quarterly, 2017, 55, pp. 129–136.

126. *Le Q.-N., Robins S., Vignat C., Wakhare T.* A continuous analogue of lattice path enumeration. Electronic Journal of Combinatorics, 2019, 26, no. 3, Paper 3.57.

127. *Leander M.* Compatible polynomials and edgewise subdivisions. Preprint [arXiv:1605.05287], 2016.

128. *Li T.* A study on lexicographically shellable posets. Ph.D. Thesis. Washington University in St. Louis, 2020.

129. *Lientz B.P.* Combinatorial problems in communication networks. A survey of combinatorial theory, *J.N. Srivastava*, with the cooperation of *F. Harary, C.R. Rao, G.-C. Rota, S.S. Shrikhande* (eds.), A volume dedicated to Professor *R.C. Bose* on the occasion of his seventieth birthday, and containing the Proceedings of the International Symposium on Combinatorial Mathematics and its Applications held at the University of Colorado, Fort Collins, on September 9–11, 1971. Amsterdam–London: North-Holland Publishing Company, 1973, pp. 323–332.

130. *MacFie A.* Enumerative properties of restricted words and compositions. Preprint [arXiv:1811.10461], 2018.

131. *Pemantle R., Wilson M.C.* Analytic combinatorics in several variables. Cambridge Studies in Advanced Mathematics, 140. Cambridge: Cambridge University Press, 2013.

132. *Pemantle R., Wilson M.C.* Twenty combinatorial examples of asymptotics derived from multivariate generating functions. SIAM Review, 2008, 50, no. 2, pp. 199–272.

133. *Ramirez J.L., Shattuck M.* Generalized Jacobsthal numbers and restricted k-ary words. Pure Mathematics and Applications (PU.M.A.), 2019, 28, no. 1, pp. 91–108.

134. *Remmel J.B., LoBue Tiefenbruck J.* Q-analogues of convolutions of Fibonacci numbers. Australasian Journal of Combinatorics, 2016, 64(1), pp. 166–193.

135. *Sarmanov O.V., Zakharov V.K.* A combinatorial problem of N.V. Smirnov. Dokl. Akad. Nauk SSSR, 1967, 176, no. 3, pp. 530–532.

136. *Shareshian J., Wachs M.L.* Chromatic quasisymmetric functions. Advances in Mathematics, 2016, 295, pp. 497–551.

137. *Shareshian J., Wachs M.L.* Chromatic quasisymmetric functions and Hessenberg varieties. Configuration spaces. Geometry, Combinatorics and Topology, *A. Björner, F. Cohen, C. De Concini, C. Procesi* and *M. Salvetti* (eds.). CRM Series, 14. Pisa: Edizioni Della Normale, 2012, pp. 433–460.

138. *Sundaram S.* The reflection representation in the homology of subword order. Algebraic Combinatorics, 2021, 4, no. 5, pp. 879–907.

139. *Taylor J.* Counting words with Laguerre series. Electronic Journal of Combinatorics, 2014, 21, no. 2, Paper 2.1.

140. 18th Austrian Mathematics Olympiad, Final round. Crux Mathematicorum, 1989, 15, pp. 264–265.

141. *Bruns W., Herzog J.* Cohen–Macaulay rings. Second edition. Cambridge Studies in Advanced Mathematics, 39. Cambridge: Cambridge University Press, 1998.

142. *McMullen P.* The numbers of faces of simplicial polytopes. Israel Journal of Mathematics, 1971, 9, 559–570.

143. *Petersen T.K.* Eulerian numbers. With a foreword by *R.P. Stanley*. Birkhäuser Advanced Texts: Basler Lehrbücher. New York: Birkhäuser/Springer, 2015.

144. *Stanley R.P.* Combinatorics and commutative algebra. Second edition. Progress in Mathematics, 41. Boston, MA: Birkhuser, 1996.

145. *McMullen P., Shephard G.C.* Convex polytopes and the upper bound conjecture. Prepared in collaboration with *J.E. Reeve* and *A.A. Ball*. London Mathematical Society Lecture Note Series, vol. 3. London: Cambridge University Press, 1971.

146. *McMullen P., Walkup D.W.* A generalized lower-bound conjecture for simplicial polytopes. Mathematika, 1971, 18, pp. 264–273.

147. *Anderson I.* Combinatorics of finite sets. Corrected reprint of the 1989 edition. Mineola, NY: Dover Publications, Inc., 2002.

148. *Bollobás B.* Combinatorics. Set systems, hypergraphs, families of vectors and combinatorial probability. Cambridge: Cambridge University Press, 1986.

149. *Frankl P., Tokushige N.* Extremal problems for finite sets. Student Mathematical Library, 86. Providence, RI: American Mathematical Society, 2018.

150. *Gerbner D., Patkós B.* Extremal finite set theory. Discrete Mathematics and its Applications (Boca Raton). Boca Raton, FL: CRC Press, 2019.

151. *West D.B.* Combinatorial mathematics. Cambridge: Cambridge University Press, 2021.

List of Notation

- $:=$ – equals by definition
- $\delta(s, t)$ – Kronecker delta, equal to 1 when $s = t$, and 0 otherwise
- $\mathbb{N}$ – nonnegative integers
- $\mathbb{P}$ – positive integers
- $\mathbb{R}$ – real numbers
- $\binom{j}{i}$ – binomial coefficient, equal to $\frac{j(j-1)\cdots(j-i+1)}{i(i-1)\cdots 1}$
- $\lfloor x \rfloor$ – greatest integer less than or equal to $x \in \mathbb{R}$ (the floor of x)
- $\lceil x \rceil$ – least integer greater than or equal to $x \in \mathbb{R}$ (the ceiling of x)
- $\mathrm{sign}(x)$ – sign 1, 0, or -1, of a real number $x \in \mathbb{R}$
- $\mathrm{c}(m; n)$ total number $\binom{n-1}{m-1}$ of compositions of a positive integer n with m positive parts

Sets and families

- $[t] =: E_t$ – set of consecutive integers $\{1, 2, \ldots, t\}$
- $[s, t]$ – set of consecutive positive integers $\{s, s+1, \ldots, t\}$
- $A \triangle B$ – symmetric difference of sets A and B
- $\hat{0}$ – empty set
- $\emptyset$ – empty family
- $\mathbf{2}^{[t]}$ – power set $\{A\colon A \subseteq E_t\}$ of the set E_t
- $\mathrm{V}(\mathcal{A})$ – vertex set $\bigcup_{i=1}^{\alpha} A_i$ of a family $\mathcal{A} := \{A_1, \ldots, A_\alpha\}$
- $\binom{E_t}{s}$ – complete s-uniform clutter $\{F \subseteq E_t\colon |F| = s\}$ on its vertex set E_t
- $|A|$ – cardinality (number of elements) of a set A
- $\#\mathcal{A}$ – number α of sets in a family $\mathcal{A} := \{A_1, \ldots, A_\alpha\}$
- $\partial(A, B)$ – measure $s - |A \cap B|$ of (dis)similarity between s-sets A and B

- $\boldsymbol{f}(\mathcal{F};t)$ – long f-vector of a family $\mathcal{F}$
- $\boldsymbol{h}(\mathcal{F};t)$ – long h-vector of a family $\mathcal{F}$
- $\mathbf{min}\,\mathcal{F}$ and $\mathbf{max}\,\mathcal{F}$ – subfamilies of all inclusion-minimal and inclusion-maximal sets of a family $\mathcal{F}$, respectively
- $\mathcal{C}^{\triangledown}$ – increasing family generated by a family $\mathcal{C}$
- $\mathcal{D}^{\vartriangle}$ – decreasing family generated by a family $\mathcal{D}$
- $F^{\complement}$ – complement $E_t - F$ of a set F
- $\mathcal{F}^{\complement}$ – family of complements $\{F^{\complement} :\ F \in \mathcal{F}\}$
- $\boldsymbol{\chi}(A)$ – characteristic vector of a subset $A \subseteq E_t$
- $\boldsymbol{\gamma}(\mathcal{F}')\,\boldsymbol{.}\,\boldsymbol{\gamma}(\mathcal{F}'')$ – concatenation of the characteristic vectors of families $\mathcal{F}'$ and $\mathcal{F}''$
- $\boldsymbol{\gamma}(\mathcal{F}') * \boldsymbol{\gamma}(\mathcal{F}'')$ – componentwise product of the characteristic vectors of families $\mathcal{F}'$ and $\mathcal{F}''$
- $\prod^{*}$ – componentwise product of several characteristic vectors
- $\widetilde{\mathfrak{a}}(a)$ and $\mathfrak{a}(a)$ – characteristic vector and the characteristic covector, respectively, of the principal increasing family $\{\{a\}\}^{\triangledown}$
- $\widetilde{\mathfrak{c}}(a)$ and $\mathfrak{c}(a)$ – characteristic vector and the characteristic covector, respectively, of the principal decreasing family $\{E_t - \{a\}\}^{\vartriangle}$

Graphs and complexes

- $\boldsymbol{H}(t, 2)$ – hypercube graph on its vertex set $\{1, -1\}^t$
- $\widetilde{\boldsymbol{H}}(t, 2)$ – hypercube graph on its vertex set $\{0, 1\}^t$
- $\mathrm{V}(\boldsymbol{D})$ – vertex set $\{D^0, D^1, \ldots, D^{2t-1}\}$ of a symmetric cycle $(D^0, D^1, \ldots, D^{2t-1}, D^0)$ in the hypercube graph $\boldsymbol{H}(t, 2)$
- $\vec{\mathrm{V}}(\boldsymbol{D})$ – vertex sequence $(D^0, D^1, \ldots, D^{2t-1})$ of a symmetric cycle $(D^0, D^1, \ldots, D^{2t-1}, D^0)$ in the hypercube graph $\boldsymbol{H}(t, 2)$
- $\boldsymbol{Q}(T, \boldsymbol{D})$ – unique inclusion-minimal subset of the vertex set $\mathrm{V}(\boldsymbol{D})$ of a symmetric cycle $\boldsymbol{D}$ in a hypercube graph $\boldsymbol{H}(t, 2)$ such that $T = \sum_{Q \in \boldsymbol{Q}(T, \boldsymbol{R})} Q$ for a vertex T
- $\mathfrak{q}(T) := \mathfrak{q}(T, \boldsymbol{D})$ – cardinality of the decomposition set $\boldsymbol{Q}(T, \boldsymbol{D})$
- $\mathfrak{q}(A) := \mathfrak{q}(A, \boldsymbol{D})$ – quantity $\mathfrak{q}({}_{-A}\mathrm{T}^{(+)}) := \mathfrak{q}({}_{-A}\mathrm{T}^{(+)}, \boldsymbol{D})$, where A is a subset of the ground set

- $\overline{\mathbf{2}^{[k]}}$ – boundary complex $\mathbf{2}^{[k]} - \{[k]\}$ of a simplex $\mathbf{2}^{[k]}$
- $\Delta^{\vee}$ – Alexander dual of a complex Δ

Sign tuples

- $\mathrm{T}^{(+)}$ – positive tuple $(1, \ldots, 1) \in \mathbb{R}^t$
- $\mathrm{T}^{(-)}$ – negative tuple $(-1, \ldots, -1) \in \mathbb{R}^t$
- $\mathrm{T}^{(+)}_{2^t}$ – positive tuple $(1, \ldots, 1) \in \mathbb{R}^{2^t}$
- $X(e)$ – eth component of a sign tuple X
- $\mathbf{S}(X, Y)$ – separation set of sign tuples X and Y
- $d(X, Y)$ – Hamming distance between tuples X and Y
- $\mathfrak{n}(X) := X^-$ and $\mathfrak{p}(X) := X^+$ – negative and positive parts of a sign tuple X, respectively
- ${}_{-A}X$ – sign tuple whose eth component $({}_{-A}X)(e)$ is 1, if $e \in A$, and $X(e) = -1$; $({}_{-A}X)(e) := -1$, if $e \in A$, and $X(e) = 1$; $({}_{-A}X)(e) = X(e)$ otherwise
- $-X$ – opposite ${}_{-E_t}X$ of a sign tuple X
- ${}_{-A}\mathcal{X}$ – set of sign tuples $\{{}_{-A}X : X \in \mathcal{X}\}$
- $(E_t, \mathcal{T})$ – oriented matroid, on the ground set E_t, with set of maximal covectors (topes) $\mathcal{T}$
- $\mathcal{H}_{\boldsymbol{D}}$ – rank 2 oriented matroid whose set of topes is the vertex set $\mathrm{V}(\boldsymbol{D})$ of a symmetric cycle $\boldsymbol{D}$ in a hypercube graph $\boldsymbol{H}(t, 2)$
- $\mathbf{max}^+(\mathcal{X})$ the subset of a set of sign tuples $\mathcal{X}$, with inclusion-maximal positive parts
- $\mathbf{M} := \mathbf{M}(\boldsymbol{D})$ – matrix $\begin{pmatrix} D^0 \\ \vdots \\ D^{t-1} \end{pmatrix} \in \mathbb{R}^{t \times t}$ whose rows are the vertices $D^0, \ldots, D^{t-1}$ of a symmetric cycle $\boldsymbol{D} := (D^0, D^1, \ldots, D^{2t-1}, D^0)$ in the hypercube graph $\boldsymbol{H}(t, 2)$
- $\boldsymbol{x} := \boldsymbol{x}(T) := \boldsymbol{x}(T, \boldsymbol{D})$ – row '$\boldsymbol{x}$-vector' $T\mathbf{M}(\boldsymbol{D})^{-1}$ associated with a vertex T and with a symmetric cycle $\boldsymbol{D}$ of a hypercube graph $\boldsymbol{H}(t, 2)$
- $\boldsymbol{x}(A) := \boldsymbol{x}(A, \boldsymbol{D})$ – vector $\boldsymbol{x}({}_{-A}\mathrm{T}^{(+)}, \boldsymbol{D})$, where A is a subset of the ground set
- $\boldsymbol{y}$ – row '$\boldsymbol{y}$-vector' ${}_{-s}\mathrm{T}^{(+)}\mathbf{M}(\boldsymbol{R})^{-1} =: \boldsymbol{x}(\{s\})$ associated with a vertex ${}_{-s}\mathrm{T}^{(+)}$ and with the distinguished symmetric cycle $\boldsymbol{R}$ of a hypercube graph $\boldsymbol{H}(t, 2)$

- $\mathtt{ro}(T)$ – relabeled opposite of a vertex T in a hypercube graph $\boldsymbol{H}(t, 2)$
- $\mathtt{rn}(T)$ – relabeled negation of a vertex $\widetilde{T}$ in a hypercube graph $\widetilde{\boldsymbol{H}}(t, 2)$

Vectors and matrices

- $\boldsymbol{w}^\top$ – transpose of a vector $\boldsymbol{w}$
- $\langle \boldsymbol{v}, \boldsymbol{w} \rangle$ – standard scalar product $\sum_e v_e w_e$
- $\|\boldsymbol{v}\| := \sqrt{\langle \boldsymbol{v}, \boldsymbol{v} \rangle}$ – Euclidean norm of a real vector $\boldsymbol{v}$
- $\boldsymbol{\sigma}(i)$ – ith unit vector $(0, \ldots, \underset{\overset{\uparrow}{i}}{1}, \ldots, 0)$ of the standard basis of a real Euclidean space
- $\mathbf{U}(t)$ – square backward identity matrix of order $t + 1$, whose (i, j)th entry is $\delta_{i+j,t}$; the rows and columns are indexed staring with 0
- $\mathbf{T}(t)$ – square forward shift matrix of order $t + 1$, whose (i, j)th entry is $\delta_{j-i,1}$; the rows and columns are indexed staring with 0
- $\overline{\mathbf{U}}(t)$ – square backward identity matrix of order t, whose (i, j)th entry is $\delta_{i+j,t+1}$; the rows and columns indexed starting with 1
- $\overline{\mathbf{T}}(t)$ – square forward shift matrix of order t, whose (i, j)th entry is $\delta_{j-i,1}$; the rows and columns indexed starting with 1
- $\mathbb{1}$ and $\mathbb{2}$ – α-dimensional column vectors $\begin{pmatrix} 1 \\ \vdots \\ 1 \end{pmatrix}$ and $\begin{pmatrix} 2 \\ \vdots \\ 2 \end{pmatrix}$, respectively

Blocking constructions, deletion and contraction

- $\mathfrak{B}(\mathcal{A})$ – blocker of a family $\mathcal{A}$
- $\tau(\mathcal{A})$ – transversal number of a clutter $\mathcal{A}$
- $\widetilde{\boldsymbol{\mathcal{S}}}$ – set covering $\{0, 1\}$-collection
- $\boldsymbol{\mathcal{S}}$ – set covering $\{1, -1\}$-collection
- $\mathcal{A} \backslash X$ – deletion of a clutter $\mathcal{A}$
- $\mathcal{A}/X$ – contraction of a clutter $\mathcal{A}$

Words

- $\mathfrak{T}(\mathfrak{s}', \mathfrak{s}''; n(\theta), n(\alpha), n(\beta))$ - number of Smirnov words (that start with $\mathfrak{s}'$ and end with $\mathfrak{s}''$), over the alphabet (θ, α, β), with exactly $n(\theta)$ letters θ, with $n(\alpha)$ letters α, and with $n(\beta)$ letters β
- $\mathfrak{F}(\mathfrak{s}', \mathfrak{s}''; n(\theta), n(\alpha), n(\beta), n(\gamma))$ - number of Smirnov words (that start with $\mathfrak{s}'$ and end with $\mathfrak{s}''$), over the alphabet $(\theta, \alpha, \beta, \gamma)$, with exactly $n(\theta)$ letters θ, with $n(\alpha)$ letters α, with $n(\beta)$ letters β, and with $n(\gamma)$ letters γ

Partially ordered sets (posets)

- $a \preceq b$ - elements a and b are comparable in a poset
- $\wedge$ - operation of meet in a lattice
- $\mathbf{min}\, V$ and $\mathbf{max}\, V$ - sets of minimal and maximal elements of a (sub)poset V, respectively
- $\mathbb{B}(t)$ - Boolean lattice of rank t

Index

Printed in the United States
by Baker & Taylor Publisher Services